Gulf Studies

Volume 9

Series Editor

This series is dedicated to serving growing academic interests in the dynamic, complex and strategically important Gulf region, offering a publication platform to scholars in the region and globally. The series takes an interdisciplinary approach to documenting the changes taking place in the Gulf societies, and examines the evolving relationship between the Gulf and other regions. The series advances non-Western perspectives in studying the Gulf societies, and their interactions with the rest of the world. The series includes work being done on the member states of the Gulf Cooperation Council (GCC), namely: Saudi Arabia, Oman, the United Arab Emirates, Qatar, Bahrain and Kuwait, in addition to Iran, Iraq, and Yemen.

The series accepts proposals for monographs, edited volumes and handbooks that provide an understanding of the Gulf societies' contemporary social, economic, and political landscapes. The series covers a wide range of topics within four broad themes, as follows:

Social and Cultural Issues in the Gulf:

Demography, migration, citizenship, gender, culture and identity, urbanization, education, new media, unemployment, youth, family, women empowerment, leadership, aging, human rights, sports, Islamic ethics, and Islam and society

Politics and Security in the Gulf:

Gulf defence and security, the GCC integration, democracy and political reform, Gulf politics and political systems, the GCC's foreign policy

Energy and Economics in the Gulf:

The political economy of the Gulf, economic diversification, international investments, international trade, knowledge economy, energy security and geopolitics of energy, renewable energy development, environmental regulations, infrastructure and urban planning, Islamic finance

The GCC and the World:

The GCC and Iran, the GCC and Africa, the GCC and South Asia, the GCC and East Asia, the GCC and Southeast Asia, and the GCC and Western Powers

Logan Cochrane · Reem Al-Hababi
Editors

Sustainable Qatar

Social, Political and Environmental Perspectives

Springer

Editors
Logan Cochrane
Hamad Bin Khalifa University
Doha, Qatar

Reem Al-Hababi
Gulf Studies Program
Qatar University
Doha, Qatar

ISSN 2662-4494 ISSN 2662-4508 (electronic)
Gulf Studies
ISBN 978-981-19-7397-0 ISBN 978-981-19-7398-7 (eBook)
https://doi.org/10.1007/978-981-19-7398-7

This Springer imprint is published by the registered company Springer Nature Singapore Pte Ltd.
The registered company address is: 152 Beach Road, #21-01/04 Gateway East, Singapore 189721, Singapore

Foreword

أولت دولة قطر اهتماما بالغاً في البحار والموارد البحرية والمحميات الطبيعية،وفي هذا الشأن نذكر على سبيل المثال لا الحصر الأنشطة التالية:

Qatar has paid great attention to the seas, marine resources and natural reserves. In this regard, and to name a few of the efforts made, we mention the following:

في مجال المحميات الطبيعة تعتبر وزارة البيئة والتغير المناخي هي الجهة الرسمية المسؤولة عن إدارة وحماية 35000 كيلومتر مربع من البيئة البحرية و11500 كيلومتر مربع من البيئة البرية وما يترتب من اجراءات تشريعية ومراقبة لضمان الحفاظ عليها وديمومتها. تغطى المساحة المحمية من البيئة البرية حوالي 24 بالمئة من جملة مساحة الدولة، في حين تمثل المساحة المحمية من البيئة البحرية ما يزيد عن 2.5 بالمئة فقط من جملة مساحة الدولة بما ينسجم ومبادرة الأمم المتحدة لنطاق عمل التنوع البيولوجي العالمي لما بعد 2020 في الحفاظ على الأنواع البحرية والموائل.

In terms of natural reserves, the Ministry of Environment and Climate Change is the official body responsible for managing and protecting 35,000 square kilometers of the marine environment and 11,500 square kilometers of the terrestrial environment, and the ensuing legislative and monitoring measures to ensure their preservation and sustainability. The protected area of terrestrial environment covers about 24 percent of the total area of the country, while the protected area of the marine environment covers more than 2.5 percent, which is in line with the initiative of the International Union for Conservation of Nature (IUCN) for the Post-2020 Global Biodiversity Framework, in preserving marine species and habitats.

تعمل دولة قطر حاليا على إعداد خطة العمل الوطنية القطرية للمحافظة على الموارد البحرية وإدارتها وذلك من خلال استقطاب ذوي الخبرة والمهارة في البيئة البحرية وما يرتبط بها من تنمية واستدامة التنوع البيولوجي.

The State is currently working on the preparation of the Qatar National Action Plan for the conservation and management of sea resources by attracting expertise and skills in the marine environment and in relevant aspects of biodiversity development and sustainability.

ولاستمرارية واستدامة العمل في الحفاظ على البيئة البحرية قامت دولة قطر بأعداد خرائط المناطق الساحلية وخرائط الموائل القاعية ذات الأهمية البيئية بالإضافة الى إعداد خطط إدارة متكاملة من أجل المراقبة ومتابعة التطورات الحاصلة على مر الزمن وتمكن الجهات المعنية لاتخاذ اللازم في التصحيح والتخفيف. ولا يفوتني أن أذكر أن دولة قطر تولي أهمية قصوى لبناء الانسان وتحسين قدراته وادائه للمشاركة الفعالة لتنمية وحماية البيئة والمحميات البحرية.

Within its continued and sustained efforts to preserve the marine environment, the State of Qatar has conducted a benthic habitat mapping of Qatar's coastal zone of environmental importance, in addition to developing integrated management plans to monitor developments over time and enable respective authorities to undertake the necessary mitigation and remedial measures. I should highlight the utmost importance Qatar attaches to building the human dimensions and enhancing their potential and performance to actively participate in developing and preserving the marine environment and reserves.

وتبرز أهمية اسهامات المجتمعات المحلية والساحلية وأصحاب المصلحة في القيام بدورهم في المحافظة على بيئة بحرية خالية من التلوث حيث تعتبر هذه الإسهامات مهمة في تعزيز السلوكيات والعقلية السليمة لتنمية وحماية البيئة البحرية والمحميات البحرية،وكما وضعت برامجاً للحد من التلوث وخصوصاً التلوث بالبلاستيك الذي يشكل هاجساً لجميع الدول المطلة على المحيط.

Contributions of the local, coastal communities and stakeholders are of great importance in maintaining the marine environment free of pollution, as they promote positive behaviours and perceptions toward the development and protection of marine ecosystems and reserves. Qatar has developed programs to restrain pollution, especially plastic pollution, which represents a concern to all oceanfront states.

كما أولت دولة قطر اهتماما كبيرا في الحفاظ على البيئة بشكل عام والبيئة البحرية بشكل خاص وحمايتها والمحافظة على مواردها الطبيعية واستغلالها للتنمية المستدامة. وللتأكيد على أولويات الدولة في حماية البيئة والمحافظة على مواردها الطبيعية وتنميتها استحدثت وزارة البيئة والتغير المناخي لضمان الاهتمام المستمر والمنهجي لتحقيق رؤية قطر 2030. اعتمدت دولة قطر نهجا استراتيجيا للوصول الى التكامل بين البيئة والحوكمة الاجتماعية والاقتصادية آخذة بعين الاعتبار رؤية قطر الوطنية 2030، التشريعات البيئية الوطنية والعالمية ذات العلاقة والمعاهدات والاتفاقيات والبروتوكولات الموقعة من قبل دولة قطر.

Qatar has also paid significant attention to preserving the environment in general and the marine environment, in particular, protecting their natural resources and utilizing them for a sustainable development. Emphasizing the State's prioritization of environment protection and preserving and developing natural resources, the Ministry of Environment and Climate Change has been established to ensure continued and systematic efforts to achieve the Qatar Vision 2030. The State of Qatar has adopted a strategic approach to reach integration between the environment and social and economic governance, that is aligned with Qatar National Vision 2030, the relevant national and global environmental legislation, treaties, and ratified agreements and protocols.

وتفضلوا فائق الاحترام والتقدير
فالح بن ناصر آل ثاني
وزير
وزارة البيئة والتغير المناخي

With utmost respect and appreciation,
Faleh bin Nasser Al Thani
Minister
Ministry of Environment and Climate Change

Contents

Editors and Contributors

About the Editors

Logan Cochrane is an Associate Professor in the College of Public Policy at HBKU and is Adjunct Associate Professor at Hawassa University. He has worked in non-governmental organizations internationally, including in Afghanistan, Benin, Burundi, the Democratic Republic of the Congo, Egypt, Ethiopia, Qatar, Saudi Arabia, South Africa, South Sudan, Tanzania, and Uganda.

Reem Al-Hababi is a Ph.D. student in the Gulf Studies Program at Qatar University. She holds a Master of Public Policy from HBKU.

Contributors

Sana Abusin Social and Economic Survey Research Institute, Qatar University, Doha, Qatar

Reem Al-Hababi Gulf Studies Program, Qatar University, Doha, Qatar

Dhabia M. Al-Mohannadi Chemical Engineering Program, Texas A&M University at Qatar, Doha, Qatar

Sara A. Al-Mohannadi School of Oriental and African Studies (SOAS), University of London, London, UK

Fahad Al-Musalmani Qatar Foundation, Doha, Qatar

Nihal Al-Saleh Environmental Management, University of London, SOAS, London, UK

Sara M. Alshuiael Environmental Science Program, Department of Biological and Environmental Sciences, College of Arts and Sciences, Qatar University, Doha, Qatar

Maryam F. Al-Thani Social and Economic Survey Research Institute, Qatar University, Doha, Qatar

Elena I. Athwal College of Law, Hamad Bin Khalifa University, Doha, Qatar

Radhouane Ben-Hamadou Environmental Science Program, Department of Biological and Environmental Sciences, College of Arts and Sciences, Qatar University, Doha, Qatar

Tarek Ben Hassen Program of Policy, Planning, and Development, Department of International Affairs, College of Arts and Sciences, Qatar University, Doha, Qatar

Logan Cochrane College of Public Policy, Hamad Bin Khalifa University, Doha, Qatar

April Torres Conkey Texas A&M University, College Station, TX, USA

Marcello Contestabile Qatar Environment & Energy Research Institute (QEERI), Hamad Bin Khalifa University, Doha, Qatar

Cristina D'Alessandro University of Ottawa, Ottawa, Canada

Sarra N. Dimassi Environmental Science Program, Department of Biological and Environmental Sciences, College of Arts and Sciences, Qatar University, Doha, Qatar

Ebaidalla M. Ebaidalla Ibn Khaldon Center for Humanities and Social Sciences, Qatar University, Doha, Qatar

Jonathan Gichuru Alliance of Biodiversity and International Center for Tropical Agriculture (CIAT), Nairobi, Kenya

Laurent A. Lambert Doha Institute for Graduate Studies, Doha, Qatar

Jenny Lawler Qatar Environment and Energy Institute, Hamad Bin Khalifa University, Doha, Qatar

Orjan Lundberg Supreme Committee for Delivery and Legacy, Doha, Qatar

Sylvie Maalouf Qatar Foundation, Doha, Qatar

Annamaria Mazzoni Qatar Environment and Energy Institute, Hamad Bin Khalifa University, Doha, Qatar

Velina Mirincheva University College London, Greater London, UK

Ahmad M. D. Mohamed Environmental Science Program, Department of Biological and Environmental Sciences, College of Arts and Sciences, Qatar University, Doha, Qatar

Damilola S. Olawuyi College of Law, Hamad Bin Khalifa University, Doha, Qatar

Leslie A. Pal College of Public Policy, Hamad Bin Khalifa University, Doha, Qatar

Cromwell Purchase Doha, Qatar

Mariam M. Razavi Environmental Science Program, Department of Biological and Environmental Sciences, College of Arts and Sciences, Qatar University, Doha, Qatar

Renee Richer Duke Kunshan University, Kunshan, China

S. Duygu Sever Hamad Bin Khalifa University, Doha, Qatar

Sa'd Shannak Qatar Environment and Energy Institute, Hamad Bin Khalifa University, Doha, Qatar

Muhammad O. Sulaiman Environmental Science Program, Department of Biological and Environmental Sciences, College of Arts and Sciences, Qatar University, Doha, Qatar

Husameldin M. Talballa Qatar University, Doha, Qatar

M. Evren Tok Hamad Bin Khalifa University, Doha, Qatar

Jason Twill Urban Apostles, Minnesota, USA

Steven Wright College of Humanities and Social Sciences, Hamad Bin Khalifa University, Doha, Qatar

Nobuyuki Yamaguchi Qatar University, Doha, Qatar

Esmat Zaidan College of Public Policy, Hamad Bin Khalifa University, Doha, Qatar

Chapter 1
Sustainable Qatar

Logan Cochrane and Reem Al-Hababi

Abstract The Qatar National Vision of 2030 has identified bold and transformational goals for the country. As envisioned by the State of Qatar, the transition to sustainable Qatar weaves together four pillars of economic, social, human, and environmental development. Yet, the country faces significant challenges, and with these challenges a range of options for future pathways. With the National Vision 2030 being launched in 2008, this volume provides an update on the key sustainability issues, focusing on environmental sustainability from a socio-political perspective.

Keywords Qatar · Sustainability · Development

1.1 Why Sustainability?

Global considerations of sustainability have taken diverse forms throughout human history. Indigenous knowledge systems commonly value sustainability as more important than other economic gains, be that in nutrient management practices in agricultural systems, water management practices in arid environments, or fishing practices that maintained ecosystem balances. Each of these decision-making processes recognized that the longevity of the system required careful and wise utilization. Exemplary of this is the Seventh Generation Principle, which guides people to make decisions that benefit people seven generations into the future (a philosophy rooted in the North American Haudenosaunee Confederacy, called the Iroquois by the French). The Islamic tradition provides principles similarly; the Qur'an explains that people are to act as custodians, caretakers, or trustees (6: 165) on the earth, avoiding excessiveness and wastefulness, walking gently on the earth (25: 63).

The original version of this chapter was revised: On pages 6 and 9, the running text has been changed. The correction to this chapter is available at https://doi.org/10.1007/978-981-19-7398-7_20

L. Cochrane (✉)
College of Public Policy, Hamad Bin Khalifa University, Doha, Qatar
e-mail: lcochrane@hbku.edu.qa

R. Al-Hababi
Gulf Studies Program, Qatar University, Doha, Qatar

L. Cochrane and R. Al-Hababi (eds.), *Sustainable Qatar*, Gulf Studies 9,
https://doi.org/10.1007/978-981-19-7398-7_1

The contemporary concept of sustainability builds on these diverse traditions, seeking to find a balance between the economic, social, and environmental needs of the present while considering the needs of the future. In defining what "sustainable development" is, the Brundtland Report of 1983 and the subsequent 1987 *Our Common Future* report defined the concept as meeting "the needs of the present without compromising the ability of future generations to meet their own needs" (United Nations, 1987, p. 54). In the last five decades, global evidence about environmental challenges has expanded, not only identifying the issues (e.g., biodiversity loss, climate change, deforestation, pollution) but also their interconnectedness. As the evidence has accrued, agreements and commitments at the international level have attempted to ensure sustainability is a global priority, a few include Agenda 21 and the 1992 Earth Summit, the Millennium Development Goals 2000–2015, the Johannesburg Declaration of 2002, the 2030 Agenda for Sustainable Development and the Sustainable Development Goals 2015–2030, the 2015 Paris Agreement on Climate Change, among many others. Despite the evidence and the enthusiasm, the global community is not on track to meet those commitments.

The State of Qatar faces a wide range of challenges related to sustainability, detailed in the following section. In 2008, a strategy was launched to embark on a transformation toward sustainability, as outlined in the Qatar National Vision 2030 (this is detailed in Chapters 2 and 3). The vision weaves together four pillars of economic, social, human, and environmental development. Despite the rapid expansion of country-specific evidence, to our knowledge, no book has attempted to bring together the challenges related to sustainability, in order to integrate the interconnected issues and put forward options that enable positive synergies across sectors and dimensions (economic, social, human, and environmental). This book contributes to this process by bringing together experts working on issues related to sustainability so that students, researchers, and decision-makers can have a broader, more holistic vision. The chapters also provide specific options for the transition toward sustainability, some being pragmatic and others calling for deep transformation. The options provided herein allow us, readers of this book, to assess not only the pathways available for us, but also provide indicators for our progress along those pathways.

1.2 Qatar in Context

The State of Qatar is a peninsula on the eastern side of the Arabian Peninsula that is geographically (11,521 sq. km) and demographically (2.7 million) a relatively small nation, in the global context. It only shares a land border with Saudi Arabia, however, within a couple of hundred kilometers of sea other neighbors include Bahrain, Iran, and the United Arab Emirates (UAE). It is beyond the scope of this brief introduction to present the deep history of the country, nor give justice to the periods of time when the lands of contemporary Qatar were ruled by foreign powers. Notable, however, is that it was only in 1971 that Qatar gained full independence. Of all subject areas, history and contemporary politics are of the most analyzed in academic literature, with a wealth of publications available for further reading (e.g., Al-Ejli, 2015; Alkhateeb, 2019; Al-Thani, 2012; Crystal, 1990; Fromherz, 2012; Gray, 2013; Kamrava, 2013; Rahman, 2005; Tok et al., 2016; Ulrichsen, 2014; Zahlan, 1979; Zweiri & Al Qawasmi, 2021).

Drawing upon hydrocarbon resources, particularly natural gas since the 1990s, the country has rapidly transformed into having one of the highest per capita GDPs and has attained a "very high" rank in the human development index. The chapters of this book provide a wealth of context on the contemporary political system; however, we feel that two areas warrant particular attention for this contextualization that complement what the chapters contribute. The first is demographics and the second is the economy, the latter being addressed by some authors as it relates to specific aspects of the economy (e.g., liquefied natural gas, energy); we offer a broader overview that situates those chapters.

Several chapters make mention of population figures; however, given the unique demographic trends within Qatar, we felt this warranted some context. Historically, the land that is now Qatar was not home to large populations, although trading and maritime towns emerged with populations in the thousands and tens of thousands. In 1960, the total population was less than 50,000 people, and today approaches 3 million (World Bank, 2022). Exploring demographic trends situates the current context as well as supports an understanding of potential future trends. There are no official statistics on the number of citizens in Qatar, although estimates suggest citizens represent only 10–12% of the population. The vast majority of the population are expatriates, and of that population, there is a large working-age male population (resulting in one of the world's most gendered population structures).

For those not familiar with the State of Qatar, it may have emerged on the radar when it was announced as the country was selected to host the FIFA World Cup in 2022. The resultant construction (e.g., of the stadiums and related infrastructure) might be assumed to be a major driver of this large expatriate workforce, and while this has been a reason for some of the population growth, the majority of the demographic expansion took place in the decade previous when liquefied natural gas (LNG) revenues enabled rapid economic growth. The population more than tripled between 2000 and 2010 (in 2000 the population was 592,467, in 2010 the population was 1,856,329, a 213% increase, see Fig. 1.1).

In December of 2010, it was announced that Qatar had won the bid to host FIFA 2022, at which time the total population was 1,637,443. More than a decade later, as of March 2022, the population had risen to 2,773,598 (a 69% increase, Fig. 1.2). The population growth rate has been slowing, and in particular since 2018, whereafter the total population has fluctuated around 2.7 million. United Nations forecasts suggest that the total population of Qatar will continue to rise, reaching 3.2 million in 2030, 3.7 million in 2050, and 3.9 million in 2100 (2017). However, these projections tend to rely upon more typical population growth rates (fertility, birth and death rates, population age and sex profiles), and in most countries, migration plays a more minor role. Given the unique status of the current population (with a majority being expatriates with temporary residence), as well as the other socioeconomic drivers that may increase or decrease migration trends, these future scenarios may provide some insight but have far more uncertainties when compared to other countries.

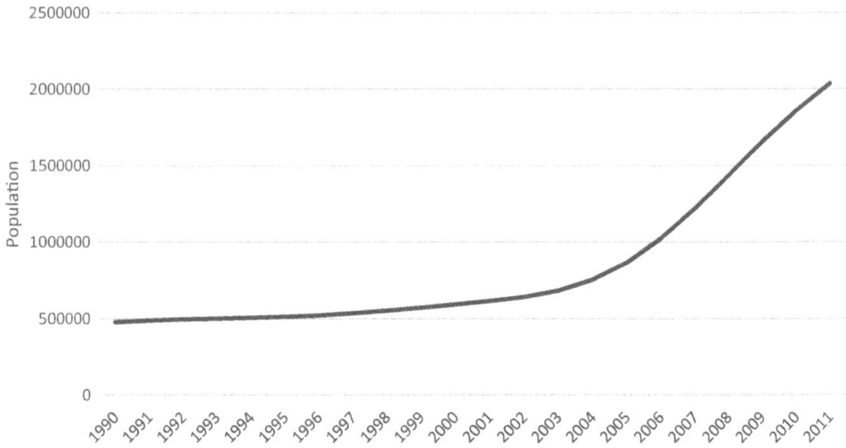

Fig. 1.1 Population of Qatar (total), from 1990 to 2011 (World Bank, 2022)

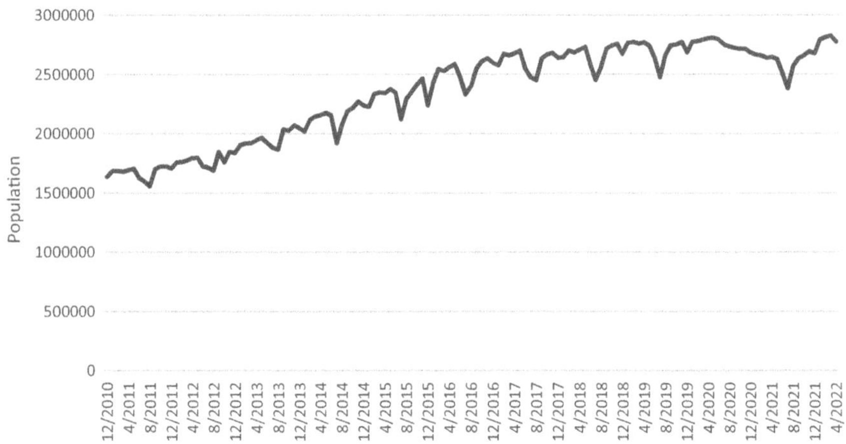

Fig. 1.2 Population of Qatar (total), monthly data from December 2010 to March 2022 (Planning and Statistics Authority, 2022)

Demographics matter when considering sustainability because many of the decisions we make today, particularly infrastructural, are intended to serve future populations for decades. This brief overview of demographics has only touched on the total population, but as Qatar aims to transition toward a knowledge-based economy, the types of expatriates living in Qatar will change, as will their needs and expectations. Ensuring that systems (e.g., transport, energy) are designed to meet those needs, while also transitioning to more sustainable ways of living, requires in-depth analysis. Indeed, this is one topic area we hope to see additional research on, particularly as it relates to the policy recommendations put forth in this collection (in the following section we also note topics that are missing from this collection that we encourage more research to be conducted on).

The second area we highlight in this Introduction is the broader economy. Hydrocarbons are often focal to discussions of the economy, and rightly so, as they account for the majority of government revenue and export earnings. Several chapters in this collection focus on hydrocarbons (see chapter description in the section that follows), what we highlight here are other areas or aspects of the economy as we feel these are critical for the sustainability transition.

Connected to the hydrocarbon resources of the country, but less often covered, are domestic industries, including producing steel, fertilizer, and a range of petrochemicals (directly using hydrocarbons or enabled by them). The largest export, by value, other than petroleum gas, crude petroleum, and refined petroleum, is ethylene polymers, or plastic, followed by fertilizer (OEC, 2022). Other than natural resources-related industries (including the subsidiary industries noted above), the largest companies in Qatar (by value), according to Forbes (2021), are from sectors including finance (Qatar National Bank, Qatar Islamic Bank, Masraf Al Rayan, Commercial Bank, Qatar International Islamic Bank, Ahlibank, Doha Bank, Qatar Insurance Company, Qatar Development Bank, Dukhan Bank), construction and real estate (Ezdan, Barwa, Aamal, United Development Company, Qatar National Cement, Qatar Investors Group, Qatari Diar), telecommunications (Ooredoo, Vodafone Qatar), retail and consumer goods (Zad, Al Meera, Baladna, Mannai Corporation, Al Mana), healthcare (Medicare Group, Hamad Medical Corporation), hospitality (Katara, Al Fardan), and aviation (Qatar Airways). Qatar also has one of the world's largest sovereign wealth funds (Qatar Investment Authority). With the expansion of these companies domestically and internationally, the economy of Qatar is diversifying (albeit from a point of low diversification two decades ago to one relatively more diversified today). According to the Observatory of Economic Complexity (OEC) (2022), economic complexity dropped following the 2017 blockade of Qatar (for additional details on the blockade, see Chapters 2 through 6 in Alkhateeb, 2019, amongst others), before which it had a higher economic complexity than the UAE.

The State of Qatar has made major investments in infrastructure over the last two decades, including in ports, airports, subway and light rail, roads and highways, major city developments (e.g., Lusail), water desalination and distribution, all of which are estimated to be worth US$ 300 billion by the Secretary-General of the Committee for Delivery and Legacy, Hassan Al Thawadi. Much of these investments have supported the growth of the above-mentioned companies and the diversification of the economy, utilizing hydrocarbon wealth to diversify the economy and enable the development of new industries. Qatar has established or has been working to develop niche industries, including in Islamic finance, tourism, sport, education, and healthcare. Such efforts, along with other government initiatives and policies to strengthen the economy (recent ones include enhancements to labor laws, the introduction of a minimum wage, changes to residency regulations, and foreign direct investment laws), appear to have yielded some progress in non-hydrocarbon sectors' performance (World Bank, 2021). The contribution of non-hydrocarbon sectors, notably the services and construction sectors, has been significant to the GDP growth rate (e.g., between 2012 and 2021) when compared to hydrocarbons. Moreover, non-hydrocarbon economic activities' share in GDP has increased from 45% in 2013 to 63% in 2020 (Planning and Statistics Authority, 2021).

Aside from the above notes on economic diversification efforts and outcomes, future sustainability pathways require more than reducing the reliance on hydrocarbons and expanding the size of the remaining sectors in the economy. Future pathways call for more attention to enhancing and deepening adaptation and mitigation. Progress has been witnessed in several sectors, as highlighted by some chapters, such as in the construction, transportation, electricity supply, and waste management activities, in which a number of sustainable or eco-friendly reforms and practices have been introduced and implemented. Progress can also be noted in the academic realm that concerns the environmental impacts and recommendations or innovations to address them. Climate change's impact on the agriculture sector in Qatar, for instance, is increasingly highlighted in research (e.g., Ben Hassen et al., 2020; Karanisa et al., 2021, among many others). However, more effort is needed to understand how environmental degradation and climate change will affect the growth patterns of non-hydrocarbon economic activities in Qatar from one end and how these activities contribute to such challenges (their environmental footprint) from the other—and accordingly, interventions to address both. A sustainable transition, in many sectors, will require significant transformation, requiring flexibility and adaptability from the public and private sectors.

1.3 The Book

With the National Vision 2030 being launched in 2008, this volume provides an update on the key sustainability issues, focusing on environmental sustainability from a socio-political perspective. *Sustainable Qatar* has nineteen chapters, to which 39 authors contributed. One of the motivations for putting this book together was a recognition of experts within Qatar working on areas of sustainability, some of whom we knew before embarking on this book journey and others we met along the way. Of the 39 authors, 34 were based in Qatar at the time of writing, with many of the remainder previously having been based in the country. This is a testament to the commitment to research and sustainability in Qatar, as experts contributing to this book are affiliated with a wide range of institutions, including Hamad Bin Khalifa University (HBKU), Qatar University, Qatar Foundation, Texas A&M Qatar, Qatar Environment and Energy Research Institute, the Doha Institute for Graduate Studies, and the Supreme Committee for Delivery and Legacy.

When envisioning this project, we sought funding to make this book fully Open Access, which was provided by the Program on Governance, Resilience, and Sustainability, a research initiative within the College of Public Policy at HBKU. Being Open Access means that this entire book is freely available for students, researchers, decision-makers, and anyone else interested. As each chapter offers policy options and recommendations, this was of fundamental importance since most policymakers often do not have access to costly academic publications. Being Open Access ensures that access to evidence is not a barrier in the decision-making process.

While we are excited that this book came together with nineteen chapters, we also recognize that there are some issues that are not covered, or not sufficiently covered. For a book offering policy implications, and one that emphasizes the interconnectedness of the issues covered, these omissions are important to make note of. From our perspective, some of the key missing topic areas include domestic transportation, aviation, international shipping and global supply chains, the built environment, marine biodiversity, cooling systems, desalination processes, demographics, inequality, gender, and other forms of social differentiation with regard to impacts of sustainable transitions, migration, health, and tipping point emergency planning. We hope that researchers will continue to make contributions to advance the evidence base supporting the sustainability transition in these, and other, areas.

1.4 Chapters

The final section of this Introduction presents a chapter-by-chapter review of the book, highlighting some key components that stood out for us as editors. The book begins with two chapters on the strategic and legal aspects of sustainability. In Chapter 2, Reem Al-Hababi tracks how environmental sustainability policy has developed in Qatar through surveying the key guiding policy documents, regulations, and institutions that concern environmental affairs. The chapter argues that while progress toward Qatar's environmental sustainability targets has been modest—many of the targets are quite ambitious—the policies have improved. Pending, however, is a detailed assessment of the institutions and laws/policies to meet and uphold these targets. To date, assessment and public reporting have been limited. On this basis, the chapter concludes that future efforts should focus on the microlevel in order to enhance institutional and technical capacities. The chapter also argues that more attention should be paid to evaluation and monitoring so as to address the persistent challenge of limited data about ecological systems. Transparency of such data would alleviate the ambiguity of the strategies' actual outcomes.

Chapter 3, by Damilola Olawuyi and Elena Athwal, narrows the focus from vision and strategy to laws and regulations. The chapter discusses the legal and governance frameworks that are in place to address sustainability and energy transition challenges. Damilola and Elena highlight how engaging in regional and international sustainability frameworks has contributed to the integration of environmental considerations in Qatar's national policies and how environmental preservation and energy transition have been at the core of a wide range of primary and secondary laws. Based upon this analysis, the authors note limitations within existing frameworks on sustainability due to the governance approach, particularly in terms of coordination and cooperation among relevant actors in delivering sustainability projects and programs, as well as the lack of knowledge and information sharing between them. The chapter suggests that there is a need for a nexus governance approach and dynamic legal innovations to enhance the coordination and coherent implementation of Qatar's rapidly increasing number of sustainability policies, laws, and organizational bodies. It also suggests the need to create an oversight body, considering the increase in spending on sustainability initiatives in Qatar, to monitor and evaluate the implementation of these initiatives in a measurable way in order to ensure real progress is made. In line with Chapter 2, Damilola and Elena highlight the critical nature of data sharing.

FIFA 2022 is a major event for a country the size of Qatar (geographically and demographically), and this international sporting event had the potential to be a catalyst for change, given how substantial the changes would be to prepare for it. In Chapter 4, Orjan Lundberg discusses how winning the bid to host the international tournament has acted as a motivator for Qatar toward achieving environmental sustainability throughout the preparation and delivery stages of the event, and the legacy that it intends to leave behind. The chapter highlights the environmental issues that usually accompany hosting such sports mega-event and the responses that the organizing entities have identified to address them. Orjan argues that such sports mega-events can serve as a platform to shed light on global environmental challenges and that hosting such events can provide an opportunity for the hosting country to catalyze a positive, sustainable development with the infrastructure, technology, skills, and supply chains that develop toward and throughout the event. Akin to Chapter 3, this chapter calls for an overarching entity that ensures cohesion across all activities and institutionalization for the future.

The Supreme Committee for Delivery and Legacy was established by the State of Qatar and leads all FIFA-related activities. While wielding significant influence between 2010 and 2022, Chapter 5 looks at a non-governmental actor working over the long term to push for sustainability innovation. Authored by Fahad Al-Musalmani and Sylvie Maalouf, the chapter presents a detailed analysis of Qatar Foundation's experience and journey as a non-state actor contributing toward environmental sustainability and the various initiatives and programs that it has developed and carried out. Fahad and Sylvie show the role non-governmental actors can play in promoting sustainability through formal and informal means, such as via education, research, advocacy, and piloting innovative solutions, thereby setting the example for other entities to undertake similar paths. This chapter concludes saying that Qatar, due to its being on the receiving end of many of the adverse effects of climate change and its influence on the global energy sector, is uniquely positioned to lead the global sustainability movement, something the authors argue the Qatar Foundation is at the forefront of as a result of its integration of research, education, and civil society.

Shifting to the key economic sector of Qatar, Steven Wright looks at the energy sector as Qatar engages with a global transition toward renewable and carbon–neutral energy futures. Chapter 6 evaluates Qatar's sustainability challenge, as a country reliant on the gas sector. The assessment also takes into account geopolitical shifts, which do not all necessarily push for decarbonization, such as the rise of demand for LNG from Qatar following Russia's invasion of Ukraine. Natural gas is viewed as low-carbon energy (transitional), which means that global demand is expected to last for some time. Domestically, gas sector revenues are needed to carry out the development plans and projects (including those intended to assist diversifying the economy and enhancing other sectors' contributions), which means that the gas sector must be able to adapt to the challenges and changes in geopolitics. Steven argues that the way forward toward sustainability requires enhancing the gas sector's ability to adapt to challenges and changes in energy markets by "maximising the relevancy, innovation, and economic returns" within and beyond the sector. The sector's future growth and development rely on progress at several forefronts, including reducing the gas sector's environmental footprint (through, for example, the deployment of carbon capture and storage technologies), increasing renewable energy capacity, and investing in "blue" and "green" hydrogen and ammonia sectors.

Turning to Chapter 7, Sara Al-Mohannadi and Dhabia Al-Mohannadi outline the challenges and opportunities of transitioning to a low-carbon economy, and particularly within the energy sector. The transition away from fossil fuels will ultimately expose Qatar to market vulnerabilities, but outline that there are ways to make this transition successfully. Building on the chapter by Steven Wright, these authors explore the implications for Qatar's energy sector in the medium and long term and the need to enhance the hydrocarbon sector's adaptability to gradual changes in energy demand. The chapter suggests that the adaptability and sustainability of the energy sector can be improved through the deployment of carbon capture storage and renewable energy technologies. Sara and Dhabia note the potentials that might exist in emerging markets, such as liquified hydrogen, and the option of investing in such sub-sectors, especially since Qatar has an advantageous position being one of the leading gas exporters worldwide.

In Chapter 8, Marcello Contestabile highlights the "systemic nature of sustainability policy problems faced by Qatar." The main systemic challenges mentioned by the chapter include moving away from the rentier-state economic model and the requirements, implications, and risks of such economic transition from social, environmental, and economic respects. The chapter argues that the multidimensional nature of sustainability challenges requires a systems perspective. Types of systems analysis tools are presented in the chapter along with a description of their key purpose and characteristics. Though there is a wide range of system analysis' frameworks and tools, the chapter highlights that the complexity of sustainability issues can be better addressed through a combination of them in which one tool or framework complements the others. The role of Qatar Environment and Energy Research Institute in establishing systems models is discussed along with the Institute's current key project, the Integrated MARKAL-EFOM System, which is intended to support policy analysis and making in Qatar. The chapter concludes that future efforts to support sustainability should include raising awareness among policymakers and various stakeholders about the potencies of the systems approach and the necessity of establishing its relevant tools in guiding policymaking and the prospects of such evidence-based policies that concern the complex challenges of sustainability.

Chapter 9 begins to look at the impacts of climate change on Qatar, and specifically how sea-level rise will affect the region in the coming decades, highlighting how this poses a national security risk that needs to be on top of policy agendas across the region. Laurent Lambert and Cristina D'Alessandro warn of the risks of the poorly documented hazards of sea-level rise on coastal cities in the Middle East and North Africa. Notably, the risks associated with sea-level rise are not limited to sea-level rise itself but are also associated with increases in extreme weather events, storm surges, and other coastal hazards. Coastal cities in the region are characterized by high population density and their significance to the states' economy. The authors suggest that technological solutions must be selected and appropriately used on a case-by-case basis; one solution will not fit all countries facing the challenges of sea-level rise. Laurent and Cristina argue that state and city leadership, alongside diverse stakeholders, are needed to ensure that the measures selected to address the challenges posed by sea-level rise are appropriate to the cities and their communities. The chapter examines the efforts made to adapt to sea-level rise and their relevant challenges in Doha as a study case. The main challenges highlighted in the chapter include the lack of transparency with stakeholders in terms of communicating the necessity to maintain a balance between mitigating risks and stimulating economic activities in coastal cities and the limited institutional capacity for sustainable urban planning (also highlighted in Chap. 3).

Chapter 10 is concerned with Qatar's marine biodiversity, particularly the impact of rig-to-reef processes—throughout the decommissioning of offshore oil platforms—on marine ecosystems. To investigate this, Radhouane Ben-Hamadou, Ahmad Mohamed, Sarra Dimassi, Mariem Razavi, Sara Alshuiael, and Muhammad Sulaiman apply a systems analysis (Driver-Pressure-State-Impact-Response) to understand the implications of such an approach. The chapter highlights that offshore oil and gas platforms have actually provided a habitat for many marine species and have assisted in increasing marine biodiversity. Accordingly, such manmade structures have been useful to marine systems during their "design life" and, therefore, should be preserved by "converting them into artificial reefs under rig-to-reef programs." Various methods of rig-to-reef are indicated along with the benefits of them, which include creating, restoring, and preserving marine habitats. On the other hand, the success of these methods, as outlined by the chapter, is associated with numerous aspects that are divided into technical factors (which relate to, e.g., water depth, geographic location, geological conditions) and non-technical factors (which relate to climate change impact, i.e., increase in temperature), in addition to legal and technological factors. Future pathways for successful rig-to-reef application and protected marine environment following the decommissioning of offshore oil and gas platforms, as per the authors, revolve around establishing "a consistent policy and regulatory framework" concerning the marine habitats surrounding oil and gas sea facilities, "clear and defined liability and compensation schemes" for relevant stakeholders, and implementation of assessment and monitoring programs.

Jenny Lawler, Annamaria Mazzoni, and Sa'd Shannak turn to another aspect of water: the domestic water system. Given that Qatar has almost no freshwater resources (limited to groundwater resources), the water sector is of critical importance for the country. In Chapter 11, the authors argue that as Qatar continues to develop and become self-sufficient, the population grows there will be increased strain on water resources. As a result, they suggest that sustainable solutions need to be implemented that include all key stakeholders in the water sector. They additionally suggest that "water management can be approached from an allocation efficiency perspective, as well as from a societal point of view in terms of quantity and access" and highlight the importance of monitoring and regulation enforcement. Analyzing supply- and demand-side factors, the chapter highlights that the strategies that concern future water sustainability pathways should integrate water quality and quantity monitoring and assessment systems and should align with food security strategies to maximize the outcomes of both.

In Chapter 12, Sana Abusin, Ebaidalla Ebaidalla, and Maryam Al-Thani draw upon their previous research on food loss and food waste in the State of Qatar and highlight one of the emerging options to reduce loss and waste, specifically focusing on the role of non-governmental organizations in reallocating these resources. The authors argue that there needs to be greater cooperation between all key stakeholders (government, non-governmental organizations, charities, academics, private companies) in order to strengthen food insecurity while in tandem reducing food loss and waste. Sana, Ebaidalla, and Maryam suggest that this includes further involving non-profit organizations to engage in waste management and food management, through which resources may be redistributed to those in need of them, preventing forms of food loss and waste while addressing food insecurity. For foods that cannot be redistributed, the authors argue it is critical to segregate organic waste so that it can be composted and create new value (and not sent to the landfill, creating costs).

April Torres Conkey, Cromwell Purchase, Renee Richer, and Nobuyuki Yamaguchi turn our attention to terrestrial biodiversity in Chapter 13. Although not a place known for its biodiversity, given its arid climate, the authors argue that Qatar's arid environment has great biodiversity despite its largely unstudied/understudied nature; a biodiversity they believe will provide the nation with immense biological resources essential to enhancing resilience to climate change. The chapter provides a readable overview of the various plants and species that are specific to Qatar's hyper-arid ecosystems and highlights their importance for climate change resilience, and notes that although Qatar's environment has "a wide range of communities and organisms from microbials to plants and animals," they remain understudied. The authors' recommendations include developing databases and conducting periodic surveys to monitor the ecosystems and their inhabitants' status and establishing guidelines for land use and agricultural or grazing practices to reduce their negative impacts on natural habitats.

Qatar is an urban country, with an estimated 99% of its population living in urban areas (primary the capital city and surrounding areas; World Bank, 2022). Chapter 14, written by Velina Mirincheva, Jason Twill, and Nihal Al-Saleh, presents an argument for thinking about 15-min cities, taking Doha as an example (and highlighting emergent examples of this form of urban development). The chapter highlights how the 15-min city concept revolves around "fulfilling the essential social urban functions" and addressing urban challenges that relate to "(a) density, (b) proximity, (c) diversity, and (d) digitalization." The chapter provides an examination of Doha City and whether it can be transformed into a 15-min or "compact, density-balanced city." For a city that is quite dependent upon personal vehicles (despite a new subway system, multiple tram systems, and a public bus system), Velina, Jason, and Nihal argue for a reconceptualization of lifestyles and urban living, one which enhances well-being and sustainability.

Chapter 15, by Esmat Zaidan, also concerns sustainable urban planning. The chapter sheds light on the concept of smart sustainable cities in the Gulf's "Post-Oil" context and the drivers and requirements for their creation. Establishing an effective zero-carbon strategy, as per the chapter, is crucial for achieving smart sustainable cities and requires assessing the socioeconomic factors that contribute to the transformation to a zero-carbon community comprehensively. The challenges of smart sustainable cities' strategies and projects in the Gulf region, as a whole, and in Qatar that are highlighted by the chapter include the orientation of past and existing infrastructure planning, which has lacked consideration for the interdependencies between various sectors and long-term needs. Other challenges identified include the resistance to implement smart sustainable cities' strategies and solutions by relevant stakeholders due to a lack of awareness about the interlinkage and interplay between their sectors. The way forward, according to the chapter, involves a comprehensive national strategy for urban planning that is articulated based on a thorough understanding for the society in Qatar and inclusion of socioeconomic and behavioral dimensions, effective engagement and collaboration of stakeholders, and sound regulations and legal framework for the information, communication, and technology sector.

While noted by other authors in brief, Husameldin Mohamed Talballa Elshaikh and Jonathan Gichuru turn our attention to an aspect of sustainability often forgotten: waste management. In Qatar, this is particularly important, as the country has high per capita waste generation, and expectations for the waste generation to rise with population growth and continued economic development. Chapter 16 discusses the strengths and weaknesses of the waste management sector and how non-hazardous solid waste is handled in Qatar. The chapter sheds light on various stakeholders' efforts toward sustainable waste management, including a case study of Qatar University's zero-waste initiative. The case study showcases a holistic approach with a variety of policy tools, active engagement of actors and stakeholders, and public participation, which are among the recommendations the chapter has identified to reduce generated waste and improve waste management processes. Drawing on the example, the way forward for Qatar includes establishing an integrated approach (top-down and bottom-up) with a mix of policy instruments, engaging actors and stakeholders, public–private partnerships and collaborations, the establishment of zero-waste frameworks, and behavioral change (which can be achieved through centralizing public participation in the efforts made).

In Chapter 17, S. Duygu Sever and M. Evren Tok explore how broader societal attitudes can change, in looking at education for sustainable development in Qatar. The authors exlore how sustainable development is integrated into the educational system in Qatar, whether in the formal or informal means, by examining the relevant policy documents, initiatives, and actors. The chapter highlights the important role the education system plays in building knowledge, skills, competencies, and mindsets of future generations to carry out the sustainability efforts further. The way forward, as per the chapter, includes a defined, explicit action plan that integrates education for sustainable development into all educational levels and the whole curricula, as well as improved access to sustainability development-specific teaching methods and innovations for educators and students.

The second last chapter of this collection, Chapter 18 by Tarek Ben Hassen, discusses the transition from a rentier-state economic model to a more sustainable, knowledge-based resilient economic model with a special focus on the entrepreneurial ecosystem in Qatar and how it can be enhanced to stimulate such a shift. The chapter argues that Qatar's knowledge-based economic sector growth faces many challenges due to its status as a rentier state. These challenges are (1) as in a rentier state, the government is the primary employer many people are not currently considering a future in STEM, hindering entrepreneurship abilities; (2) the passive entrepreneur is the most popular type of entrepreneur, with many having companies on the side of their full-time job in the public sector to provide additional income while still holding a steady job; (3) there is a lack of access to finance for entrepreneurs due to the culture surrounding Qatari banks. Tarek concludes by arguing that Qatar's transition toward a more knowledge-based economy will require participation from all stakeholders and a national plan that avoids a silo mindset. He also suggests that export and activity diversification could help along with the implementation of sector-specific policies.

Chapter 19, the Conclusion by Leslie Pal, Logan Cochrane and Reem Al-Hababi, attempts to offer a reflection on the book chapters by providing a brief on the sustainability context in Qatar, the policy frameworks, their implementation, and the key recommendations for future pathways. The chapter also highlights sustainability topics that are missing from this volume.

We hope that you, the readers of this collection, find this book a useful source of diverse issues and perspectives. We believe the range of opinions found within this collection reflect the wide range of options and pathways available to the State of Qatar, and it is, for this reason, we encouraged authors to write as they felt best appropriate, rather than according to a pre-determined framework that all issues ought to fit within. We greatly look forward to ideas and discussions that emerge following the publication of this book.

References

Al-Ejli, O. (2015). *Sheikh Jassim Al-Thani, founder of Qatar: A historical study of a nineteenth-century Gulf and the Arabian Peninsula*. Page Publishing.

Alkhateeb, H. M. (Ed.). (2019). *Qatar: Political, economic and social issues*. Nova Science Publishers.

Al-Shelek, A. Z., Oqail, M. M., & Ibrahim, A. A. Y. (2015). *Qatar's modern and contemporary development: Chapters of political, social and economic development*. Independent.

Al-Thani, M. (2012). *Jassim the leader: Founder of Qatar*. Profile Books.

Ben Hassen, T., El Bilali, H., & Al-Maadeed, M. (2020). Agri-Food markets in Qatar: Drivers, trends, and policy responses. *Sustainability, 12*, 3643.

Crystal, J. (1990). *Coalitions in oil monarchies: Rulers and merchants in Kuwait and Qatar*. Cambridge University Press.
Fromherz, A. J. (2012). *Qatar: Rise to power and influence*. Bloomsbury.
Forbes. (2021, December 9). *The 30 most valuable companies in Qatar*. https://www.forbesmiddleeast.com/lists/the-30-most-valuable-companies-in-qatar/
Gray, M. (2013). *Qatar: Politics and the challenges of development*. Lynne Rienner Publishers.
Kamrava, M. (2013). *Qatar: Small state, big politics*. Cornell University Press.
Karanisa, T., Amato, A., Richer, R., Sara, A. M., Skelhorn, C., & Sayadi, S. (2021). Agricultural production in Qatar's hot arid climate. *Sustainability, 13*(7), 4059.
Observatory of Economic Complexity. (2022). *Qatar*. https://oec.world/en/profile/country/qat
Planning and Statistics Authority. (2021). *Qatar economic outlook 2021–2023*. https://www.psa.gov.qa/en/knowledge/Doc/QEO/English_QEO_2021-2023.pdf
Planning and Statistics Authority. (2022). *Monthly figures on total population* [Data File]. Retrieved from https://www.psa.gov.qa/en/statistics1/StatisticsSite/pages/population.aspx
Rahman, H. (2005). *The emergence of Qatar: The turbulent years, 1627–1916*. Routledge.
Tok, M. E., Alkhater, L. R., & Pal, L. A. (Eds.). (2016). *Policy-making in a transformative state: The case of Qatar*. Palgrave Macmillan.
Ulrichsen, K. C. (2014). *Qatar and the Arab Spring*. Oxford University Press.
United Nations. (1987). *Report of the world commission on environment and development: Our common future*. United Nations General Assembly.
United Nations. (2017). *World population prospects*. United Nations.
World Bank. (2021). *Macro poverty outlook: Country-by-country analysis and projections for the developing world*. https://thedocs.worldbank.org/en/doc/77351105a334213c64122e44c2efe523-0500072021/related/mpo-am21.pdf
World Bank. (2022). *Population, total—Qatar* [data file]. Retrieved from https://data.worldbank.org/indicator/SP.POP.TOTL?locations=QA
Zahlan, R. S. (1979). *The creation of Qatar*. Routledge.
Zweiri, M., & Al Qawasmi, F. (Eds.). (2021). *Contemporary Qatar: Examining state and society* (Vol. 4). Springer Nature.

Logan Cochrane is an Associate Professor in the College of Public Policy at HBKU and is Adjunct Associate Professor at Hawassa University. He has worked in non-governmental organizations internationally, including in Afghanistan, Benin, Burundi, the Democratic Republic of the Congo, Egypt, Ethiopia, Qatar, Saudi Arabia, South Africa, South Sudan, Tanzania, and Uganda

Reem Al-Hababi is a PhD student in the Gulf Studies Program at Qatar University. She holds a Master of Public Policy from HBKU

Chapter 2
The Evolvement of Qatar's Environmental Sustainability Policy: The Strategies, Regulations, and Institutions

Reem Al-Hababi

Abstract Environmental sustainability has been increasingly present on Qatar's policy agenda since the early 2000s, with the ratification of the Permanent Constitution, the issuance of several environmental laws, and the launch of Qatar National Vision 2030, and its supplementary strategies. This chapter tracks the development of environmental policy through surveying the country's key 'guiding' policy documents; Qatar National Vision 2030, the first and second National Development Strategies, sectoral strategies, and through reviewing laws and regulations that have been issued so far to safeguard the ecological systems and their inhabitants. This chapter also investigates the respective government institutions and how their role has evolved over the years.

Keywords Environmental Policy · Sustainability · Environment · Qatar

2.1 Introduction

Environmental sustainability has been one of the most pressing topics on Qatar's policy agenda over the past two decades, considering the steady economic growth, massive population expansion, and the rapid urbanization the country has been experiencing and the implications of such growth patterns on the degradation of the environment and the latter's impact, in addition to climate change, on the society and the state as a whole.

With all the transformations in various respects that Qatar has been going through, environmental policy, too, has been subject to several changes during this time period. Several strategies, laws, and institutions have been established at various levels to address the environmental challenges. Whether or not they have been successful in doing so, this chapter intends to provide an overview of environmental sustainability

R. Al-Hababi (✉)
Gulf Studies Program, Qatar University, Doha, Qatar
e-mail: reem.alhababi@outlook.com

L. Cochrane and R. Al-Hababi (eds.), *Sustainable Qatar*, Gulf Studies 9,
https://doi.org/10.1007/978-981-19-7398-7_2

policy evolvement and tries to identify the gaps that might exist. The chapter first investigates related national policies and strategies, then goes over the laws and regulations issued, and lastly, the relevant government institutions.

2.2 National Policies and Strategies

Although the 2000s were significant to environmental sustainability policy planning in Qatar since they have witnessed the establishment of the first medium- and long-term nationwide strategies,[1] environment has been the concern of a number of policy initiatives throughout the 1970s, 1980s, and 1990s. These policy initiatives came in various—yet scattered—forms, whether laws (such as Law No. 8 of 1974 on General or Public Cleanliness, see Sect. 2.3), institutions (such as the establishment of Standing or Permanent Environmental Protection Committee in 1981, see Sect. 2.4), or the ratification of regional and international environmental treaties (such as accessing the 1954 International Convention for the Prevention of Pollution of the Sea by Oil in 1980) (United Nations Environment Program 1984).

The state's strategies concerning environmental affairs have been developing gradually in terms of scope and level since the start of the new millennium. One of the early medium-term strategies that were developed at the sub-sectoral level was the National Biodiversity Strategy and Actions Plan (NBSAP), launched in 2004. It was developed at a time in which the subject of biodiversity has been the focus of several international and regional treaties, such as the Convention on Biological Diversity in 1996, the Convention to Combat Desertification in 1999, the Convention on International Trade in Endangered Species of Wild Fauna and Flora in 2001, and the Convention on the Conservation of Wildlife and their Natural Habitats in the Countries of Gulf Cooperation Council in 2004—which Qatar has ratified the accession to. The NBSAP intended to provide a 10 years roadmap for sustaining biodiversity, marine ecosystems, and aquatic resources. The second NBSAP was launched in 2015 and aims to achieve the objectives set in the first strategy along with a few additional by 2025.

The years that followed the launch of NBSAP witnessed the development of more sophisticated long- and medium-term strategies with wider scopes, including the Qatar National Vision, National Development Strategies, the National Environment and Climate Change Strategy, as well as sectoral and institutional strategies, respectively, detailed in the following subsections.

[1] In 1995, the General Secretariat for Development Planning prepared the 'Public Strategy of Social and Economic Development in Qatar', but the policy document was never implemented or executed due to the dissolution of the General Secretariat soon after (Al-Kuwari, 2012).

2.2.1 Qatar National Vision 2030

Launched in 2008, Qatar National Vision 2030 (QNV 2030) was the first of its kind in the state. The launch of the Vision came during a time period wherein there were repeated calls for establishing an integrated approach to sustainable development at global, regional, and national levels, from the Earth Summit in 1992 to the General Assembly Special Session on the Environment in 1997 and the World Summit on Sustainable Development in 2002. The idea of developing the Vision perhaps has emerged due to the deliberations and conclusions of such conferences. QNV 2030 was the first policy document attempting to provide a long-term comprehensive development strategy for the State of Qatar, and it appears to be largely influenced by the United Nations' (UN) manifestation of sustainable development (i.e., 1987's Brundtland Report and subsequent UN publications).

The sustainable development approach of QNV 2030 is similar to what Peterson (1997) described as the competing objectives approach, in which a strategy is designed to reconcile social, economic, and ecological goals. The Vision is constituted of four 'pillars': social, human, economic, and environmental development, and for each, national challenges and objectives are identified. In addition to the challenges that are specific to each pillar, five major challenges were recognized by QNV 2030 and are supposed to be at the core of the efforts made toward the four pillars. Three of these challenges concern environmental sustainability; (a) attending to current needs without compromising future generations' needs, (b) managing economic growth and population, and (c) sustaining the social and economic development that the state has been experiencing while protecting the environment.

The environmental development pillar of the Vision is concerned with environmental management and addressing the aforesaid three challenges along with challenges that are specific to Qatar's ecosystems, which are (a) the impact of climate change on sea levels, (b) the impacts of declining water and hydrocarbon resources, (c) pollution, and (d) environmental degradation.

QNV 2030 also identifies the means that need to be established for environmental sustainability, which include (a) an environmentally aware population, (b) a comprehensive legal system, (c) effective environmental institutions, (d) a comprehensive sustainable urban development planning, and (e) regional and international cooperation.

While the aforementioned guiding environmental principles and considerations in QNV 2030 are 'nice-sounding,' they appear to be too general (Al-Kuwari, 2012) and could apply in many temporal and spatial contexts, which is an issue that perhaps partially originates from the definition of sustainable development itself and any policy planning that concerns it due to its complexity and ambiguity at the interpretation (i.e., defining actions) and application levels (Karki, 2004). QNV 2030 does not provide a defined starting point for the environmental challenges it has highlighted (e.g., statistics on the status of water and hydrocarbon resources, level of pollution), nor does it give a defined desired 'destination'; it leaves both to the next layer of policy guiding documents which is the National Strategy.

2.2.2 National Development Strategy

Guided by QNV 2030, two National Development Strategies (NDS) were launched (NDS-1 in 2011 and NDS-2 in 2018) to define and prioritize the national development goals along with time-bound targets for a period of five years. The NDS intends to provide a medium-term structure for various sectoral and enterprise strategies to ensure that they are coordinated and aligned. Moreover, NDS is supposed to form the basis for the national policy framework and be complemented further by the sectoral strategies.

In its attempt to provide an integrated approach, NDS-1—as well as NDS-2—sought to integrate environmental sustainability in the various sectoral strategies wherever possible. For instance, throughout the implementation of the Strategy the healthcare sector was intended to be enabled by targets set under environmental sustainability—e.g., the latter's target to raise the public awareness and encourage healthier living and working environments—and was also intended to act as an enabler for environmental sustainability, within the healthcare sector's target to establish monitoring and health standards. Similar notes of intended alignments among the sectoral targets were indicated throughout the Strategy document. Regardless of the actual outcomes of NDS-1, such cross-sectoral planning concerning the environment is recent to the state's policymaking and has developed gradually following the launch of QNV 2030.

NDS-1 specified four themes: water, air quality, waste management, and biodiversity as the environmental development priorities for the period from 2011 to 2016 and identified a set of targets for each. While the Strategy document provided some insights on the drivers of degradation in these four areas and the risks of such, it missed providing adequate baseline statistics on their status at the time of the Strategy's launch. On one hand, statistics is a prerequisite for the formulation of 'meaningful' environmental policy. On the other, it is essential throughout the monitoring and progress assessment stages (Rao, 1983). The absence or inadequacy of statistics at the planning stage raises concerns about the basis upon which the environmental targets (detailed in the following paragraph) were determined in NDS-1 and also about the ability to conduct progress assessment activities throughout the implementation. Also, the appropriateness of the identified environmental policies in Qatar's context is in question considering that they seem to have been developed somewhat 'blindly' due to the lack of data about many of the targeted respects.

Fourteen environmental targets were identified in NDS-1. In terms of water preservation, a comprehensive National Water Act was to be developed and enacted. For air quality, a national policy was to be developed and launched to manage air pollution, greenhouse gas (GHG) emissions, and climate change-related challenges. The specified targets toward this end included eliminating instances of excess ozone levels and halving gas flaring volumes that result from energy production. For waste management, a multidimensional strategy was to be developed and initiated to reduce the waste levels that are generated by households and various commercial and industrial activities and increase the volumes of recycled waste. A target that was identified here was to keep the generated domestic waste at 1.6 kg per capita per day. For biodiversity, a central database that conducts detailed surveys on Qatar's biodiversity and provides baseline data to inform policymaking was to be established and operated. The other target set was to expand the nature reserves and conservation areas. With regard to environmental management, the plan was to develop urban planning to include a green dimension. The first planned action was to create a network of tree-lined green areas in Doha instead of large, open parks. The second was to carry out a nationwide campaign to raise the population's awareness of Qatar's ecological endowment and their role in supporting environmental sustainability. Other activities that were highlighted by the Strategy in relation to environmental sustainability were building more regional and international partnerships and establishing environmental performance reporting requirements for the industrial sector.

The environmental sustainability medium-term targets set in NDS-1 indeed were ambitious. By 2016, however, only two targets were fully met—according to NDS-2. Efforts were successful in reducing gas flaring that results from energy production by around 30 percent. Also, the target to maintain domestic waste generation per capita per day at the pre-Strategy level was met and even reduced to 1.3 kg. In terms of the factors underpinning the failure to meet remaining targets, NDS-2 highlighted that they relate to policy planning and implementation gaps, such as unclear prioritization of targets, poor coordination at sectoral and cross-sectoral levels, and lack of planning and implementation capabilities at the institutions' level.

NDS-2 was launched in 2018, with targets to be met by the year 2022. The second Strategy's identified concerns include the urban expansion's pressure on ecological systems, waste, air pollution, surface groundwater levels, and the lack of sanitation network availability in all residential areas and its stress on the environment. NDS-2, too, attempts to approach environmental sustainability in an integrated manner. For instance, economic infrastructure sector targets include environmental dimensions. One of the set targets in this regard is to increase the proportion of the recycled materials that are used in infrastructural projects. Such targets, if met, are meant to contribute to both economic and environmental development.

A total of eight targets specific to environmental development were identified in NDS-2, (a) reducing levels of air pollutants, (b) improving coastal and marine water quality, (c) keeping the generated domestic waste under the rate of 1.6 kg, (d) creating a green belt around Doha, (e) raising public awareness on biodiversity status and establishing a national database about the status of ecological systems and their inhabitants, (f) managing nature reserves and ecosystems in a sustainable manner, (g) establishing a database on Qatari environment within the Ministry of Environment, and (h) building an environmentally aware and supportive society. The programs or projects that were identified to assist in achieving these eight targets revolve around establishing sub-sectoral plans (e.g., air quality management plan, coastal and marine water quality control plan, waste management plan, climate change mitigation and adaptation plan, nature reserves management plan), creating databases (e.g., national biodiversity database), and carrying out awareness campaigns.

It is worth noting also that the targets that are identified under other pillars of NDS-2 (e.g., sustainable economic prosperity) are intended to, directly and indirectly, contribute toward environmental sustainability. One of the main targets that are identified for economic infrastructure development, for instance, is to increase the power production capacities through increasing the solar energy share in the energy mix to meet the increasing demand for electricity. Deployment of renewable energy would contribute—along with relevant programs and projects—to national targets regarding air quality and GHG emissions. So is the case for the target to increase the proportion of recycled materials that are used in public and private projects. Such a target, if met, would contribute to the state's targets on waste management and the overall goal of environmental preservation. This complementary approach, again, is a unique feature of the post-QNV 2030 policy planning.

Unlike NDS-1, no mid-term review report was published for NDS-2. As of the first quarter of 2021, it remained unclear whether NDS-2 has been successful—compared to the first at least—in meeting the environmental development targets. Assessing and reporting progress has been lagging behind throughout the duration of both strategies. Moreover, it seems that the respective governmental entities are more active in sharing the progress toward sustainability with external, international bodies, while locally, they are less active. Three national voluntary reviews on the implementation of the 2030 Agenda for Sustainable Development—in 2017, 2018, and 2021—have been prepared and submitted. While such reporting is necessary, similar efforts at the national level are important as well with regard to progress toward NDS-2 targets.

It is worth mentioning that the development of the third NDS (or NDS-3) for the period 2023 to 2027 is underway following the Prime Minister's Resolution No. 1 of 2022 on establishing the National Steering Committee, which will supervise the preparation of the Strategy document and follow up on its implementation.

The fourth quarter of the year 2021 has witnessed the unveiling of the first National Environment and Climate Change Strategy 2021 to 2030 and the National Climate Change Plan 2030 (see Sect. 2.2.3 and 2.2.4). Both documents are supposedly significant to environmental sustainability policy in Qatar since they intend to pave the way toward meeting the state's targets of not only QNV 2030 but also those that it has committed to accomplishing at the international level (i.e., the 2030 Agenda for Sustainable Development and the Paris Agreement). It is worth mentioning as well that, unlike the NDS and relevant sectoral strategy 5-years duration, the timeframe that has been set in both policy documents extends from 2021 to 2030. What is unique also about both policy documents is that they are similar to the NDS in terms of level and scope. Although the Ministry of Environment and Climate Change (formerly the Ministry of Municipality and Environment) is the principal owner of the two policy documents, numerous entities (governmental and non-governmental) are responsible for carrying out the necessary measures and programs to meet the identified targets.

This perhaps would promote the Ministry's supervisory and advisory role after a period of slight passiveness pre the recent government reshuffle, through which the Ministry seemed to be more active and focused on municipal affairs (Al-Kuwari & Rahaman, 2018).

2.2.3 National Environment and Climate Change Strategy

The National Environment and Climate Change Strategy (QNE) was announced in 2021. The Strategy has identified five priority areas for action (a) GHG emissions and air quality, (b) biodiversity, (c) water, (d) circular economy and waste management, and (e) land use, and several objectives were set toward each area. The objectives are aligned with those of QNV 2030 and the NDS, but perhaps are more defined.

The QNE targets toward GHG emissions and air quality are (a) GHG emissions to be reduced by 25 percent by 2030, (b) ambient air quality standards to be enhanced by 2024, (c) indoor air quality standards to be enhanced, too, by 2024 in accordance with World Health Organization guidelines, (d) national network for air quality monitoring to be established by 2023, and (e) a national air emissions inventory to be established by 2023. The pathway toward QNE's GHG emissions and air quality targets include increasing carbon capture and storage application, increasing renewable energy capacity, and investing in blue carbon technologies.

The targets toward biodiversity are (a) nature reserve areas' proportion to be increased to 25 percent and effectively managed in line with the post-2020 Global Biodiversity Framework, (b) endangered species to be protected and recovered, (c) fishing activities to follow sustainable practices, (d) biodiversity awareness, participation, and capabilities to be enhanced, and (e) biodiversity conservation to be incorporated national and institutional planning.

For Water, the targets identified are (a) all water resources to be effectively monitored, (b) reverse osmosis (RO) and sustainable technologies to be used to produce more than 55% of desalinated water, (c) groundwater extraction to be reduced by 60%, (d) total water network loss to be maintained under 8% and real loss under 5%, (e) per capita household water consumption to be reduced by 33% from 2019 levels, and (f) 100% of recycled water to be reused.

Targets for circular economy and waste management include (a) 15% of municipal waste to be recycled, (b) 100% of unsanitary landfills to be closed and rehabilitated, (c) recycled materials to account for 35% of used materials in construction projects, (d) circular public procurement in public infrastructure to be achieved at 30%, (e) per capita food waste to be reduced by 50% at consumer and retail levels, and (f) circular industry principles to be applied across industrial areas.

In terms of land use, targets are (a) farmland productivity to be improved by 50% compared to 2019 levels, (b) water consumption of crops produced to be improved by 40% in comparison to 2019 levels, (c) agrochemicals use to be monitored and reduced, (d) soil quality to be monitored and improved, (e) public transport modal share to be increased to 16%, (f) open space and recreation strategy to be developed by 2022, and, lastly, (g) green building requirements to be established and mandated for new development projects.

The Strategy document does not indicate specific programs or initiatives, nor does it list out key stakeholders or actors. It does, however, outline the enablers of achieving the QNE targets, which include (a) establishing a committee at the Cabinet-level to review and follow up on the implementation process, (b) establishing a supporting team from various ministries to assist in monitoring and following up processes, (c) increasing the cooperation with the international and scientific communities, (d) deployment of state-of-art technologies for environmental assessment and monitoring, (e) providing financial resources, (f) developing necessary legislative frameworks, (g) building human capacities and national talents, and (h) raising public awareness. Moreover, it does specify baselines for some of the identified targets (e.g., 2019 levels for water and land use targets), which would allow assessing the progress toward them throughout the implementation. Such specifications were missing in former policy documents.

2.2.4 National Climate Change Action Plan

The National Climate Change Action Plan (NCCAP), too, was approved and launched in 2021. The NCCAP policy document intends to combine climate change implemented and planned adaptation and mitigation measures 'under a single framework' (p. 10). The most significant goal of NCCAP is to reduce GHG emissions by 25%, which is aligned with the QNE target. On the mitigation side, key sectors for action include (a) oil and gas, (b) power and water, (c) transportation, and (d) building, construction, and industry. On the adaptation side, sectors for action include (a) economy, (b) infrastructure, (c) water management, (d) health care, (e) biodiversity, and (f) food security.

Mitigation measures in oil and gas sectors involve (a) deployment of carbon capture and storage technologies, (b) reducing methane emissions, (c) establishing flare reduction projects, and (d) establishing energy efficacy programs. Mitigation measures in the power and water sector include (a) establishing water conservation regulations, (b) enhancing energy conservation in buildings, (c) expanding renewable energy capacity, and (d) enhancing energy and water production efficiency. Mitigation measures in the transportation sector involve (a) increasing the use of public transportation, (b) improving shipping and aviation sectors, (c) improving efficiency standards for vehicles, and (d) electrification of cars and buses. Mitigation measures in the building, construction, and industry sectors include (a) establishing green building standards and (b) increasing the recycling of construction and other wastes.

Adaptation measures as per the NCCAP document are more general under some sectors and more defined under others. For instance, adaptation measures involve establishing a diversified, knowledge-based and resilient economy and resilient infrastructure. For water management, adaptation measures include reducing water consumption and increasing the use of recycled water. For healthcare, adaptation measures include conducting studies to understand climate change's impacts on public health and raising awareness of them. Adaptation measures in biodiversity include improving scientific knowledge of the state's biodiversity, conserving biodiversity, raising awareness, and building necessary capacities. Lastly, adaptation measures in food security involve diversifying trade partners, increasing self-sufficiency in perishable commodities, establishing strategic reserves for storable commodities, and establishing an efficient local food supply chain.

Unlike QNE, NCCAP lists out relevant stakeholders and the key initiatives they are to carry. Enablers as per the Action Plan document include (a) community awareness, (b) education and human capital, (c) technology and research, and (d) incentives and regulations.

2.2.5 *Sectoral and Sub-sectoral Strategies*

The Environmental Sustainability Strategy (ESS) was launched by the Ministry of Municipality and Environment in parallel with the NDS-1 and NDS-2 for the periods 2011 to 2016 and 2018 to 2022. The second ESS indicates the specific programs and projects that are supposed to assist in achieving the environmental development targets of the NDS along with the responsible department. The programs outlined in ESS include (a) establishing a waste management manual, (b) developing and implementing a management plan for solid waste, (c) raising awareness of environmental protection, (d) developing and implementing a climate change mitigation and adaptation plan in addition to stimulating strategic partnerships at various levels, (e) establishing nature reserves management plan, and (f) creating a database within the Ministry. The aforementioned programs are linked to NDS-2 time target, meaning they should be completed by 2022. This is highlighted because the second ESS also lists many ambitious strategic initiatives of the Environmental Affairs Department that would contribute to the targets of NDS-2, but it is not clear if they are intended to be met by 2022 as well. This perhaps suggests that the issue of unclear prioritization of targets that has been faced once before with NDS-1 might reoccur. It could also highlight the planning challenges faced at the institutional level.

On another note, while NDS-2 does not refer to Qatar's Intended Nationally Determined Contributions (INDCs), even though the state has submitted them in 2015 and has ratified Paris Agreement in June 2017, the ESS indicates that one of the strategic initiatives of the Ministry's Environmental Affairs Department is to develop and implement INDCs' plan. It is worth mentioning also that the elements of the ESS 2011 to 2016 differ from those of ESS 2018 to 2022. The water sector is not included in ESS 2018 to 2022 since the water-related activities and projects were transferred to the Ministry of Energy and Industry, which has been dissolved in November 2018 following the government reshuffle.

At the sub-sector level, the strategy of the Ministry of Municipality's urban planning and urban development sector has identified the threats of climate change (e.g., sea-level rise and flooding, increasing temperatures, and harm to biodiversity) and a number of initiatives to develop the sector's mitigation and adaptation capacities against each threat (see Chap. 9). They are included in a dedicated policy document established for the sector (Climate Change Strategy for Urban Planning and Urban Development Sector in the State of Qatar, launched in 2018). In addition to QNV 2030, the policy document and the whole urban planning sector are supposedly guided by Qatar National Master Plan or Qatar National Development Framework 2032 (QNDF). QNDF 2032 is a long-term spatial development strategy that is intended to act as the guiding document for urban planning and urban development in the state's 8 administrative municipalities. The policies for sustainable urban planning and development seem to be comprehensively and thoroughly planned and articulated. Yet, the owner principal of the policies, i.e., the Ministry of Municipality, seems to lack control in terms of implementing them, whereas semi-governmental entities, such as Qatar Railways Company, have much influence on urban development (Al-Thani, 2019).

Also, Qatar National Research Strategy (QNRS) has specified desalination, wastewater reuse, and deployment of solar energy among the major challenges related to the environment and energy that need to be prioritized on the R&D agenda in the country. QNRS is supposed to provide the framework to drive Qatar Foundation R&D's program plans toward addressing critical national priorities in line with QNV 2030. Therefore, the aforementioned challenges have been assigned to the relevant research institute, which is Qatar Environment and Energy Research Institute (QEERI) (see Chap. 8).

2.2.6 *Institutional Level Strategies*

While the recent practice for strategies at the national and sectoral levels has been, to some extent, consistent in which policy documents are prepared and launched within a specific timeframe, at the institutions and agencies level, it is unclear whether such consistency exists. This can be understood since such documents are usually developed for internal use and not shared or accessible for public.

For the Ministry of Environment and Climate Change (formerly the Ministry of Municipality and Environment), the various departments' strategic initiatives are highlighted in the previously mentioned sector strategy, the second ESS. However, the strategy document does not indicate whether those initiatives are to be accomplished by 2022—in parallel with NDS-2.

Though the development process of environmental strategies at various levels (i.e., national, sectoral, institutional) has come a long way in Qatar, the development itself is not the determining success factor. Implementation constraints remain in question, and whether there is an adequate institutional and technical capacity (see Sect. 2.4) to achieve the desired outcomes outlined in these strategies (although recognized by QNV, NDS a challenge that has been faced from the very beginning). Stakeholders' engagement in the development process of the strategies is also in question, and whether a participatory approach has been followed throughout the preparation and drafting processes. QNV 2030 has been criticized as being a top-down approach (e.g., Al-Kuwari, 2012). This could reflect the level of endorsement and, hence, the ownership of the strategies by the respective entities, and the implications of the level of endorsement on the implementation process eventually.

2.3 Laws and Regulations

One of the unique features of the Permanent Constitution of Qatar is that it recognizes the state's obligation toward preserving the environment and pursuing a sustainable development approach, as stated in Article-No 33 of the Constitution. However, many environmental laws and regulations were issued long before the Constitution was decreed in June 2004. One of the very early laws to protect the environment was Law No 8 of 1974 on public or general cleanliness, which generally prohibited depositing and abandoning waste in any public space and provided a guideline on handling and managing waste. The law was active for around four decades and was only recently repealed by Law No 18 of 2017, which indicates similar principles.

Marine biodiversity was one of the early environmental themes to be addressed by the law. Law No 4 of 1983 was issued to regulate the exploitation and protection of living aquatic resources. Among several purposes, the Law aims to (a) protect marine biodiversity from harmful practices and overfishing and (b) detect and prevent the use of harmful substances that could impact the living aquatic resources' growth, reproduction, and migration. The issuance of the law came only a few years after Qatar's accession to the Kuwait Regional Convention for Cooperation on the Protection of the Marine Environment from Pollution in 1978, its Action Plan, and its subsequent protocols, namely the Protocol concerning Regional Cooperation in Combating Pollution by Oil and Other Harmful Substances in Cases of Emergency in 1982.

Law No 32 on preventing damage to plant ecology and ecosystems was issued in 1995. The main purpose of the law is to regulate grazing areas and seasons and indicate the prohibited practices that have an adverse impact on plants' ecology and ecosystems.

The 2000s, however, witnessed the issuance of several important laws concerning environment protection and management. This could be related to the formation of the Supreme Council of Environment and Natural Reserves in the year 2000, which has been active in proposing policies and legislation, and the accession to several regional and international organizations and agreements throughout the 1990s.

The key legislation on environment protection is Law No. 30 issued in 2002 with the aim to (a) protect and preserve the quality and balance of the environment, (b) combat all forms and sources of pollution and avoid its impacts, and (c) develop natural resources and conserve biodiversity to ensure the benefit of current and future generations. The executive bylaw for the law was issued in 2005 and covers the environment and sustainable development, development projects' environmental impact, environmental disasters' emergency plans, waste and hazardous materials, air pollution, water pollution, and marine environment protection.

Perhaps the most significant law on conserving wildlife and natural habitats is Law No 19 of 2004, which revolves around prohibiting activities that harm ecological systems and their inhabitants and undertaking necessary arrangements to rehabilitate and manage natural habitats and preserve endangered species. Several nature reserves were established following the issuance of the law, such as Lusail Reserve in 2005, Al Mashabiya, Al Eraiq, and Al Thakhira Reserves in 2006, Khor Al Adaid Reserve in 2007, Southern Area Reserve in 2018, and Al Rafaa Reserve in 2020. Prior to the 2004's Law, several laws were in force on conserving wildlife. Law No 4 of 2002 was issued to regulate hunting activities of wildlife animals, including mammals, birds, and reptiles. Moreover, hunting seasons for certain birds and wild animals are also defined by ministerial decrees—seasons were specified for the years 2002, 2003, 2006, 2007, 2008, and 2009.

Law No 5 of 2006 was issued to regulate trade activities of endangered wildlife—fauna and flora—and their products. Also, a National Biosafety Committee was established by Ministerial Decree No. 11 of 2007 to (a) propose policies, regulations, and executive bylaws for biosafety, (b) establish guidelines and procedures for licensing importing of genetically modified organisms and their products, and several other related responsibilities. The Minister of Environment issued Resolution No 37 of 2010 on conserving turtles and seabirds from extinction. The Resolution prohibits approaching, disturbing, or poaching the nests of both species in the area of Fuwairit Beach during the breeding season—from April to July.

In terms of laws to address air pollution, there is Law No 19 of 2015 on issuing the Unified Law of the Gulf Cooperation Council for the Arab States on the Control of Substances that Deplete the Ozone Layer. Also, Ministerial Decree No 310 of 2020 has been issued for establishing and operating a national air quality monitoring network, which intends to contribute to solving the persistent issue of lack of data and monitoring tools for air pollution in Qatar.

The level of these laws' enforcement, however, remains unclear considering the previously mentioned institutional capacity of respective government entities and their actual ability to make necessary arrangements to ensure environmental laws' application.

2.4 Institutions

Over the years, the number of government agencies that are concerned with environmental affairs increased, and their responsibilities—at least theoretically—expanded. Long before the establishment of a dedicated ministerial body, the first entity formed to look after environmental affairs was the Standing (or Permanent) Committee for Environmental Protection in 1981. The Committee was attached to the Prime Minister's Office and functioned within the governmental structure that was governed by the Amended Provisional Constitution. The Committee consisted of representatives from several government sectors and held its meetings in the Ministry of Public Health. Its responsibilities included (a) proposing policies for environmental protection along with implementation plans, (b) drafting environmental protection legislation and regulations and following up on their implementation, (c) acting as a coordination agency between various bodies concerned with environmental protection and research laboratories and centers, evaluate their activities, and provide recommendations, (d) monitoring activities of various sectors in areas of provision of data, measurements, and analyses, carrying out studies on ecological conditions, sources of pollution, ways to address them, evaluating those studies and provide recommendations and funding accordingly, (e) verifying the availability and adequacy of monitoring, measuring and control systems, (f) raising awareness of the environment through educational programs and media campaigns, and (g) representing the state at bodies and organizations and regional and international meetings. Another supervisory responsibility was added in 1986, by which the Committee was to review development plans and projects before their implementation. The Committee's approval was to be sought for megaprojects, whether those of the public or private sector. After nearly a decade and a half under the Prime Minister's Office, the Committee was attached to the Ministry of Municipal Affairs and Agriculture in 1994.

In 1990, the Ministry of Municipal Affairs and Agriculture was also assigned to contribute with the respective bodies in the state to protect the environment and address the issue of pollution in all its forms and sources and eliminate its effects. Within the Ministry, a Department of Environmental Protection was established and was to (a) propose and implement environmental protection programs, (b) contribute to assessing environmental analyses conducted for public and private projects, (c) monitor and follow up on pollution incidents, and (d) contribute to the environmental awareness educational programs and media campaigns in coordination with concerned entities. When the Standing Committee for Environmental Protection became under the Ministry in 1994, the Department was also assumed to act as the technical secretariat of the Committee to assist it in carrying out its responsibilities.

Both the Committee and the Department kept functioning under the Ministry of Municipal Affairs and Agriculture until the year 2000, which marks a significant year for the government structure. The law establishing the Standing Committee for Environmental Protection was repealed by Law No 11 of 2000, establishing the Supreme Council of Environment and Natural Reserves (SCENR) to undertake the same role but with a larger scope. The SCENR responsibilities included (a) establishing general policies for environmental protection, achieving sustainable development, and conserving endangered wildlife and their natural habitats, (b) developing action plans that are necessary to carry out these general policies and supervising their implementation, (c) monitoring all related activities to the aforementioned aspects and evaluating their outcomes, (d) preparing related legislation and regulations and following up the enaction, (e) creating a national database and a reference laboratory for the Qatari environment, (f) evaluating the studies concerned with preserving the environment when planning development projects and providing an opinion about their impact, (g) identifying the problems that are caused by pollution and environment degradation and cooperate with respective government agencies to address them, (h) representing the state at organizations and regional and international meetings, (i) developing plans to improving human capacities with regards to ways and means to protect the environment and natural reserves, and (j) raising public awareness about environment through educational programs and media campaigns.

Two committees were established under the umbrella of the SCENR in 2007 and later under the Ministry of Environment following the government restructure in 2008. The first was the Climate Change Committee, and the second was the Clean Development Committee (following the ratification of the Kyoto Protocol in 2005). Both committees were merged in 2011 to form the Climate Change and Clean Development Committee. The established Ministry of Environment's responsibilities did not differ much from those of the SCENR following the government restructuring. Throughout the 2010s, The Ministry of Environment has undergone several restructurings in 2014, 2016, and 2019 in which it has been merged with the Ministry of Municipality to form the Ministry of Municipality and Environment (MME).

Referring back to the Climate Change and Clean Development Committee and its structure, the Committee is chaired by the Assistant Undersecretary for Environmental Affairs, and it is constituted of representatives from MME, Ministry of Foreign Affairs (the Committee Vice-Chair), Ministry of Commerce, Qatar Petroleum (currently known as Qatar Energy), Civil Aviation Authority, and Qatar University. It is worth noting that the Ministry of Energy and Industry (established at the time) was not represented in this committee, even though this Committee's scope covers the energy and industrial sectors. The Committee's responsibilities include (a) suggesting national policies and action plans related to reducing GHG emissions, (b) ensuring that the government and non-government entities implement UNFCCC and Kyoto Protocol obligations and develop the required reports and studies, (c) developing databases based on UNFCCC and Kyoto Protocol requirements, (d) suggesting a strategy for Clean Development Mechanism (CDM), (e) spreading awareness and knowledge on Clean Development (CD) projects and their goals, (f) engaging in the activities related to the Committee's scope of work at local, regional, and international levels, and several other activities for the Committee to carry.

A Climate Change Department was established under the Environmental Affairs Department within MME. Article-No. (29) of Emiri Decree No. (11) of 2019 on the organizational structure of MME states that the Climate Change Department's responsibilities include (a) suggesting policies reduce the emissions causing climate change and following up their implementation, (b) suggesting policies and programs to encourage the deployment of renewable energy resources, clean production projects, and sustainable consumption, (c) reviewing and approving CD and renewable energy projects and programs to adapt to climate change, (d) preparing studies and reports on climate change in cooperation with respective entities, (e) monitoring the implementation of international conventions that Qatar has joined and highlighting the efforts made regarding climate change at the international level, and (f) collaborating with international organizations and benefit from their programs, activities, and support under the international conventions concerned with climate change, and several other activities for the Department to take care of.

The Climate Change and Clean Development Committee and the Climate Change Department's responsibilities seem to overlap. For example, both are responsible for proposing national policies for emissions reduction, ensuring implementation as per the commitment to international conventions, promoting awareness of climate change, and both bodies are expected to collaborate with regional and international organizations. It is unclear whether the Department is acting as the technical secretariat of the Committee or if the latter works in parallel with the former. Such overlapping reflects a lack of collaboration management and might lead to duplicated efforts, inefficient use of time and resources, and policy incoherence.

Furthermore, the current framework seems blurry regarding overseeing current projects and plans and ensuring that they are aligned with the targets and objectives of NDS, QNV 2030, and other multilateral agreements Qatar has ratified. For instance, the responsibilities of the Standing Committee of Water Resources, which was established under the Council of Ministers by Emiri Decision No. (19) of 2011, include (a) suggesting policies related to water resources in line with the overall development plans and in consideration for sustainability and environment protection, (b) implementing projects and programs related to managing and developing water resources, (c) prioritizing water resources development projects and developing proposals of their action plans, (d) following up on the implementation of projects, programs, and studies related to water resources, (e) suggesting appropriate resolutions for issues of wasteful use of water, and (f) proposing amendments to water resources protection, management, and development laws. The Climate Change and Clean Development Committee and the Standing Committee of Water Resources seem to undertake a supervisory and advisory role over specific development projects and plans, which also creates some sort of overlap of efforts between the two, especially since some development projects might fall within the scope of both committees, such as decarbonizing water desalination. It is essential to define each committee's responsibilities over the other and put in place cooperation or coordination mechanisms. The structure and how these committees, along with related government institutions, should interact under the Council of Ministers should also be clarified. Those committees are supposedly part of the same network concerning subjects related to environmental sustainability.

In 2021, the council of ministers was reshuffled one more time, and the Ministry of Environment and Climate Change (MOECC) was established. The challenge of climate change has now become a key aspect of the Ministry's responsibilities and activities, whether in terms of policies that it shall propose and implement, the necessary studies that it shall conduct, monitoring, and so on. It is worth noting that, as of the 2022s quarter, the ministry structure is yet to be announced and confirmed through a decree-law. Government restructuring is a means that is usually used to cope with fiscal stresses, especially during global financial crises and economic recessions. The most significant government restructures in Qatar were made a mid or shortly after such crises, whether in 2009 or 2014. The MOECC has indeed undergone several organizational restructures in which it has either been merged with the

Ministry of Municipality or segregated. The question here would be whether such reoccurring organizational restructuring could have limited the Ministry's ability to plan and implement the strategies' targets. Moreover, a concern also arises on where and how certain environment-related matters would fall following the recent government reshuffle and the establishment of a dedicated ministerial body for environmental affairs and the implications of their assignment, whether under MOECC or a different governmental body. Consider waste management as an example. On the one hand, waste collection and handling are the responsibility of the local municipalities under the Ministry of Municipality. On the other, waste management is significant for environmental protection and accordingly within the scope of MOECC (see Chap. 16).

2.5 Conclusion

Although the progress toward environmental sustainability targets so far remains modest, the policy has evolved gradually in terms of scope and level. On the planning side of the policy, several strategies and action plans—e.g., NDS, QNE, NCCAP, ESS, and NBSAP—are in place with ambitious targets. However, it is the implementation side of the policy that is challenging, and whether the relevant institutions have the capacity to carry the strategies through. The same can be said about the laws and the regulations that have been issued and to what extent, really, they are in force. The situation seems ambiguous also due to the lack of assessment and progress reporting activities—at least those that are accessible by the public. Future efforts perhaps should focus on the microlevels of the strategies and on developing relevant entities' institutional and technical capacities. Also, attention should be paid to the evaluation and assessment phases of the policymaking process, as they should feed into the planning phase to avoid rearticulating targets that might be beyond reach. Moreover, addressing the persistent issue of limited data and information about ecological systems and their inhabitants is essential to alleviate the ambiguity of the strategies' actual outcomes.

References

Al-Kuwari, A. (2012). The visions and strategies of the GCC countries from the perspective of reforms: The case of Qatar. *Contemporary Arab Affairs, 5*(1), 86–106.

Al-Kuwari, M., and Rahaman, K. (2018). Climate change strategy for the urban planning and development sector in Qatar. *International Society of City and Regional Planners*. https://isocarp.org/app/uploads/2021/08/ISOCARP_2018_Al-kuwari_40.pdf

Al-Thani, S. (2019). Neighborhood sustainability and livability in Qatar national development framework 2032. *International Journal of Research in Environment Studies, 6*, 47–56.

Article No. 33 of The Permanent Constitution of the State of Qatar. (2004). Retrieved from https://www.almeezan.qa/LawView.aspx?opt&LawID=2284&language=en#Section_6674

Cabinet Decision No. 4 of 1986 on Presenting Development Plans and Projects to the Standing Committee for Environmental Protection. (1986). Retrieved from https://almeezan.qa/LawView.aspx?opt&LawID=1654&language=ar

Cabinet Decision No. 15 of 2011 Establishing climate change and clean development committee and determining its functions. (2011). Retrieved from https://www.almeezan.qa/LawView.aspx?opt&LawID=3172l&language=en

Decree-Law No. 11 of 2000 Establishing the Supreme Council of Environment and Natural Reserves. (2000). Retrieved from https://almeezan.qa/LawView.aspx?opt&LawID=462&language=en

Decree-Law No. 20 of 1993 Organising the Ministry of Municipal Affairs and Agriculture and Assigning Competencies Thereof. (1993). Retrieved from https://almeezan.qa/LawPage.aspx?id=526&language=en

Decree-Law No. 24 of 1990 Organising the Ministry of Municipal Affairs and Agriculture and Assigning Competencies Thereof. (1990). Retrieved from https://almeezan.qa/LawPage.aspx?id=2703&language=en

Emiri Decision No.1 of 2008 Reshuffling the Council of Ministers. (2008). Retrieved from https://almeezan.qa/LawView.aspx?opt&LawID=4003&language=ar

Emiri Decision No. 4 of 1989 Establishing the Supreme Council for Planning. (1989). Retrieved from https://almeezan.qa/LawView.aspx?opt&LawID=4130&language=en

Emiri Decision No.4 of 2021 Reshuffling the Council of Ministers. (2021). Retrieved from https://almeezan.qa/LawView.aspx?opt&LawID=8762&language=ar

Emiri Decision No.16 of 2009 on Setting the Competencies of the Ministries. (2009). Retrieved from https://almeezan.qa/LawView.aspx?opt&LawID=2752&language=en

Emiri Decision No. 19 of 2011 on Establishing the Standing Committee of Water Resources. (2011). Retrieved from https://www.almeezan.qa/LawView.aspx?opt&LawID=2759&TYPE=PRINT&language=en

Emiri Decision No.57 of 2021 on Setting the Competencies of the Ministries. (2021). Retrieved from https://almeezan.qa/LawView.aspx?opt&LawID=8766&language=ar

Emiri Decree No. 11 of 2019 on the Organisational Structure of the Ministry of Municipality and Environment. (2019). Retrieved from https://www.almeezan.qa/LawView.aspx?opt&LawID=7932&language=ar

General Secretariat for Development Planning. (2008). *Qatar national vision 2030.* Doha. Retrieved from https://www.psa.gov.qa/en/qnv1/Documents/QNV2030_English_v2.pdf

General Secretariat for Development Planning. (2011). *Qatar national development strategy 2011–2016: Towards Qatar national vision 2030.* Doha. Retrieved from https://www.psa.gov.qa/en/knowledge/Documents/Qatar_NDS_reprint_complete_lowres_16May.pdf

Government Communications Office. (2021). *Environment and Sustainability.* https://www.gco.gov.qa/en/focus/environment-and-sustainability/

Karki, N. (2004). *The manifestation of sustainable development and issue of sustainability in development discourse: A comparative study of ideological and practical perspectives* [Capstone Project, School of International Training]. https://digitalcollections.sit.edu/capstones/815

Kuwait Regional Action Plan for the Protection and Development of the Marine Environment and the Coastal Areas of Bahrain, Iran, Iraq, Kuwait, Oman, Qatar, Saudi Arabia, and the United Arab Emirates, April 15–23, 1978, http://www.ropme.org/Uploads/Events/EBM/11-KAP.pdf

Kuwait Regional Convention for Co-operation on the Protection of the Marine Environment from Pollution, April 24, 1978, http://www2.ecolex.org/server2neu.php/libcat/docs/TRE/Full/En/TRE-000537.txt

Law No. 4 of 1981 for the Establishment of the Standing Committee for Environmental Protection. (1981). Retrieved from https://almeezan.qa/LawView.aspx?opt&LawID=323&language=en

Law No. 4 of 1983 on the Exploitation and Conservation of Living Aquatic Resources in Qatar. (1983). Retrieved from https://www.almeezan.qa/LawPage.aspx?id=3981&language=en

Law No. 4 of 2002a Regulation of the Hunting of Wild Animals including Mammals, Birds, and Reptile. (2002). Retrieved from https://www.almeezan.qa/LawPage.aspx?id=325&language=en

Law No. 5 of 2006 on the Regulation of Trade in Endangered Wildlife Fauna and Flora and their Products. (2006). Retrieved from https://www.almeezan.qa/LawPage.aspx?id=2619&language=en

Law No. 8 of 1974 on General Cleanliness. (1974). Retrieved from https://www.almeezan.qa/LawView.aspx?opt&LawID=4099&language=en

Law No. 14 of 1994 on Amending some Provisions of Law No. 4 of 1981 for the Establishment of the Standing Committee for Environmental Protection. (1994). Retrieved from https://almeezan.qa/LawPage.aspx?id=65&language=en

Law No. 18 of 2017 on General Cleanliness. (2017). Retrieved from https://www.almeezan.qa/LawView.aspx?opt&LawID=7478&language=ar

Law No. 19 of 2004b on Conservation of Wildlife and their Natural Habitat. (2004). Retrieved from https://www.almeezan.qa/LawView.aspx?opt&LawID=141&language=en

Law No. 19 of 2015 on Issuing the Unified Law of the Gulf Cooperation Council for the Arab States on the Control of Substances that Deplete the Ozone Layer. (2015). Retrieved from https://www.almeezan.qa/LawPage.aspx?id=6750&language=ar

Law No. 30 of 2002b on Promulgating the Law of the Environment Protection. (2002). Retrieved from https://www.almeezan.qa/LawPage.aspx?id=4114&language=en

Law No. 32 of 1995 on Preventing Damage to Plant Ecology and Ecosystems. (1995). Retrieved from https://www.almeezan.qa/LawPage.aspx?id=265&language=en

Ministerial Decree No. 9 of 2007a Establishing the Climate Change Committee. (2007). Retrieved from https://almeezan.qa/LawView.aspx?opt&LawID=3936&language=en

Ministerial Decree No. 10 of 2007b Establishing the Clean Development Committee. (2007). Retrieved from https://almeezan.qa/LawView.aspx?opt&LawID=3494&language=en

Ministerial Decree No. 11 of 2007c Establishing the National Biosafety Committee. (2007). Retrieved from https://www.almeezan.qa/LawPage.aspx?id=3511&language=en

Ministerial Decree No. 310 of 2020 on Air Quality. (2020). Retrieved from https://www.almeezan.qa/LawPage.aspx?id=8523&language=ar

Minister of Environment Resolution No. 37 of 2010 on the Conservation of Turtles and Seabirds from Extinction. (2010). Retrieved from https://almeezan.qa/LawView.aspx?opt&LawID=3737&language=en

Ministry of Development Planning and Statistics. (2018). *Qatar Second National Development Strategy 2018 - 2022*. Doha. Retrieved from https://www.psa.gov.qa/en/knowledge/Documents/NDS2Final.pdf

Ministry of Environment. (2014). *Qatar national biodiversity strategy and action plan 2015–2025*. Retrieved from https://www.cbd.int/doc/world/qa/qa-nbsap-v2-en.pdf

Ministry of Environment and Climate Change. (2021). *National environment and climate change strategy*. Retrieved from https://www.mecc.gov.qa/wp-content/uploads/2022/04/Environmental-And-Climate-Change-Strategy-Book-AR.pdf

Ministry of Municipality and Environment. (2016). *Qatar national master plan: Qatar national development framework*. Retrieved from https://www.mme.gov.qa/QatarMasterPlan/Downloads-qnmp/QNDF/English/English_QNDF.pdf

Ministry of Municipality and Environment. (2017). *Climate change strategy for urban planning and urban development sector in the state of Qatar*. Retrieved from https://www.mme.gov.qa/QatarMasterPlan/Downloads-qnmp/ClimateChange/Stage%200%20-%20Executive%20Summary/Stage%2000%20Executive%20Summary%20English.pdf

Ministry of Municipality and Environment. (2018a). *Climate change strategy for urban planning and urban development sector in the state of Qatar*. Retrieved from https://www.mm.gov.qa/QatarMasterplan/Downloads-qnmp/ClimateChange/Stage%200%20-%20Executive%20Summary/Stage%2000%20Executive%20Summary%20English.pdf

Ministry of Municipality and Environment. (2018b). *Environmental sustainability strategy 2018–2022*. Doha. Retrieved from https://www.mme.gov.qa/pdocs/cview?siteID=1&docID=14466&year=2018

Ministry of Municipality and Environment. (2021). *Qatar national climate change action plan 2030.* Retrieved from https://www.mme.gov.qa/pdocs/cview?siteID=2&docID=23349&year=2021

Peterson, T. (1997). Sustainable development comes of age. In T. R. Peterson (Ed.), *Sharing the Earth: The rhetoric of sustainable development* (pp. 6–33). University of South Carolina Press.

Prime Minister's Resolution No. 1 of 2022 Establishing the Third National Development Strategy National Steering Committee. (2022). Retrieved from https://almeezan.qa/LawView.aspx?opt&LawID=8841&language=ar

Protocol Concerning Regional Cooperation in Combating Pollution by Oil and Other Harmful Substances in Case of Emergency, February 14, 1982, http://www2.ecolex.org/server2neu.php/libcat/docs/TRE/Full/En/TRE-000745.txt

Qatar Foundation. (2014). *Qatar national research strategy 2014.* Retrieved from https://www.qnrf.org/Portals/0/Download/QNRS%202014.pdf

Rao, C. R. (1983). Statistics, statisticians and public policy making. *Sankhyā: The Indian Journal of Statistics, Series B (1960–2002)*, *45*(2), 151–159.

Resolution No. 4 of 2005 by the Chairperson of the Supreme Council of the Environment and Natural Reserves (SCENR) Issuing Executive Bylaw for Law No. 30 of 2002 on Environment Protection. (2005). Retrieved from https://www.almeezan.qa/LawPage.aspx?id=2108&language=en

United Nations Environment Program Regional Office for West Asia. (1984). *The state of environment in Qatar.* United Nations Environment Program. https://wedocs.unep.org/bitstream/handle/20.500.11822/28296/TSEQTR84.pdf?sequence=1&isAllowed=y

Reem Al-Hababi is a PhD student in the Gulf Studies Program at Qatar University. She holds a Master of Public Policy from HBKU.

Chapter 3
Law and Governance Innovations on Sustainability in Qatar: Current Approaches and Future Directions

Damilola S. Olawuyi and Elena I. Athwal

Abstract Over the last few decades, Qatar has developed a clear, comprehensive, and committed legal order that places sustainability squarely at the heart of its rapid economic development. In addition to the Qatar National Vision 2030 (QNV 2030), which accentuates the need for balanced and holistic social, economic, and environmental development, Qatar has been a hub of research, technology development, and public dialogue on sustainable development. More recently, Qatar has also launched a National Environment and Climate Change Strategy that outlines practical steps to achieve the environment pillar of the QNV 2030. These efforts underline Qatar's commitment and vision to promote balanced and holistic social, environment, and economic development across all key sectors. However, while the impetus and appetite for sustainability is rapidly increasing in Qatar, dynamic legal innovation is required to ensure greater coordination and coherence in the implementation of such efforts. This chapter evaluates law and governance innovations that have been put in place in Qatar to advance sustainability and low-carbon transition. It also proposes innovative legal approaches for addressing practical challenges to implementation, especially issues of institutional coordination, interoperability, and reducing overlaps in regulatory arrangements.

Keywords Law · Governance · Innovation · Sustainability · Climate change · Qatar · Environment

D. S. Olawuyi (✉) · E. I. Athwal
College of Law, Hamad Bin Khalifa University, Doha, Qatar
e-mail: dolawuyi@hbku.edu.qa

E. I. Athwal
e-mail: elat32212@hbku.edu.qa

L. Cochrane and R. Al-Hababi (eds.), *Sustainable Qatar*, Gulf Studies 9,
https://doi.org/10.1007/978-981-19-7398-7_3

3.1 Introduction

This chapter evaluates law and governance innovations that have been put in place in Qatar to advance sustainability and low-carbon transition. It also examines practical limitations to the coherent implementation of extant laws, and the legal framework for addressing those challenges.

Ever since the *1991 Arab Declaration on Environment and Development and Future Prospects* (1991 Arab Declaration) proclaimed that all Arab countries shall respect, protect, and fulfill 'the rights of peoples to sustainable development, as set out in the Rio Declaration,' policy and legislative efforts to promote sustainability have been on a rapid ascendancy across the region, including in Qatar (League of Arab States, 2001, pp. 6–7). According to the Declaration, Arab countries agree to limit the degradation of the environment and natural resources and manage them 'in a sustainable manner that ensures Arab water and food security, the preservation of ecosystems and biodiversity and the control of desertification' (League of Arab States, 2001, p. 5). Although non-binding, this Declaration provides normative guidance on how Arab countries could advance sustainability, which entails balancing and ensuring that projects and policies aimed at promoting economic development, do not result in social and environmental challenges.

Prior to 1991, Qatar had already made significant progress as one of the first few countries in the Arab region to enact laws aimed at addressing social and environmental impacts of economic development in key sectors (Olawuyi, 2022c, pp. 1–25). However, the increasing international awareness of the impacts of economic activities such as oil and gas production, infrastructure development, illegal traffic and trade in endangered plant and animal species, and transboundary movement of hazardous chemicals and wastes to developing countries, all created greater momentum on the need for sustainability focused legislative frameworks across the world, especially more so in the Arab region. Qatar, like many other Arab countries, faces a complex web of sustainability threats because of its dry and arid climate which increases vulnerability to food insecurity, water scarcity, loss of biodiversity, as well as the problem of climate change which is projected to escalate and worsen pre-existing sustainability challenges across the region (Olawuyi, 2022b). Efforts to respond to the complex and multifaceted sustainability challenges facing the world have spurred the rapid development and enforcement of sustainability frameworks in Qatar.

Over the last few decades, Qatar has developed a clear, comprehensive, and committed legal order that places sustainability squarely at the heart of its rapid economic growth and development. For example, the environment pillar of the QNV 2030 expressly recognizes the need for an 'agile and comprehensive legal system that protects all elements of the environment, responding quickly to challenges as they arise' (General Secretariat for Development Planning, 2008, p. 36). Furthermore, Qatar emerged as one of the first countries in the region to ratify the Paris Agreement on Climate Change and put in place very strong Nationally Determined Contributions on a low-carbon and sustainable future. Additionally, over the last years, the Ministry of Environment and Climate Change has issued a wide range of substantive laws and policies that cast a special spotlight on environmental protection in the country (Olawuyi, 2022c). Additionally, institutional measures have also been put in place, such as the landmark achievement of Qatar's establishment of a National Climate Change Committee, an administrative body tasked with monitoring climate change (Olawuyi, 2022c). Qatar has also launched a National Environment and Climate Change Strategy which outlines practical steps to achieve the environment pillar of the QNV 2030 (Government of Qatar, 2021a, b). Due to these efforts, Qatar has gained increased recognition as a regional leader in efforts to address climate change and advance renewable energy technologies in the Arab world (Ataullah, 2017).

However, while the impetus and appetite for sustainability are rapidly increasing in Qatar, dynamic legal innovation is required to ensure greater coordination and coherence in the implementation of such efforts. After surveying the scope of law and governance frameworks on sustainability in Qatar, this chapter assesses the potential for greater coherence in implementation of such efforts. The rest of the chapter is structured as follows. After this introduction, Sect. 3.2 discusses the scope and contours of legal frameworks on sustainability in Qatar. Section 3.3 examines the practical challenges to the implementation of sustainability frameworks in Qatar, and then develops a case on the need for greater institutional coordination, interoperability, and holistic integration in the implementation of sustainability programs. Section 3.4 is the concluding section.

3.2 Legal Frameworks on Sustainability in Qatar

The period since 1991 has seen rapid formalization and development of sustainability laws and policies in Qatar. The scope of Qatar's sustainability laws and policies can be understood under three broad categories: international and regional efforts (Sect. 3.2.1), domestic legislation (Sect. 3.2.2), and scholarly publications and local initiatives by organizations (Sect. 3.2.3).

3.2.1 International and Regional Efforts

The development of sustainability frameworks in Qatar flows directly from efforts by Qatar to implement the rapidly evolving international and regional frameworks that aim to integrate social and environmental considerations into economic development. Qatar has been an active player in international and regional efforts on sustainability. For example, ever since the United Nations Framework Convention on Climate Change (UNFCCC) was adopted in 1992 as one of the Rio Conventions, Qatar has progressively and proactively mainstreamed international law on climate change into its domestic strategies (Olawuyi, 2022b). In 2011, the Ministry of Environment of Qatar submitted its *Initial National Communication* to the UNFCCC, which expressly recognized Qatar's commitment to contribute to ongoing global efforts to address the problem of climate change (UNFCCC, 2012). In light of this initial communication, and in order to support global climate change negotiations, Qatar, in 2012, hosted the 18th Conference of the Parties (COP) to the UNCFCCC (2012). The importance of the 18th COP Conference cannot be understated, as it helped secure the future of international climate action by setting the stage for the Paris Agreement to come to fruition (United Nations, 1994). Qatar was also one of the first few countries in the Arab world to ratify the Paris Agreement, a treaty which sets a clear goal to reduce emissions and keep the rise in global temperature under 2 °C (3.6°F) and limit it to 1.5 °C above pre-industrial levels starting in the year 2020. Qatar ratified the Kyoto Protocol in January 2005 and the Paris Agreement in June 2017 (United Nations, 2015a, 2015b, 2015c). Furthermore, the country accepted the Doha Amendment (the second commitment round of the Kyoto Protocol) in 2020 (United Nations, 2012).

Qatar's commitment to the UNFCCC and the Paris Agreement has resulted in the adoption of national plans, policies, and strategies that recognize the need to lower current levels of GHG emissions (State of Qatar, 2021). Qatar's most recent submission, the *Ministry of Municipality and Environment's Nationally Determined Contribution of August 2021*, builds upon the 2015 Intended Nationally Determined Contribution (INDC), by reflecting a balanced approach between adaptation, economic diversification, and mitigation. Qatar also sets a more ambitious targets of achieving 25% reduction in greenhouse gas emissions (GHGs), relative to its baseline scenario (Business-As-Usual, or BAU) (State of Qatar, 2021). Furthermore, Qatar has mobilized financial support for global climate action. For example, Qatar made a contribution of $100 million to support small island states and least developed countries to deal with climate change (Al Sharq Newspaper, 2021). Similarly, the Qatar Fund for Development was directed to continue mobilizing support to address climate change and promote green growth in these countries (Al Sharq Newspaper, 2021). These efforts reflect Qatar's increasing role at the international level as a key enabler of global efforts to address sustainability threats posed by climate change.

Regionally, Qatar subscribes to a wide range of sustainability declarations and instruments in the Arab region, as outlined in Table 3.1 below. Though not legally binding, these soft law instruments lay the foundation for all Arab countries, including Qatar, to ensure the balanced implementation of economic development programs in a manner that does not result in social and environmental challenges. They also emphasize the need to integrate low-carbon emission planning into national policies, strategies and planning along with the United Nations Sustainable Development Goals (UN SDGs). For example, the *Pan-Arab Renewable Energy Strategy, 2030* of the League of Arab States sets a target of increasing installed renewable energy power generation capacity across the region by the year 2030. In line with this, Qatar has outlined plans to generate 20% of its electricity from solar systems by 2030 (General Secretariat for Development Planning, 2008; Government of Qatar, 2011a). In line with the 2016–2020 development plan, Qatar has set a $30 billion infrastructure investment plan, which includes the design and delivery of more than 700 legacy projects over the next five to seven years in solar energy, technology, and other crosscutting infrastructure projects (Olawuyi, 2018a). A number of solar projects are already under construction in Qatar, including a 200 megawatt solar farm project that will consist of approximately 800,000 square meters of photovoltaic (PV) panels (Olawuyi, 2018a).

Qatar is aware that it cannot achieve these goals alone. In addition to its participation in regional soft law instruments, Qatar has also taken a leadership role on sustainability in the region by hosting key regional meetings and conferences on sustainability (Meltzer et al., 2014).

3.2.2 *Domestic Legislation*

In addition to international and regional efforts, Qatar's domestic laws include comprehensive provisions on environment and sustainable development.

3.2.2.1 Qatar's Constitution

Qatar is one of the countries in the world with a clear constitutional provision on environmental protection. Article 33 of Qatar's Constitution of 2003 provides that the state 'endeavours to protect the environment and its natural balance, to achieve comprehensive and sustainable development for all generations' (Olawuyi, 2018b, p. 487). Thus, not only is the protection of the environment expressly stated in the highest legal source of the nation, but in particular, goes beyond to emphasize protection of the natural balance, such as the ecosystem and its biodiversity, while tying that into comprehensive, meaning considering broadly the impacts of, sustainable development on future generations.

Furthermore, as noted by the Constitution, Qatar's principal source of law is Shariah law, as Islam is the primary religion of the country. Therefore, Qatar, as a state upholding Islamic principles, through Shariah law, emphasizes that humankind is a mere steward or trustee (*khalifa*) of the earth and not a proprietor nor a disposer or one who ordains; and that all Muslims have the solemn duty to maintain and preserve the natural environment from disequilibrium or damage (IUCN, 1983). Some innovative ways that the protection of the environment has been addressed legally is through this idea of trusteeship. For example, the *khalifa* is responsible for governing resources for the benefit of future generations (Olawuyi, 2022c). Closely related to this idea of *khalifa*, is the principle of *waqf*, which essentially reinforces the fiduciary duty of the *khalifa*, with regard to trusteeship over natural resources, insofar as the beneficiaries, such as civil society, can ask the *khalifa* to fulfill the duty of care in terms of environmental matters. This *khalifa* principle provides a faith-based imperative and foundation on the need for clear and coherent efforts to advance sustainability and resource conservation in Qatar. Similarly, the idea of *waqf* can be used as a tool for green finance (Idllalène, 2022). These initiatives from a faith-based perspective mirror some of the efforts under the United Nations 'Faith for Earth Initiatives' frameworks (UN Environmental Programme, 2019a).

Table 3.1 Key declarations and instruments relating to climate change by regional bodies in the Middle East and North African (MENA) region (Olawuyi, 2022a)

Instrument	Highlights
Arab Declaration on Environment and Development and Future Perspectives, Cairo, 1991	The Declaration called on all Arab countries to limit the degradation of the environment and natural resources and manage both in a sustainable manner
Arab Declaration to the World Summit on Sustainable Development, 2001	It renewed the commitment of Arab countries to work together to advance environmental protection and sustainable development, especially by addressing the vulnerabilities of Arab countries to water scarcity, food insecurity, and climate change
Abu Dhabi Declaration on the Future of Environmental Action in the Arab World, 2001	It identifies the priority of environmental problems in the Arab world as: 'the acute shortage and deteriorating quality of sources of water; the paucity and deteriorating quality of exploitable land; the imprudent consumption of natural resources; urban sprawl and its associated problems; the degradation of marine, coastal and watered areas.'
	It also identifies the need for clean production methods and technologies to reduce emissions
Arab Charter on Human Rights, 2004	Article 38 states that 'every person has the right to an adequate standard of living for himself and his family, which ensures their well-being and a decent life, including food, clothing, housing, services and the right to a healthy environment. The States parties shall take the necessary measures commensurate with their resources to guarantee these rights.'
The Arab Ministerial Declaration on Climate Change of December 6, Cairo, 2007	It contains regional aspirations by all Ministers responsible for environment and climate change to include 'policies to deal with climate change issues in all sectors within national and regional policies for sustainable development.'
	The Declaration also provides that: 'Adaptation to measures that address climate change shall be fully consistent with the economic and social development and in such a way so as to achieve sustainable economic growth and eradication of poverty.'
Cairo Declaration on Development Challenges and Population Dynamics in a Changing Arab World, 2013, *(paras 66, 85, and 87)*	It called on Arab countries to develop regional and local climate change response measures that take into account the distribution, vulnerability, and resilience of the targeted populations

(continued)

Table 3.1 (continued)

Instrument	Highlights
Rabat Islamic Declaration on Environment Protection and Achieving Sustainable Development Goals, 2017	Called on Islamic countries to advance climate change mitigation and adaptation in all aspects of planning at national levels. It also affirms the importance of a green economy transition
Arab Framework Action Plan on Climate Change, 2010–2020	It creates a ten-year master plan for climate mitigation and adaptation in the Arab region
Pan-Arab Renewable Energy Strategy, 2030 (League of Arab States)	It identifies the integration of renewable energy as a tool for addressing climate change vulnerabilities in the Arab region. It sets a target of increasing installed renewable energy power generation capacity across the region by the year 2030

New models of green financing have been introduced to assist in the modernization of *waqf* and *zakat*, or almsgiving in Qatar, in particular, based on these fundamental Islamic principles (Aassouli, 2022).

3.2.2.2 Primary Legislation

Qatar's Law No. 30 of 2002 on Environmental Protection outlines general provisions for the protection of the environment, including ensuring environmental quality, avoiding pollution and damage from development projects, protection of the public health in addition to the local flora and fauna, and lastly provides for environmental awareness under Article 7 (Government of Qatar, 2022a). Pursuant to this law, the main authority in charge of environmental matters is the Ministry of Environment and Climate Change, which oversees regulations and restrictions maintaining the environment, and implementing conditions and criteria for the protection of the environment. One article of note is Article 6, which states that all public and private bodies must include an environmental protection and pollution control clause in local and international agreements and contracts which may be detrimental to the environment, and these agreements and contracts shall include applicable penalties and the obligation to bear the costs of repairing the environmental degradation and harm. According to Article 8, the Ministry oversees all approvals, control, and supervision of private and public development projects, including setting standards and measures related to environmental impact assessment of projects, and the procedures and requirements for issuing authorizations. Furthermore, Article 34 states that anybody working in the exploration, extraction, production, and refining of oil must comply with the executive bylaw for this law and other applicable best practices. Moreover, in Part 2, there is the legislation that establishes the basis for the governmental authorities to penalize and sanction all forms of air pollution, i.e., any chemical, physical, or biological change or modification of the natural characteristics of the atmosphere, in proportions that could be harmful to human life and nature, including contributions to climate change.

3.2.2.3 Secondary Legislation

In addition to primary legislation that aim to enhance balanced social, environmental, and economic development, Qatar has put in place a wide range of regulations, action plans, directives, and national vision that address various aspects of sustainability. For example, while the social development pillar of the QNV 2030 aims to promote the development of a just society based on rule of law and active engagement of citizens in public life, the economic development pillar focuses on promoting economic diversification, entrepreneurship, and a high standard of living for current and future generation of the people of Qatar (General Secretariat for Development Planning, 2008, pp. 3, 8–9, 24). Furthermore, the environmental pillar emphasizes the management of the environment to ensure harmony between economic growth, social development, and environmental protection. In order to balance these needs, the QNV expresses Qatar's aim to achieve 'a diversified economy that gradually reduces its dependence on hydrocarbon industries' by the year 2030 (Government of Qatar, 2011a, p. 11). The environment pillar also identifies 'the need for an agile and comprehensive legal system that protects all elements of the environment, responding quickly to challenges as they arise' (General Secretariat for Development Planning, 2008; Government of Qatar, 2011a). Evidently, Qatar has recognized the need for integrated economic development, built on an entrepreneurial, low-carbon, and green economy.

Similarly, the *Executive By-Law for the Environment Protection Law* contains elaborate provisions on the reduction of all sources of air pollution, especially the emission of several GHGs that cause climate change (Government of Qatar, 2002). For example, Article 65 states the requirements that any party working in the oil sector must comply with the procedures determined by this bylaw, such as (i) installing the separation and flaring systems required to produce and transport petroleum products; (ii) taking the required measures to avoid leaks of gas during drilling tests; managing storage tanks emissions according to international standards; (iii) using of compressed air in instruments tools instead of natural gas whenever is possible; and iv) getting rid of associated gas that cannot be utilized, in a safe manner according to the international standards (Government of Qatar, 2002). The regulatory focus on flare minimization is yielding results as Qatar has recorded significant reductions in levels of gas flaring and methane emissions (Bawazir et al., 2014).

In addition to the Executive By-Law, *Qatar's Resolution of the Council of Ministers* also establishes a Committee on Climate Change and Clean Development Mechanism (CDM) that plays similar roles in laying down strategies, policies, and communications on climate change in the country (Government of Qatar, 2011b). Furthermore, it outlines ways to ensure that governmental and non-governmental organizations in Qatar implement the provisions of the UNFCCC and the Kyoto Protocol and provides the means to prepare studies and reports accordingly for the creation of databases, national assessments, reviews, and evaluations (Government of Qatar, 2011b).

Similarly, in 2017, the Government of Qatar established the *National Committee on Climate Change* which is responsible for monitoring the progress on climate change, as well as approving low-carbon projects that can lead to the reduction of GHG emissions. Moreover, in 2021, the *National Environmental and Climate Strategy* was launched in Qatar, which aims to achieve a target of 25% reductions in GHGs by the end of 2030. The National Strategy includes national projects which adopt tangible measures to reduce GHGs, including: (i) the largest carbon capture project in the MENA region; (ii) improving the efficiency of the hydrocarbon toxicity treatment; (iii) the Tarsheed program (National Program for Conservation and Energy Efficiency) which conserves electricity and water consumption, including initiatives by Kahramaa, the national water authority, to reduce the loss of desalinated water and to encourage water recycling and reuse; and iv) the building of the Al Kharsaah solar power station with a capacity of 800 MW (FIFA 2022; QNA, 2021a). Not only are there initiatives within Qatar, but Qatar is playing a pivotal role in supporting the developing countries' fight against the impact of climate change, including donations of $100 million for small island states and least developed states to deal with the impact of climate change and natural hazards (QNA, 2021a). The government's role encouraged other institutions to get on board, such as the Qatar Investment Authority, with its One Planet Sovereign Wealth Initiative (Qatar Investment Authority, 2022).

In terms of integrating environmental considerations into investments, the Qatar Investment Authority is a founding member of the One Planet Sovereign Wealth Fund (SWF) Working Group, which aims to integrate climate change analysis and environmental considerations into investment decisions (One Planet, 2019). Through this commitment, Qatar has shown increased commitment to allocate its SWF investments to 'finance the smooth transition to a more sustainable, low-carbon economy' as envisaged by the Working Group (One Planet, 2019). By integrating climate change considerations into the design, financing, and implementation of SWF investments, Qatar aims to leverage savings from the extractive industries to invest in assets and projects worldwide that contribute to decarbonization and sustainable development (Olawuyi, 2020a).

In October 2017, Qatar also launched the initiative to establish the Global Drylands Alliance (GDA), announced by His Highness Sheikh Tamim bin Hamad Al Thani the Amir of the State of Qatar at the 68th United Nations General Assembly (UNGA), 17 September, 2013. The GDA is one of the international mechanisms aimed at confronting the broad consequences of climate change, such as food security and dryland ecosystems (Global Dryland Alliance, 2022aa). The GDA implements research and development programs, including technological innovation, and strives to improve members' food security policies and planning ecosystems (Global Dryland Alliance, 2022b).

The establishment of the Ministry of the Environment and Climate Change in 2021, also underscores the increasing focus and emphasis on environmental sustainability in Qatar (QNA, 2021b). The target of 25% emission reduction by 2030 means that each sector must take action and integrate their efforts together to reduce and encourage efficiency. Reducing food waste is also included in the country's National Food Security Strategy 2018–2022 (see Chap. 12). Consequently, Qatar has established a household solid waste treatment center in Mesaieed, which is the first integrated solid waste treatment facility in the Middle East (QNA, 2021b). In 2020, the center produced more than 30,000 tons of organic fertilizer, and generated about 269,000 megawatts/hour of electrical energy, and more than 33 million cubic meters of biogas, in addition to treating more than 30,000 tons of car tires in each of the Umm al-Afaei landfill and Rawdat Rashed landfill, and recycling about 420,000 tons of construction waste (QNA, 2021b; see Chap. 16).

3.2.3 Scholarly Publications and Local Initiatives by Organizations

In addition to primary sources of law, such as the constitution, acts and codes, and international treaties, the development of sustainability law and policy in Qatar is highly influenced by secondary sources of law, such as scholarly publications, reports, and local community awareness spearheaded by local non-governmental organizations. While such community-led efforts are legally non-binding, they provide strong influence for the increasing local awareness on sustainability in Qatar. For example, the Arab Youth Climate Movement (AYCM) has emerged as one of the largest youth-led climate education, capacity development, and advocacy groups in the MENA region with branches in more than 15 countries, including Qatar. Similarly, the Qatar Green Building Council provides important reports, information, and best practices that support informed policy making on low-carbon, energy-efficient, and environmentally sustainable practices in building design and construction. As a result of this and other similar efforts, Qatar ranked second in terms of the number of green and environment-friendly buildings in the MENA region, with 1,406 sustainability certified buildings with Global Sustainability Assessment System (GSAS) green building standards (Ataullah, 2021). Similarly, the National Museum of Qatar recently became the first national museum in the world to have received a LEED (Leadership in Energy and Environmental Design) Gold and a 4 Star GSAS sustainability rating (Qatar Museums Authority, 2021). Ahead of the FIFA World Cup™ 2022, a wide range of initiatives have been launched by the Government of Qatar, as well as sustainability institutes such as the Josoor Institute, which aim to deliver a carbon–neutral tournament (see Chap. 4). Increased government funding for research, especially through the Qatar National Research Fund, has also fostered scientific cooperation and research on sustainability. For example, in 2018, the Association of Environmental Law Lecturers in Middle East and North African Universities (ASSELLMU) was established at a conference hosted by Hamad Bin Khalifa University to serve as a professional network for all MENA environmental law academics (Mrema &

Smagadi, 2022). ASSELLMU has since been at the forefront of creating awareness on sustainability in Qatar and across the region (ASSELLMU 2019).

The wide range of law and governance efforts ongoing in Qatar which aim to accelerate energy transition, waste management, economic diversification, and green growth in key sectors, reflect Qatar's important role as a regional leader and hub for sustainability in the Arab world. However, to effectively achieve the wide range of ambitious sustainability targets stipulated in different laws and policies, there is a need to address gaps that may hinder coherent implementation of the various sustainability programs. The next section explores key legal issues and gaps that must be addressed in order to advance law and governance frameworks on sustainability in Qatar.

3.3 Advancing the Coherent Implementation of Sustainability Policies and Programs in Qatar: The Need for an Integrated Approach

Given the rapid profusion of sustainability policies and laws in Qatar, dynamic legal innovation is required to continually enhance their coordinated and coherent implementation. First, the rapidly increasing volume of ongoing programs and initiatives, as well as the number of administrative and governance bodies with supervisory functions in economic, social, and environment spheres can result in implementation challenges if not holistically monitored. In addition to the challenges of coordinating the diverse functions and mandates of the various bodies, there has also been a tendency to implement sectoral and piecemeal responses in various sectors (Olawuyi, 2020b). This raises the need to promote greater coordination among the various agencies and ministries in order to enhance greater cooperation, interoperability, and resource sharing in the implementation of sustainability programs. Over the last several years, a nexus governance approach has been offered as a solution to this problem (Olawuyi, 2022b; Salam et al., 2017). The nexus governance approach allows policymakers to have a systemic understanding and view of the trade-offs and synergies between the various sustainability programs and implementing bodies to avoid overlapping and duplicative sector-specific actions and programs (Olawuyi, 2022b; Salam et al., 2017). The interconnections between the SDGs underscore the need for a nexus approach to ensure the delivery of sustainability efforts in a coherent and integrated manner. An integrative governance framework will provide appropriate incentives such as linked and coordinated regulation, structural integration of expertise, knowledge, and information, as well as holistic programming by actors in diverse domains that can help move toward successful transformation (Olawuyi, 2022b; Salam et al., 2017).

Second, with increased spending on sustainability initiatives in Qatar comes the question of how such programs are to be evaluated and measured to ensure real and measurable progress. To ensure the coherent and integrated implementation of sustainability programs at all levels of government, a number of countries, such as Hungary, New Zealand, Canada, United Kingdom among others have established focal institutions that will evaluate, monitor, and assess the performance of government departments and agencies against sustainability targets and threshold (Szabó and Cordonier Segger, 2021). For example, Hungary established the position of Ombudsman for Future Generations in 2008, with the responsibility to evaluate and measure how all government agencies and bodies are implementing sustainability programs (Szabó and Cordonier Segger, 2021; Olawuyi, 2021). Similarly, Canada established the Office of the Commissioner of the Environment and Sustainable Development, an independent unit, housed within the Office of the Auditor General of Canada, which oversees, reviews, and appraises sustainable development programs across all government departments (Government of Canada, 2022). The Commissioner supports the work of government ministries and assists them in harmonizing their working methods and reporting requirements on sustainable development. Appointing an Ombudsperson or establishing a coordination agency or committee to spearhead and evaluate the implementation of sustainability programs in Qatar can further enable important stakeholders in key sectors to cooperatively explore co-benefits in the design and delivery of sustainability programs thereby reducing program or role duplication and enhancing overall effectiveness. Such a coordinating entity can also spearhead sustainability training and education to deepen the capacity of officials and stakeholders to implement sustainability programs.

Similar to the need for institutional coherence is the need for greater data and information sharing on sustainability programs. For example, in order to address gaps in data sharing and limited access to statistical information on the impact of sustainability programs, the Arab Strategy for Housing and Sustainable Urban Development 2030 calls on all Arab countries to develop regional observatory systems for sharing information on housing and urban development. It also encourages integrated approaches that promote interoperability, exchange of expertise as well as 'capacity building and human development in the field of housing, and urban development at the local, national and regional levels' (League of Arab States, 2016). Sustainability programs in Qatar can be enhanced when clear and accessible statistical information on the scope of ongoing programs are made readily available in publicly accessible platforms. Not only will such information dissemination allow members of the public to measure progress, but it will also enhance information access and planning in various entities and institutions, including private sector institutions that rely on such sustainability information for resource planning and investment analyses.

The rapid ascendancy of sustainability law and policy in Qatar has also placed greater emphasis on the need for business enterprises and private actors to integrate sustainability into their operations and value chains. For example, QatarEnergy (formerly Qatar Petroleum) has played a leading role in this regard through its sustainability reports that show efforts to align its operations, procurements, and programs with all of the SDGs (Qatar Energy, 2022). Similarly, Qatar Foundation has increasingly promoted sustainability initiatives, such as its Car Free Day program, among others (see Chap. 5). There is a need for other stakeholders in private and public sectors to further develop similar programs, including green procurements, sustainability screening and due diligence, research and development as well as public awareness drives. There is a strong business case, in terms of cost, reputation, and effectiveness, for doing so. Not only will such efforts increase the corporate and ethical responsibility of business enterprises, but they could also significantly help enterprises to minimize, anticipate, and mitigate legal risks that may result from the adverse impacts of their operations and projects on Qatar's short- and long-term sustainability.

3.4 Conclusion

Qatar has made significant progress in placing sustainability at the heart of economic, social, and investment decision-making. Through active international and regional engagement, as well as the enactment of domestic legislation, policies, and programs on sustainability, Qatar has set the right tone with innovative national targets and action plans that align themselves with mainstream international efforts to protect the environment and promote sustainability.

However, while Qatar has increasingly developed greater capacity for coherent coordination of sustainability programs and responses; there remains still a great deal of fragmentation between social, environment, and economic actors. A clear and comprehensive integration of sustainability programs across diverse sectors can be enhanced through a nexus governance approach. Furthermore, it is necessary to coordinate training and promote the cooperation in the delivery of sustainability projects and programs, particularly in terms of sharing best practices, so that barriers to information sharing and knowledge exchange can be removed for a more interconnected and open system among ministries and various agencies. Likewise, increased financing for research and sustainability networks can significantly cement Qatar's growing reputation as a regional hub for sustainability innovations.

As clear sustainability legislation and rules continue to emerge in Qatar, business enterprises, investors, lenders, insurance companies, and lawyers can reduce legal liability and risks arising from the direct and indirect impacts of their activities by also integrating sustainability considerations into business decision-making processes and planning.

References

Aassouli, D. (2022). Mobilizing Islamic climate finance. In D. Olawuyi (Ed.), *Climate change law and policy in the Middle East and North Africa region* (pp. 204–230). Routledge.

Al Sharq Newspaper. (2021, November 11). Minister of environment and climate change: Qatar gives great attention to climate change risks. https://hukoomi.gov.qa/en/news/minister-of-environment-and-climate-change-qatar-gives-great-attention-to-climate-change-risks. Accessed 28 March 2022.

ASSELLMU, Association of Environmental Law Lecturers in Middle East and North African Universities. https://assellmu.org. Accessed 20 March 2022.
Ataullah, S. (2017, November 26). Qatar regional leader in addressing climate change. *The Peninsula*. https://thepeninsulaqatar.com/article/26/11/2017/Qatar-regional-leader-in-addressing-climate-change
Ataullah, S. (2021, November 7). Qatar has highest number of GSAS-certified green buildings. *The Peninsula*. https://thepeninsulaqatar.com/article/07/11/2021/qatar-has-highest-number-of-gsas-certified-green-buildings. Accessed 8 April 2022.
Bawazir, I. , Raja, M. , and I. Abdelmohsen. *Qatargas Flare reduction program*. Paper presented at the International Petroleum Technology Conference, Doha, Qatar, January 2014. doi: https://doi.org/10.2523/IPTC-17273-MS
Beard, J. (2019, July 11). Eight asset managers unite for One Planet initiative. *City Wire*. https://citywireselector.com/news/eight-asset-managers-unite-for-one-planet-initiative/a1249346
Decree-Law No. 11 of 2000 Establishing the Supreme Council of Environment and Natural Reserves (Repealed) which repeals and replaces the functions of the Supreme Council.
FIFA World Cup Qatar 2022. (n.d.). Sustainability at the FIFA World Cup: Our sustainability framework. https://publications.fifa.com/en/sustainability-report/sustainability-at-the-fifa-world-cup/our-sustainability-framework/. Accessed 8 April 2022.
General Secretariat for Development Planning. (2008). Qatar national vision 2030. www.gco.gov.qa/wp-content/uploads/2016/09/GCO-QNV- English.pdf.
Global Dryland Alliance: Making Drylands Food Secure (2022a). https://www.globaldrylandalliance.org/en/
Global Dryland Alliance: Making Drylands Food Secure (2022b). Vision and Mission. https://www.globaldrylandalliance.org/en/about-us/vision-mission/
Government of Canada. (2022). Sustainable development strategy for the office of the auditor general of Canada—2020–2023 (February 2022 Update). https://www.oag-bvg.gc.ca/internet/English/acc_rpt_e_43626.html. Accessed 10 April 2022.
Government of Kuwait (2016). Kuwait's National Vision 2035, Kuwait Third National Master Plan, https://andp.unescwa.org/sites/default/files/2020-09/New%20Kuwait%202035-Kuwait%20National%20Development%20Plan.pdf Accessed 10 April 2022.
Government of Qatar. (2002). Decision No. (30) of 2002 of the President of the Supreme Council of environment and natural protection concerning the issuance of the executive by-law for the environment protection law, Qatar
Government of Qatar. (2003). Article 7 of Qatar's law 10/2003 of the judiciary act.
Government of Qatar. (2018). State of Qatar second voluntary national review 2018, Submitted to the High-Level Political Forum, 42–433.
Government of Qatar. (2021a). Ministry of municipality and environment. Nationally Determined Contribution. https://www4.unfccc.int/sites/ndcstaging/PublishedDocuments/Qatar%20First/Qatar%20NDC.pdf. Accessed 8 April 2022.
Government of Qatar. (2021b). PM launches Qatar's national environment. *Climate Strategy* (31 October, 2021b) https://hukoomi.gov.qa/en/news/pm-launches-qatar-s-national-environment-climate-strategy
Government of Qatar. (2022a). Law No. 30 of 2002 promulgating the law of the environment protection. https://almeezan.qa/LawPage.aspx?id=4114&language=en. Accessed 1 April 2022a.
Government of Qatar. (2022b). Resolution No. 4 of 2005 - executive bylaw for the environmental protection law.
Government of Qatar (2020). The future is ours: Sustainability report 2018 & 2019. https://km.qa/Tarsheed/SavingWithTarsheed/Kahramaa%20sustainability%20report_2020_.pdf. Accessed 8 April 2022.
Government of Qatar. (2011b). Qatar's resolution of the council of ministers No. 15 of 2011 establishing the committee on climate change and clean development at the Ministry of Environment. https://www.almeezan.qa/LawView.aspx?opt&LawID=3172&TYPE=PRINT&language=en. Accessed 1 April 2022.

Idllalène, S. (2022). The role of environmental Waqf. In D. Olawuyi (Ed.), *Climate change law and policy in the Middle East and North Africa region* (pp. 43–62). Routledge.

Intergovernmental Panel on Climate Change. (2006). 2006 IPCC guidelines for national greenhouse gas inventories. Task Force on National Greenhouse Gases. https://www.ipcc-nggip.iges.or.jp/public/2006gl/. Accessed 1 April 2022.

International Union for Conservation of Nature. (1983). Islamic principles for the conservation of the natural environment, IUCN Environmental Policy and Law Paper 20.

League of Arab States. (2001). First and Third Agenda, Arab declaration to the world summit on sustainable development (adopted on 24 October 2001 by the Arab Ministers responsible for development affairs, planning, and environment, meeting at the headquarters of the League of Arab States in Cairo, Egypt). https://www.hlrn.org/img/documents/Arab_Declaration_Sustainable_Dev.pdf. Accessed 1 April 2022.

League of Arab States. (2016). Arab Strategy for Housing and Sustainable Urban Development 2030. https://unhabitat.org/arab-strategy-for-housing-and-sustainable-urban-development. (Arab Strategy).

Meltzer, J., Hultman, N., and Langley, C. (2014). Low-carbon energy transitions in Qatar and the Gulf cooperation council region global economy and development program. The Brookings Institution. https://www.brookings.edu/wp-content/uploads/2016/07/low-carbon-energy-transitions-qatar-meltzer-hultman-full.pdf. Accessed 1 April 2022.

Ministry of Municipality and Environment of Qatar. (n.d.) Qatar's climate change strategy (CCS) for the urban planning and development sector (MME Qatar). www.mme.gov.qa/QatarMasterPlan/English/strategicplans.aspx?panel=ccs. Accessed 6 January 2021.

Mrema, E. M., & Smagadi, A. (2022). The United Nations environment programme—promoting climate law education in the Middle East and North Africa. In D. Olawuyi (Ed.), *Climate change law and policy in the Middle East and North Africa region* (pp. 21–42). Routledge.

Olawuyi, D. (2018a). Advancing innovations in renewable energy technologies as alternatives to Fossil fuel use in the Middle East: Trends, limitations, and ways forward. In D. Zillman, M. Roggenkamp, L. Paddock, & L. Godden (Eds.), *Innovation in energy law and technology: Dynamic solutions for energy transitions* (pp. 354–370). Oxford University Press.

Olawuyi, D. (2018b). Human Rights and the environment in Middle East and North African (MENA) Region: Trends, limitations and opportunities. In J. May & E. Daly (Eds.), *Encyclopedia of Human Rights and the environment, indivisibility, dignity and legality* (pp. 483–493). Edward Elgar.

Olawuyi, D. (2020a). *Can MENA extractive industries support the global energy transition?* Extractive Industries and Society Journal. https://doi.org/10.1016/j.exis.2020.02.003

Olawuyi, D. (2020b). Sustainable development and water-energy-food Nexus: Legal challenges and emerging solutions. 103(1) Environmental Science and Policy, 103(2022), 1–9. https://doi.org/10.1016/j.envsci.2019.10.009.

Olawuyi, D. (2021). Embedding intergenerational justice across government: Regional trends in Africa. In M. Szabó and M.C. Cordonier Segger (Eds.), *Sustainable development: Intergenerational justice and treaty implementation* (pp. 638–655). Cambridge: Cambridge University Press.

Olawuyi, D. (2022a). Nature and sources of climate change law and policy in the MENA region. In D. Olawuyi (Ed.), *Climate change law and policy in the Middle East and North Africa region* (pp. 3–20). Routledge.

Olawuyi, D. (2022b). *Climate change law and policy in the Middle East and North Africa region.* Routledge.

Olawuyi, D. (2022c). *Environmental law in Arab States.* Oxford University Press.

One Planet. (2019). Asset managers endorse the One Planet SWF framework eight leading asset managers endose the One Planet SWF framework and announce the formation of the One Planet Asset Managers Initiative. https://oneplanetswfs.org. Accessed 1 April 2022.

Qatar Green Building Council. https://www.usgbc.org/organizations/qatar-green-building-council?view=overview. Accessed 1 April 2022.

Qatar Investment Authority. (n.d.). Climate. https://www.qia.qa/en/About/Pages/Climate.aspx. Accessed 8 April 2022.

Qatar Museums Authority. (2021). Sustainable from the start: NMoQs award-winning design. https://qm.org.qa/en/experience/stories/sustainable-nmoq/. Accessed 8 April 2022.

Qatar Resolution of the Council of Ministers No. 15 of 2011 establishing the Committee on Climate Change and Clean Development at the Ministry of Environment. https://www.almeezan.qa/LawView.aspx?opt&LawID=3172&TYPE=PRINT&language=en. Accessed 1 April 2022.

Qatar's Law No. 3 of 1975 on the Commercial, Industrial and Public Premises, and amending laws thereof which focused on maintaining public hygiene and sanitary conditions in commercial, industrial, and public premises.

Qatar's Law No. 8 of 1974 on Public Hygiene (replaced by Public Hygiene Law No. 18 of 2017).

Qatar Energy. (2022). Climate Change Mitigation. https://www.qatarenergy.qa/en/Sustainability/Pages/ClimateChangeMitigation.aspx. Accessed 1 April 2022.

QNA. (2021a). Cabinet okays national climate change plan. *Gulf Times*. https://www.gulf-times.com/story/699568/Cabinet-okays-National-Climate-Change-Plan

QNA. (2021b). QNE critical to addressing climate change: Minister. *Gulf Times*. https://m.gulftimes.com/content/pdf/Dailynewspaper/Main2021b_10_29413615.PDF. Accessed 8 April 2022.

Reuters/The Qatar Peninsula. (2020). Qatar to Build 800 MW Solar Power Plant on 10 sqkm Plot. *The Peninsula*. https://thepeninsulaqatar.com/article/19/01/2020/Qatar-to-build-800-MW-solar-power-plant-on-10-sqkm-plot

Salam, P.A., Pandey, V.P., Shrestha S., and Anal, A.K. (2017). The need for the nexus approach', in Water-Energy-Food Nexus: Principles and Practices. In P. Abdul Salam, S. Shrestha, V. P. Pandey, and A.K. Anal (Eds.), *Water-Energy-Food Nexus: Principles and Practices* (pp. 3–10). Washington, DC: John Wiley and Sons, Inc.

Shafi, N. (2019). The Arab world's best weapon against climate change? Its young people. World Economic Forum. www.weforum.org/agenda/2019/01/the-arab-worlds-best-weapon-against-climate-change-its-youth/

State of Qatar. (2021). Ministry of Municipality and Environment, Nationally Determined Contribution (NDC), August 2021. https://www.mme.gov.qa/pdocs/cview?siteID=2&docID=23348&year=2021

The Peninsula. (2018, December 5). National museum of Qatar gets LEED gold. www.thepeninsulaqatar.com/article/05/12/2018/NMoQ-gets-LEED-Gold,-GSAS-4-Star-Sustainability-ratings.

United Nations (n.d.). Paris agreement—status of ratification https://unfccc.int/process/the-paris-agreement/status-of-ratification.

United Nations. (2012). Doha Amendment to the Kyoto Protocol. U.N. Doc. FCCC/KP/CMP/2012/13/Add.1, Decision 1/CMP.8. https://unfccc.int/process/the-kyoto-protocol/the-doha-amendment. Accessed 1 April 2022.

United Nations. (2015a). Conference of the parties, adoption of the Paris Agreement, Dec. 12, 2015a, UN Doc FCCC/CP/2015a/L.9/Rev/1. https://unfccc.int/sites/default/files/resource/parisagreement_publication.pdf. Accessed 1 April 2020.

United Nations. (2015b). Islamic declaration on global climate change, August 18, 2015b. https://unfccc.int/news/islamic-declaration-on-climate-change.

United Nations. (2015c). Transforming our world: The 2030 agenda for sustainable development. UN Doc. A/RES/70/1. https://sustainabledevelopment.un.org/content/documents/21252030%20Agenda%20for%20Sustainable%20Development%20web.pdf.

United Nations Development Programme. (2013). Regional Bureau for Arab States, water governance in the Arab region: Managing scarcity and securing the future, United Nations Publications.

United Nations Environment Programme. (2019a). Faith for Earth dialogue synthesis report, 8 May 2019a. https://www.unep.org/resources/synthesis-reports/faith-earth-dialogue-synthesis-report. Accessed 20 March 2022.

United Nations Environment Programme. (2019b). Using green and digital technologies to reduce food waste at the consumer level. https://wedocs.unep.org/handle/20.500.11822/36802. Accessed 20 March 2022.

United Nations Environment Programme. (2019c). Case study: Doha, Qatar https://wedocs.unep.org/bitstream/handle/20.500.11822/36802/CSDA.pdf. Accessed 8 April 2022.

United Nations Educational Scientific and Cultural Organization. (n.d.). Qatar: Climate change communication and education. https://education-profiles.org/northern-africa-and-western-asia/qatar/~climate-change-communication-and-education. Accessed 1 April 2022.

United Nations Framework Convention on Climate Change. (1992). 1771 U.N.T.S. 107, 165; S. Treaty Doc No. 102–38 (1992); U.N. Doc. A/AC.237/18 (Part II)/Add.1; 31 I.L.M. 849 (1992). https://unfccc.int/node/61146. Accessed 1 April 2022.

United Nations Framework Convention on Climate Change. (1998). Kyoto Protocol to the United Nations framework convention on climate change. https://unfccc.int/resource/docs/convkp/kpeng.pdf. Accessed 1 April 2020.

United Nations Framework Convention on Climate Change. (2015). Paris agreement under the United Nations framework convention on climate change. Conference of the Parties Twenty-First Session Paris, 30 November to 11 December 2015, FCCC/CP/2015/L.9, Articles 2(1), 4, 6, and 7.

United Nations General Assembly. (1994). United Nations framework convention on climate change (UNFCCC), 20 January 1994, UN Doc. A/RES/48/189.

UNHABITAT. (n.d.). Arab strategy: For housing and sustainable urban development 2030. https://unhabitat.org/sites/default/files/2017/05/Arab-Strategy-English.pdf

US Green Building Council. (2022). *Benefits of green building*. www.usgbc.org/press/benefits-of-green-building. Accessed 1 April 2022.

Vikør, K.S. (2014). Sharīʿah. In E. El-Din Shahin (Ed.), *The Oxford encyclopedia of Islam and politics*. Oxford: Oxford University Press.

World Bank. (2018). *Beyond scarcity: Water security in the Middle East and North Africa*. MENA Development Report World Bank.

World Commission on Environment and Development. (1987). *Our common future*. Oxford University Press.

World Green Building Council. (n.d.). *MENA regional network*. https://www.worldgbc.org/our-regional-networks/middle-east-north-africa . Accessed 6 January 2021.

Damilola S. Olawuyi is Professor and UNESCO Chair on Environmental Law and Sustainable Development at the Hamad Bin Khalifa University (HBKU) Doha, Qatar, and chancellor's Fellow at the Institute for Oil, Gas, Energy, Environment, and Sustainable Development (OGEES Institute), Nigeria. A prolific and highly regarded scholar, he has published several influential peer-reviewed articles, books, and reports on climate finance, energy infrastructure, and extractive resource governance. His most recent book publications are Environmental Law in Arab States (Oxford University Press, 2022), Climate Change Law and Policy in the Middle East and North Africa Region (Routledge, 2021), Local Content and Sustainable Development in Global Energy Markets (Cambridge University Press, 2021), the Human Rights-Based Approach to Carbon Finance (Cambridge University Press, 2016), and Extractives Industry Law in Africa (Springer, 2018), among many others. A Senior Advocate of Nigeria, he has been admitted as Barrister and Solicitor in Alberta, Canada; Ontario, Canada; and Nigeria.

Elena I. Athwal is a Research Fellow at the UNESCO Chair on Environmental Law and Sustainable Development, Hamad Bin Khalifa University (HBKU) Doha, Qatar. She is a former Assistant Professor in the Art History Department at Virginia Commonwealth University School of the Arts in Qatar. She is a specialist in the urban development of several Arab cities and in urban public policy. She is currently studying in the JD program of the HBKU College of Law, in Doha, Qatar.

Chapter 4
FIFA World Cup 2022 as a Catalyst for Environmental Sustainability in Qatar

Orjan Lundberg

Abstract Hosting the FIFA World Cup, an international football competition for national men's teams, in Qatar has acted as a catalyst for environmental sustainability. A mega sporting event such as the FIFA World Cup requires concerted efforts from nearly all sectors of a country. A strong sustainability program with a focus on preparation, event time operations, and post-event legacy was required to capitalize on the opportunities associated with hosting such an event. A robust and inclusive process to design such a program for the event laid the foundation for positive legacy outcomes. The legacy outcomes are categorized as physical, knowledge, and institutional, all being important contributors towards the realization of the Qatar National Vision 2030. This chapter describes the sustainability program development process, provides examples of how strategic initiatives for Sustainable Buildings, Greenhouse gas Emissions, and Waste Minimization catalysed positive developments leading to strong and tangible legacy outcomes. It is plausible that future events may look to this program to seek inspiration and lessons learnt for how to design and benchmark their sustainability programs. Qatar will also host many large events in the future, including the Asian Games 2030, and will capitalize on the developments that the FIFA World Cup brought about.

Keywords Mega-events · Sports-event · Sustainability · Legacy · Sustainable buildings · Greenhouse gas emissions · Climate action · Recycling · Circularity · Sustainable development

4.1 Introduction

The FIFA World Cup, an international football competition for national men's teams, is arguably the biggest and most exciting sporting event in the world. It is held every four years in different locations with the winner crowned the world champion of football. Over 3 billion people watched the FIFA World Cup live in 2018, either on

O. Lundberg (✉)
Supreme Committee for Delivery and Legacy, Doha, Qatar
e-mail: o.lundberg@sc.qa

L. Cochrane and R. Al-Hababi (eds.), *Sustainable Qatar*, Gulf Studies 9,
https://doi.org/10.1007/978-981-19-7398-7_4

TV, via online platforms, or in stadiums (FIFA World Cup Russia, 2018). This event has an enormous global reach.

In 2009, the State of Qatar set out on a journey bidding for the hosting rights of the FIFA World Cup. The hosting concept presented was a compact tournament with nearly all facilities in the capital city of Doha and remaining tournament sites within 50 km. The bid included world-class facilities, such as new hotels and stadiums. Environmental sustainability was a cross-cutting theme running through the bid, with the promise of stadiums certified to green building standards and a fully carbon neutral hosting of the event itself. In fact, the bid saw the design and construction of a 500-seat showcase stadium with the purpose of demonstrating that it is possible to provide a comfortable outdoor football stadium environment in the middle of summer powered by renewable energy. This showcase stadium is located with the Al Thumama Stadium precinct.

On the 2nd December 2010, Qatar was announced as the host for the 22nd edition of the FIFA World Cup to be held in the year 2022. The then bid committee was converted to a delivery committee, which eventually evolved into the Supreme Committee for Delivery and Legacy (SC) with a mission to:

> To share Qatar's unique identity with the world through amazing FIFA World Cup™ experiences that will connect people like never before. We celebrate football and our region, build intercultural understanding and develop opportunities. Hosting the tournament advances Qatar's ongoing mission for a sustainable future and a lasting legacy for our country, the Middle East, Asia and the world (FIFA World Cup Qatar, 2022d, para. 1).

Although one of the main tasks for the SC was ensuring the delivery of all preparations required for hosting the FIFA World Cup in Qatar, there was always a focus and an emphasis on the sustainable legacy of the tournament and how it could catalyse positive change across the country and the region. While the SC was the coordinating Committee in Qatar, numerous, if not all, other government, semi-government, and private sector entities in Qatar played major roles in delivering the FIFA World Cup preparations. This chapter focuses on the role that the FIFA World Cup has played in Qatar as a catalyst for environmental sustainability, in the lead up to the event as well as its legacy afterward.

4.2 Setting the Direction

Qatar made bold sustainability commitments in the bid using the Qatar National Vision 2030 (QNV 2030), published in 2008, as a guiding document (see Chap. 2). The QNV 2030 is setting the direction for the country's development and is in many ways a blueprint for sustainable development. With 2022 as an approximate midpoint for the QNV 2030, the FIFA World Cup is the perfect milestone to expedite and catalyse sustainable development and major sustainability-oriented projects.

In parallel to the ambitions of the State of Qatar, FIFA was also increasing their environmental ambitions for their flagship event, the FIFA World Cup. FIFA's sustainability plans, strategies, and reports became increasingly sophisticated for the Germany 2006, South Africa 2010, Brazil 2014, and Russia 2018 editions of the event. The Olympic Games had also continuously increased their environmental ambitions, with a starting point at the 1994 Lillehammer winter Olympics, via Sydney 2000, and the groundbreaking London 2012 games. Talavera et al. (2019) provide a comprehensive historical timeline of sustainability and sports mega-events, demonstrating an increasing emphasis on the topic over time across all aspects of mega-events.

Qatar and FIFA formed a joint venture, the FIFA World Cup Qatar, 2022 LLC (Q22), which also held high sustainability ambitions. This was demonstrated by appointing a Sustainability Manager as one of their first new positions. FIFA, the SC, and Q22 quickly formed a strong sustainability partnership and formalized both working and steering committees that held regular meetings throughout the event preparations. This resulted in a historic decision, where the FIFA World Cup Sustainability Strategy would be a joint strategy between FIFA and the host country for the first time. The strategy was jointly developed, and it was agreed it would also be jointly delivered. The strategy was developed using a comprehensive process with extensive consultation, involving over 100 stakeholders through surveys, focus groups, and interviews (FIFA World Cup Qatar, 2022a). The development process can be summarized in three steps: (1) thorough analysis of context and existing strategies and requirements, (2) identification of strategic priorities through a materiality assessment and a human rights salience analysis, with both results eventually combined through an innovative process, and (3) definition of objectives, initiatives, and action plans. The document was an internal working document for some time before it was published in January 2020.

The publication of the strategy was complemented with a detailed document describing the development process (FIFA World Cup Qatar, 2022a). The intention of this document was to communicate the process used to stakeholders, but also to contribute to the growing body of knowledge around sustainability strategy development for major sporting events. The strategy covers all three phases of tournament delivery—the preparation, staging, and post-tournament activities—from the time the strategy was developed to the completion of post-tournament activities in 2023, when all temporary structures will have been dismantled and the final tournament sustainability report published.

The FIFA World Cup 2022 Sustainability Strategy covers five pillars, has 22 objectives and over 70 different initiatives designed to deliver the objective of the strategy (FIFA World Cup Qatar, 2022c). This strategy is by far the most comprehensive sustainability strategy for any FIFA event and sets a new benchmark for future sustainability strategies for FIFA World Cups (Fig. 4.1).

This joint approach to sustainability strategy development was a first for any FIFA event. Previous sustainability strategies were developed by FIFA and their local organizing committee without much involvement of the host nation. The new

Fig. 4.1 The FIFA World Cup 2022 sustainability framework's five strategic pillars

joint approach enabled a stronger focus on the event preparation period and the legacy of the event.

The sustainability program for the FIFA World Cup 2022 in Qatar was unique in many ways. With Qatar being a comparatively small host country for the FIFA World Cup, most sectors in the country were involved in the delivery of the preparations for the event. This presented an opportunity for extensive partnerships with national stakeholders to deliver sustainability programs. The SC, FIFA, and Q22 were the immediate tournament organizers but relied heavily on partnerships with other public sector organizations and the local, regional, and international private sector. Partnerships for delivery and for maximum legacy benefit was very important for the organizers and served as a guiding principle to deliver the sustainability program.

The focus of this chapter lies with the environmental sustainability priorities in the Strategy, which are:

Sustainable Buildings (Sect. 4.3)
• Design, construct, and operate FIFA World Cup 2022™ sites to limit environmental impacts, while building local sustainable building expertise, supply chains, and standards.
Greenhouse Gas Emissions (Sect. 4.4)
• Measure, mitigate, and offset all FIFA World Cup 2022™ GHG emissions, while advancing low-carbon solutions in Qatar and the region.
Air Quality
• Minimize local air pollution from FIFA World Cup 2022™ stadiums, training sites, overlay infrastructure and transport services, promoting access to cleaner technologies in Qatar.
Waste Minimization (Sect. 4.5)
• Minimize waste sent to landfill from FIFA World Cup 2022™ sites, and promote waste management and recycling solutions in Qatar.
Water Conservation
• Minimize water use during the construction and operation of FIFA World Cup 2022™ sites and promote water conservation in related sectors.

Three priorities are discussed in detail in this chapter: Sustainable Buildings, Greenhouse Gas Emissions, and Waste Minimization.

To deliver initiatives in these priority areas, sustainable procurement and extensive training and capacity building were used as important tools. This chapter aims to summarize the key environmental outcomes of the programs and illustrate with examples how the program acted as a catalyst for further development in the same direction in Qatar. The chapter is not a comprehensive report on all aspects of the sustainability program and its outcomes, the final FIFA World Cup 2022 Sustainability Report will fulfil that function.

The legacy impact within the focus areas of the chapter can be categorized in three domains: physical aspects, knowledge aspects, and institutional aspects:

- *Physical aspects* refer to new infrastructure and material supply, such as parks, buildings, and rail systems.
- *Knowledge aspects* refer to new skills, experiences, and innovations and are often delivered through training courses, on-the-job learning, conferences, articles, film clips, and books.
- *Institutional aspects* refer to new or improved organizations that regulate or enable sustainable development, for example an upskilled government department.

4.3 Sustainable Buildings

In a global context, buildings and infrastructure are major contributors to overall greenhouse gas emissions. Their designs have long-term impacts, as they normally stand to serve their purpose for several decades. As a result, improved water or energy efficiency of their design can lead to very large cumulative savings.

The design and construction of stadiums for the tournament started well before the publication of the Sustainability Strategy, however, based on the commitments in Qatar's hosting bid and FIFA's green building requirements, the SC set stringent green building requirements for all eight stadiums. The requirements included certification to the independent green building certification scheme, Global Sustainability Assessment System (GSAS), and for the stadium contractor to appoint a sustainability manager and team with certain sustainability credentials. Green building certification requirements for such large infrastructure projects created a ripple effect of sustainability throughout the supply chains of materials, equipment, and services. For example, all electrical equipment, such as Air-conditioners, fans, pumps, lights, etc., installed in the buildings are assessed for energy efficiency. All plumbing equipment such as taps, toilets, and showers were assessed for water efficiency. All chemicals used, such as adhesives and paints were assessed for their chemical content. To meet the criteria for all these types of material and equipment, these criteria had to be included in the requirements used for procurement. When sustainability criteria are a consideration to award winners of commercial tenders, suppliers will make sure they are able to provide products that meet the sustainability requirements as it affects their bottom line. With suppliers responding to sustainability criteria, sustainable products become more readily available in the market at better prices.

Large projects such as stadiums have a long construction process that span over several years. The environmental impact from construction activities was managed in several ways. First and foremost all the activities had to go through the environmental through compliance with the national environmental legislation and permitting process governed by law 30 of 2002. One key component of that process was to develop and implement a comprehensive Construction Environmental Management Plan (CEMP). In addition to the environmental permitting and the CEMP, all stadium construction projects implemented the GSAS Construction Management certification system. This system added an additional layer of control and assessment of the environmental impact from construction practices. These processes enabled clear processes and strict controls on environmental impacts such as groundwater, dust, noise, waste, energy and water use, chemical and fuel handling, etc.

The major projects of Msherieb Downtown Doha and Doha Metro were under construction in Qatar when the construction for the stadiums commenced. These two projects also included green building certification requirements and targets. These major projects required sustainable products, sustainable certification, and sustainability professionals around the same time and created a notable change and uptake of green practices in the local construction sector. This also resulted in a steep increase in the number of GSAS projects in recent years both in Qatar and in neighbouring countries such as Kuwait and Oman (Planning & Statistics Authority, 2020).

As a result of the green building requirements for the stadiums, thousands of professionals were trained in green practices, and thousands of hours were spent inspecting and auditing the stadiums for their green credentials. New documents with best practices and practical advice for future projects were published and incorporated into professional development programs. The environmental performance of all materials used in the buildings was assessed, and it was found that the energy and water demands were reduced by 40% compared to international standards and 80% of construction waste was recycled.

The green building program for the FIFA World Cup was not limited to stadiums. The headquarters of the SC and Q22—Al Bidda Tower—became the very first high rise building in Qatar to receive a GSAS operations certification. This certificate was awarded for the green operational practices implemented by the facility management team. The key learnings from certifying the headquarters were documented and shared in a 'Green Book' with other government offices. These lessons were carried forward to inform the operations of the stadiums supporting the achievement of GSAS Operations certification at the stadiums. Assessing the green aspects of the operations of buildings was completely new in Qatar and this program opened the door to improved building management practices in the country.

Hotels are another building type that play a very important role when hosting a mega sporting event. Over 1 million visitors, including officials, players, fans, and media are expected to visit Qatar over the duration of the tournament. The SC collaborated with the Qatar Green Building Council (QGBC) to consult Qatar's hospitality sector and encourage them to adopt sustainable practices in the lead up to the FIFA World Cup 2022™ and beyond. This program has supported hotels to earn certification, through initiatives like the Green Key Award. As a leading international standard of excellence in the field of environmental responsibility and sustainable operation within the tourism industry, the award assesses 13 criteria, including waste management, green activities offered to guests, and staff engagement. A comprehensive industry survey was carried out with over 100 hotels, which documented the challenges and opportunities for implementation of sustainable practices in the hospitality sector. Green Key has certified 11 hotels in Qatar since 2016. Best practice examples that were implemented in hotels across Qatar include implementing a 100% plastic free policy, reducing food waste, extending recycling programs, using green cleaning products and non-toxic building materials, and revising menus to include more locally sourced produce.

Summarizing the outcomes of the Sustainable Building program for the FIFA World Cup shows significant progress and contribution to all the three legacy themes. Below are a few key examples of how the sustainable building program acted as a catalyst for sustainable development in Qatar.

4.3.1 Physical

The new stadiums have pushed the boundaries on GSAS scoring with several 'first ever' achievements. Al Janoub Stadium was, for example, the first ever building, not just stadium, to achieve the GSAS certification for design, construction, and operations. As another example, the Education City Stadium was one of the very first buildings to reach the five star GSAS certification level.

Given the warm temperatures in Qatar, cooling was a challenge in and of itself, made more challenging with the sustainability ambitions. To achieve this, most of the stadium precincts include a district cooling system that produces chilled water for cooling on a large scale in an energy efficient manner, in some cases allowing additional future buildings to connect to this efficient source of cooling after the FIFA World Cup. These type of systems are up to 50% more efficient than conventional cooling systems (Qatar2022, 2022b). Another physical element of the stadiums includes very large new green spaces in Qatar, with over one million square metres of parklands in the precincts around the stadiums. These parks use plants suitable to the local environment grown in the SC's own plant nursery. The parks are irrigated with recycled water, allowing further development of the local recycled water network. New green spaces also have a cooling effect on the local environment.

Finally, and perhaps one of the most significant aspects of the physical legacy is the change throughout the supply chain. When major development projects, such as stadiums, set clear sustainability requirements, it impacts the supply chain and suppliers will bring new lines of products to the market to meet these requirements. The combined efforts of Doha Metro, Msherieb, and the SC Stadium projects have improved the availability of more sustainable products in the Qatari market.

Examples of regulatory changes that have been introduced recently is the minimum energy efficiency and labelling standard for air-conditioners. Green building requirements are also included in the Qatar Construction Standard. Local manufacturers and importers of sustainable materials has increased, as an example stadium seats were produced locally.

4.3.2 Knowledge

The stadium design and construction program was innovative in many ways and new solutions and technologies were tried and tested in many areas, including to advance sustainability. With over 50 sustainability professionals across the program at its peak, the amassed sustainability expertise was vast. Monthly sustainability knowledge sharing sessions were held to foster a culture of sharing, peer review, and inspiration. Case studies were documented throughout the process, which were shared across an online platform accessible to all who were working on the program for immediate implementation on respective projects. These case studies were later analysed further and summarized in best practice reports that were shared in public for anyone working on construction projects in Qatar and the region. These best practice reports covered topics such as waste management, dust control, energy efficiency, and water management. In addition to this semi-formal way of training, formal training sessions were provided specifically on the GSAS certification system. Thousands of hours of learning were provided through these processes, with summary reports of best practices published on Waste and Dust Management.

4.3.3 Institutional

GSAS certified stadiums was a commitment made during the bid to host the FIFA World Cup. This green building system provided a robust and stringent assessment of the sustainability credentials of the stadiums. The stadium buildings were not like any other building; typical space layouts, user patterns, and equipment were different to standard buildings. As with most standards, the goals and performance aspects remain the same while the application and assessment may change ever so slightly to make the assessment and the compliance practical. For example, since the stadiums were significantly larger than most other buildings, an additional site audit was introduced to be able to check all aspects. The teams on site also provided extensive review comments and suggestions for how to streamline and improve the application and user experience of the GSAS assessment system. One outcome from this was implementation of the GSAS gate online submission and assessment system.

Many of the stadium precincts had district cooling energy centres and there was a need and interest to assess the energy efficiency of these systems. The GSAS administrators developed a new calculator in response to this interest that provided a streamlined way of calculating the Seasonal Energy Efficiency Ratio (SEER) of the full district cooling system. Understanding the SEER values of district cooling system designs will contribute to better decision-making in future district cooling projects.

To summarize, the sustainable building program acted as a catalyst in many ways and contributed to advancing the wider design and construction in Qatar and the region. New solutions, upskilled professionals, valuable resources and new green supply chains, strengthened governing institutions all contributed to sustainable practices.

4.4 Greenhouse Gas Emissions

The bid commitment to deliver a carbon neutral FIFA World Cup in Qatar was very ambitious and innovative at the time. Nothing similar had been attempted on such a large scale. When Qatar was awarded the right to stage the first FIFA World Cup in the Middle East and Arab world, it not only committed to organizing an amazing tournament—but also a mega-event delivered in a sustainable manner.

Part of the strategy includes the delivery of the first carbon neutral FIFA World Cup in the history of the event. To achieve this, all infrastructure projects—including the eight stadiums that will be used for Qatar 2022—had to meet stringent sustainability benchmarks. When the SC, Q22, and FIFA set out its carbon neutral objectives, several key areas were identified in the planning. Achieving carbon neutrality is a process made up of five key components. Firstly, it required raising awareness among key stakeholders, including the general public. This was followed by creating a detailed estimation of the carbon footprint, also known as greenhouse gas emission inventory. The third step was to mitigate and limit carbon emissions, followed by investing in credits from green projects that will offset any of the remaining emissions associated with Qatar 2022. The final step is to ensure a lasting legacy of climate action.

The greenhouse gas inventory was developed by an international carbon expert consultancy in partnership with a local Qatari firm under a contract with the tournament organizers. The partnership made sure that all relevant best practices and international standards were applied, while understanding local context and data sources when estimating the carbon emissions for the event. There was also a component of capacity building within the local consulting firm that later went on to carry out similar studies in Qatar. Bringing sustainability expertise to the local market is a key enabler for ongoing sustainable development, and in this case, better understanding of greenhouse gas inventory studies and analysis of where to focus climate action for best impact.

Staging the most compact version of the World Cup in modern history was one important feature in the carbon reduction strategy. Many of the measures taken to reduce carbon emissions centred around the compact nature of the tournament. Fans, players, and officials will stay in one accommodation during the tournament. With the longest distance between stadiums being just 75 km, fans will have the opportunity to attend at least two matches in a single day during the group stage, since five of the stadiums are connected to the Doha Metro and the other three accessible by connecting bus services. Many of the buses serving the stadiums will be electric buses, since 741 new fully electric buses have been supplied and will be in use by the time of the event, making up 25% of the bus fleet. The SC has also worked with the hotel sector in Qatar to encourage green practices to reduce carbon emissions from accommodation. This project saw publication of best practice guidelines for the hotel sector to reduce energy consumption, manage waste, and provide efficient water management and reuse, as well as practical guides for how to achieve certification for green hotel operations (Qatar2022, 2022a).

The compact nature of Qatar 2022 will not only provide convenience for fans and an optimal performance environment for participating teams, but it will also eliminate the carbon footprint of domestic air travel that fans, players, and officials were required to do to attend matches in previous World Cups. With air travel being recognized as one of the world's largest sources of carbon emissions, this will have a significant impact on reducing the tournament's carbon footprint.

Reducing carbon emissions was also factored into the building of Qatar 2022 venues—most notably the eight World Cup stadiums. Careful consideration was given during the design and construction of the stadiums for lower energy and water consumption, the reusing and recycling of water and materials wherever possible, highly efficient cooling systems, the use of renewable energy for lighting in some stadiums, creating designs that allows for natural light, native trees and plants in stadium parklands, and efficient waste management strategies. The reduction in carbon emissions from introducing new green space and plants was estimated at 23,000 tonnes per annum (Spanos et al., 2021).

All eight stadiums that will be used during Qatar 2022 have been built to be 40% more water and energy efficient when compared with conventional designs. Many of the stadiums have exceeded industry ratings by implementing non-mandated requirements for sustainability. One such example is Stadium 974. The 40,000-capacity venue will be the first fully demountable stadium in FIFA World Cup™ history, with its components being used to create other sporting projects during legacy mode. This pioneering approach will set new benchmarks in waste reduction. The carbon benefits of this innovative design approach were studied in detail, and it was concluded that the demountable approach was beneficial, provided that the new location was not too far away (FIFA World Cup Qatar, 2022b).

Another key component to achieve carbon neutrality will be the use of solar energy to help power Qatar during the tournament. Qatar is currently developing a large-scale 800 MW solar energy plant on a $10km^2$ plot in Al Kharsa'ah, in the central part of Qatar. The voluntary carbon market will be used to procure carbon credits to offset the remaining carbon emissions. The carbon credits came from projects such as solar, hydro and wind energy, waste treatment, energy efficiency, landfill efficiency, and industrial efficiency. All carbon credits came from verified third-party standards and from a variety of countries, including Qatar.

These initiatives will leave an important sustainability legacy for Qatar and the region and provide an invaluable blueprint for the use of sustainable practices in future sporting events held around the world.

4.4.1 Physical

One of the main physical carbon reduction projects in Qatar has been the transformation of the public transportation system in the country, including the Doha Metro, Lusail and Education City light rail, and the new fleet of electric buses. These solutions are providing electrified mass transit options, and reducing carbon emissions per kilometre travelled per passenger significantly when compared to the more conventional modes of private transport typically used. The Al Kahara'ah large-scale solar power plant will supply renewable power for the tournament. Once the tournament concludes, the plant will continue to produce clean renewable energy for many years to come, leaving a significant legacy as a carbon reduction project. The numerous carbon offset projects in Qatar and in the region that received funding from the purchase of credits towards the FIFA World Cup will continue to operate and deliver carbon savings for many years beyond the event. The funding that came through this mechanism contributed to the realization of these projects, as they became more financially viable due to this funding.

4.4.2 *Knowledge*

The Qatar 2022 program included a large amount of outreach on the topic of carbon. This included a webinar series with Josoor Institute, an online course delivered on the Edx platform with Hamad Bin Khalifa University (HBKU), numerous conference presentations including international United Nations conferences as well as local conferences such as the Qatar Sustainability Summit and Qatar Sustainability Week. These activities have built knowledge and capacity within Qatar, and have also facilitated a sharing of lessons learned globally.

In order to assess carbon reductions, new and adopted calculation methodologies were developed. Some examples of this include the carbon assessment of the demountable Stadium 974 and the extended scope of the event carbon footprint due to the inclusion of additional activities of the host country that were excluded in assessments of previous events. Some of the projects that were the basis for the carbon credits were either using new technology or were in new locations, requiring the assessors of these credits to develop new methodologies to accurately estimate and verify the carbon reductions. The development of these new methodologies and technologies will have legacy impacts beyond Qatar and beyond 2022.

4.4.3 *Institutional*

The carbon program for the FIFA World Cup 2022 saw the formation of the Qatari institution Global Carbon Council (GCC), a voluntary carbon offsetting program. The GCC aims to assist organizations that reduce their carbon footprints, help different sectors of the economy to diversify by adopting low-carbon pathways and catalyse climate actions on ground. The key activities of the GCC are to:

- Support organizations implementing GHG reduction projects and measures;
- Develop standards for emission reduction calculations and monitoring of GHG reduction projects;
- Establish market system for carbon offsets;
- Establish a carbon offsetting program that deliver credible, cost-effective, and sustainable carbon offsets; and
- Communicate organizational and regional climate actions.

The SC supported the formation of the GCC through providing sustainability expertise and providing funding to develop the first GCC-specific GHG calculation methodologies, alongside a commitment to procure carbon credits from this scheme once operational. This allowed the GCC to approach prospective carbon reduction project owners in Qatar and in the region to establish baselines and calculate emission reductions with the goal to convert carbon reductions to carbon offsets, which in turn were cancelled for the FIFA World Cup hosts to achieve carbon neutrality. To date, nearly 200 projects have been registered with the GCC. The GCC is now a fully operational institution that has been recognized by several international governing bodies for its credibility and will remain in operation long past 2022 carrying out its mission to expedite climate action in Qatar and the region. Similarly, those carbon reduction projects that have been assessed with credits issued, will continue to generate carbon credits that can be traded for years to come generating revenue that can be re-invested in additional climate action projects.

4.5 Waste Minimization and Recycling

The Qatar National Vision and the UN Sustainable Development Goals both place a strong emphasis on reducing waste, encouraging recycling, and moving towards a circular economy. The approach for Qatar 2022 followed the standard waste hierarchy, simply put Reduce—Reuse—Recycle. The program focused on construction waste in the first phase while later focusing on event time waste.

Construction waste was managed through contractual requirements to achieve certain recycling targets on select sites and to achieve the GSAS certification requirements, which also cover construction waste. The overall recycling rate for construction waste reached 79%. Detailed tracking and review of the waste management systems and its performance drove a systematic change to how construction waste was handled on the stadium sites, which contractors will carry forward to future projects. The SC collected data, best practices, and lessons learned throughout the construction period. This information was summarized and published in a free to access resource that guides any contractor on how to achieve high recycling rates during construction (GORD, 2021). This resource is used by contractors across Qatar and examples from the resource are used in GSAS training courses. Practical examples and case studies of effective methods implemented locally are very valuable for local contractors. A lot of information about this topic comes from practices implemented abroad in a different context, which in many cases can prevent implementation in Qatar. This new resource with local examples has been received very well by professionals working in Qatar.

Construction waste and event time waste are two quite separate issues to tackle. Although the principles of reduce, reuse, recycle are the same, the type of waste and the operational environment is very different. The FIFA Club World Cup 2020 and FIFA Arab Cup 2021 were the two main test events ahead of the FIFA World Cup. These events were also used to refine the waste and recycling operations. The testing included new operational models, communication and awareness campaigns, waste bin design, upstream reduction measures, and downstream treatment options. Waste streams were also measured, which resulted in the insight that organic waste was by far the biggest contributor to downstream waste. The organic waste was mainly made up of food leftovers, compostable cutlery, and grass clippings. Cardboard and plastics were also important contributors.

During the event time, most spectators are focused on the event that they have come to attend and may not pay much attention to recycling and source segregation, hence the waste in the bins was mixed and not very well segregated. To increase the segregation of waste streams, manual on-site segregation was carried out. This system allowed cleaner waste going to each recycling facility. et al. Bayt Stadium during the FIFA Arab Cup, the recycling rate for event time waste reached 70%. On-site composting facilities were used to process the organic waste. The compost was then used at one of the largest farms in Qatar.

Additionally, a significant portion of single-use items were sourced as compostable rather than plastic, which was the standard practice. Workforce food service was carried out in a buffet style, which eliminated a large amount of food packaging and created a nicer atmosphere for the workforce. Reducing single-use food packaging and replacing single-use plastics with compostable alternatives were key strategies to reduce the overall waste and increase the recycling rates.

During the two test events, it became obvious that the general awareness of recycling among spectators and workforce was very low. In response to this problem, the SC created the OneTide community campaign (One Tide n.d.). Using awareness messages online and at in-person events, the OneTide campaign aims to raise awareness and understanding of recycling in general and plastic waste recycling in particular.

Reflecting on how the FIFA World Cup program acted as a catalyst in the area of waste management and recycling, there are legacy benefits in several areas, as discussed in the following sections.

4.5.1 Physical

The physical legacy of the waste and recycling program is the material that was not sent to landfill, but rather diverted to other treatment processes. This diverts a negative environmental impact and a cost benefit as a potential new revenue stream for stakeholders in Qatar. National capabilities for organic waste treatment through large-scale treatment and smaller-scale composting have developed significantly during the years of Qatar's FIFA World Cup preparations. Local capabilities now amount to several options for local composting services, local manufacturing of certain composting machines, and of course availability of locally produced compost. Qatar's arid environment combined with its ambitions to plant a million trees and to build new green parks makes the shift to locally produced compost even more important. Another systems shift is the availability of compostable single-use cutlery, food packaging, and plates during the lead up to the tournament. The supply and local manufacturing of these types of products enables reduction of plastic waste and increases the composting rates recovering much more valuable material. Although single-use plastics are still used in many places across Qatar, the lead up to the FIFA World Cup has seen a significant shift in the update of compostable alternatives. These physical changes are contributing to a broader societal shift towards more sustainable consumption.

4.5.2 Knowledge

Following the two test events and the construction projects, it is now proven that it is possible to achieve high recycling rates in Qatar in many sectors, however a concerted effort is required. It is also well understood what is required to reach even higher rates and close the loop by achieving zero waste to landfill. These experiences have been documented and shared either in public or directly with relevant stakeholders. New operating procedures, a consolidated list of recyclers, and even new businesses have been established as a result. The consolidated list of all known recycling companies in Qatar is a valuable resource in delivering waste and recycling projects. At the start of the program, the providers were unknown to most and it was very challenging to find recyclers for any material. Through the collective effort of the professionals working on the program, a long list of recycling companies in Qatar covering nearly all materials was available for all to use. This was a contributing factor to achieving an average recycling rate at nearly 80% for construction waste and 70% of event time waste et al. Bayt Stadium during the FIFA Arab Cup.

4.5.3 Institutional

The OneTide awareness program was a direct outcome of FIFA World Cup 2022 preparations and a response to the low awareness of recycling in general and plastic waste in particular. OneTide has been using football stars to engage with a wide audience on these topics. OneTide has also produced short clips highlighting some local waste and recycling champions that are contributing to improved waste and recycling in Qatar. In addition, several targeted training courses were provided, for example on how to reduce plastic waste from merchandize and gifts. The OneTide program is intended to carry on after the tournament and continue to deliver education and awareness programs. With Qatar's recycling sector growing fast, new laws, regulations, and standards have been drafted or passed to improve the processing of materials, encourage source segregation of waste, and allow recycled materials in more applications (see Chap. 12 and Chap. 16).

4.6 Future Options and Conclusions

Sustainability in sports and mega-events is a rapidly growing field receiving more emphasis and attention. This is partially because of the opportunities associated with the spotlight that a sports club or event can provide to the issue of sustainability, but also the growing sense of responsibility among the sports industry. The response to these issues by organizers of events such as Olympics and FIFA World Cups has become increasingly sophisticated over the years, and it is expected that this will continue to grow over time. Beyond the mega-events, other large events are also taking on the sustainability challenge.

Hosting the FIFA World Cup in Qatar is a unique opportunity to catalyse positive development in the country in many different areas, including environmental sustainability. This FIFA World Cup has already had profound impact on the country as a whole. This short book chapter gave some brief insights and examples of how the event contributed to improved environmental sustainability. The event catalysed physical improvements, enhanced knowledge, and institutional developments. The sustainability approach at the FIFA World Cup in Qatar was the most comprehensive to date. The strategy development, the human resources, the programs, the reach, and the depth of the impact were beyond what had been seen before in the history of the FIFA World Cup.

At the time of writing, the FIFA World Cup 2022 has not yet started, consequently it is not possible to conduct a comprehensive assessment of all plans and interventions. Although public progress reporting is carried out at fifa.com, this reporting is mainly focused on progress against plans rather than impact. Following all recent FIFA World Cups, a sustainability report has been published with comprehensive data sets provided and with assessment of impacts. These reports have been prepared in compliance with international best practice standards. The sustainability report for FIFA World Cup 2022 Qatar will be published in 2023.

It seems reasonable and plausible therefore that future events may look to this program to seek inspiration and lessons learnt for how to design and benchmark their programs. Qatar will also host many large events in the future, including Asian Games 2030, and should capitalize on the developments that the FIFA World Cup has brought about: new infrastructure, new operating models, new skills, and new supply chains. With sustainability becoming increasingly more important in selection of event host nations, Qatar can have a sustainability edge on many other hosting cities and countries. Its approach to sustainability has provided a blueprint future mega-event host nations can follow. The continuation of this requires a systematic approach to capturing best practices and lessons learnt across the sustainability value chain, and to keeping up with new developments and ideas. The physical, knowledge, and institutional gains catalysed by the FIFA World Cup 2022 set a strong foundation for the continuation and advancement of these sustainability gains.

References

FIFA World Cup Russia 2018. (2018). 2018 FIFA World Cup Russia™: Global Broadcast and Audience Summary. https://digitalhub.fifa.com/m/2589b77c20849beb/original/njqsntrvdvqv8ho1dag5-pdf.pdf

FIFA World Cup Qatar 2022. (2022a). Development of the Sustainability Strategy. https://www.qatar2022.qa/sites/default/files/FWC-2022-Strategy-Process-Development.pdf

FIFA World Cup Qatar 2022. (2022b). Greenhouse Gas Emission Analysis of a Demountable FIFA World Cup™ Stadium. https://www.qatar2022.qa/sites/default/files/documents/greenhouse-gas-emission-analysis-of-a-demountable-fifa-world-cup-stadium.pdf

FIFA World Cup Qatar 2022. (2022c). Sustainability Strategy. https://www.qatar2022.qa/sites/default/files/documents/FWC-2022-Sustainability-Strategy.pdf

FIFA World Cup Qatar 2022. (2022d). Vision. https://www.qatar2022.qa/en/about/vision-mission-values

Gulf Organization for Research and Development (GORD). (2021). Waste Management Best Practices for Construction Sites: Case Studies from FIFA World Cup Qatar 2022™ Stadiums. https://www.qatar2022.qa/sites/default/files/documents/fifa-world-cup-2022-construction-site-waste-management-report.pdf

Qatar2022. (2022a). Green Hotels. https://www.qatar2022.qa/sites/default/files/documents/Green-Hotels-Qatar.pdf

Qatar2022. (2022b). Sustainable Stadiums. https://www.qatar2022.qa/sites/default/files/documents/FIFA-World-Cup-Qatar-2022%E2%84%A2-Sustainable-Stadiums-EN.pdf

Planning and Statistics Authority. (2020). Environment Statistics Bulletin. https://www.psa.gov.qa/en/statistics/Statistical%20Releases/Environmental/EnvironmentalStatistics/Environment_Statistics_bulletin_2020_En.pdf

One Tide. (n.d.). Home. https://www.onetide.org/

Spanos, I., Kucukvar, M., Bell, T. C., Elnimah, A., Hamdan, H., Al Meer, B., Prakash, S., Lundberg, O., Kutty, A. A., & AlKhereibi, A. H. A. (2021). How FIFA World Cup 2022™ can meet the carbon neutral commitments and the United Nations 2030 Agenda for Sustainable Development?: Reflections from the tree nursery project in Qatar. *Sustainable Development*. https://doi.org/10.1002/sd.2239

Talavera, M. A., Al-Ghamdi, S. G., & Koc, M. (2019). Sustainability in mega-events: Beyond Qatar 2022. *Sustainability, 11*(22). https://doi.org/10.3390/su11226407

Orjan Lundberg has broad sustainability experience across strategy development, implementation, communications, training, and execution. His expertise lies in strategic sustainability advisory with practical implementation at the core. Orjan has been driving the host nation's environmental program for the Qatar 2022 FIFA World Cup since 2013. Prior to that Orjan was Executive Director at the Green Building Council of Australia. Orjan thrives on creating meaningful collaborations across industry sectors to drive sustainability transformation. His work spans across diverse sectors such as construction, architecture, urban design, sports, events management, hotel, food and beverage and academia. Orjan is also an experienced conference speaker, lecturer and moderator and a citizen of the world with extensive cross-cultural work experience from three continents. He holds a Masters of Civil Engineering degree from Chalmers University of Technology in Sweden.

Chapter 5
Qatar Foundation: A Sustainability Innovator

Fahad Al-Musalmani and Sylvie Maalouf

Abstract Addressing sustainability issues requires a complex multisectoral systems shift that cuts across government and public sector to private sector and civil society. Non-state actors have long played a role in promoting sustainability; from nonprofits to youth movements, to civil activists' think tanks and policy institutions, there are several examples of how these have shaped sustainability considerations worldwide. However, it is rare to find areas of intersection for the different actors. Far too commonly, efforts are siloed and as a result miss out on potential synergies from collaboration. Nowhere in the world do research, education, and civil society come together as they do in Qatar Foundation (QF), which is what places it so well as a critical sustainability innovator. Sustainability has been at the core of QF's mission for over 25 years. Recognizing its own role as a contributor to the country's ecological footprint, QF has been striving to enhance its sustainability practices and is committed to advance education on sustainability, promote and pilot innovative solutions, and advocate for national policy change. This chapter highlights this unique case study, one of the few places where preK-12 and tertiary education are found together, alongside NGOs, research institutes, think tanks, and cultural institutions, resulting in a unique ecosystem catalyzing innovation in sustainability.

Keywords Qatar Foundation · Sustainability · Innovation · NGOs · Community · Youth · Research · Education · Pilot · Testbed · Policy · Advocacy

5.1 The Role of Non-State Actors in Promoting and Enhancing Sustainability

Non-state actors have long played a role in promoting sustainability globally. Nongovernmental Organizations (NGOs), like the World Wildlife Fund set up in 1961, work around the world to conserve nature and protect biodiversity and have millions of supporters (World Wildlife Fund n.d.). Greenpeace is possibly one of the

F. Al-Musalmani (✉) · S. Maalouf (✉)
Qatar Foundation, Doha, Qatar
e-mail: falmusalmani@qf.org.qa

L. Cochrane and R. Al-Hababi (eds.), *Sustainable Qatar*, Gulf Studies 9,
https://doi.org/10.1007/978-981-19-7398-7_5

most recognized environmental entities working on promoting awareness and action to protect the planet from environmental injustice (Greenpeace n.d.).

The list of environmental and sustainability-focused NGOs and institutes is long, but the evidence of their increasing influence in promoting change has become more visible over the years. The recent May 2021 case, "Milieudefensie et al. v Royal Dutch Shell" where a court in the Netherlands ruled against Shell, ordering it to reduce its CO_2 emissions by 45% compared to 2019 levels by 2030, serves as an example of a landmark achievement for a civil society-led environmental group (BBC, 2021). Its applications and ramifications for the oil producer outside the Netherlands are yet to be seen, but the outcome is a testament to the power and change that can be brought up by NGOs.

Civil activism regarding climate change and environmental issues has been rising around the world, and the youth movement has become more vocal, having gained recognition among global leaders with the launch of the first youth-led conference held in Montreal ahead of the Conference of the Parties (COP) 11 in 2005. Since then, the Conference of Youth (COY) takes place yearly in advance of the annual COP meetings, allowing youth to participate in the discussions and voice their concerns to leaders around the world. Most recently, the COY 2021 and pre-COP26 meetings took place in Milan. Entitled "Youth4Climate: Driving Ambition", the conference invited prominent youth activists from around the world to take part in and engage in climate change dialogues (UN, 2021).

How did these NGO and youth movements reach such prominence? How can we empower the generation of tomorrow to lead the discussions related to sustainability and ensure a prosperous future for themselves and for future generations? Education and educational systems are a key component in answering both of these questions. The United Nations has long advocated for education to provide knowledge, skills, values, and understanding required to ensure promoting sustainable behaviors through its Education for Sustainable Development (ESD) framework (UNESCO n.d.). The enablers for youth-driven transformation include formal education, but also informal, experiential, and extra-curricular learning, where independent organizations such as research institutes, policy centers, or think tanks, play a key role. These institutions have been critical in shifting thinking, promoting policies for behavioral change, advocating for greener living standards, incorporating sustainability considerations in national budgets, holding leaders accountable for protecting our natural environment, and more.

Addressing sustainability issues requires a complex multisectoral systems shift. However, it is rare to find areas of intersection between the different actors outlined above. Far too commonly, efforts are siloed and as a result miss out on potential synergies from collaboration. Nowhere in the world do research, education, and civil society come together as they do in Qatar Foundation (QF), which is what places it so well as a critical sustainability innovator. This chapter highlights this unique case study, one of the few places where preK-12 and tertiary education are found together, alongside NGOs, research institutes, think tanks, and cultural institutions.

5.2 Qatar Foundation: A Unique Ecosystem with Amplified Impact

QF and its flagship project, Education City, presents a one-of-a-kind ecosystem that brings together multiple stakeholders and segments of society. Within its over 1,270 hectares, it hosts several world-renowned universities and its homegrown research-based Hamad Bin Khalifa University (HBKU). Education City hosts schools (including for specialized education), early education centers, the largest regional library, research and policy institutes, tech start-ups—as part of the Qatar Science and Technology Park (QSTP)—and multiple cultural institutions. This unique ecosystem allows it to be an active non-state actor, promoting sustainability through research and academia, community and student activists, and its own sustainability best practices.

Sustainability has been at the core of QF's mission for over 25 years. Recognizing its role as a contributor to the country's ecological footprint, QF has been striving to enhance its sustainability practices, as detailed in this chapter.

5.2.1 Leading by Example

A key component of sustainability relates to the built environment, and in Qatar's particular context this draws attention to the energy use and efficiency of the buildings. Several of Education City's built infrastructure follow international sustainability ratings. The QF Student Housing complexes have dashboards that detail the buildings' energy and water efficiency, and, in some cases, their air quality. Within Education City, many building structures are certified as green buildings under internationally recognized schemes. This includes the headquarters, an 11-floor structure built in 2015, with a total floor area spanning 27,366sqm that accommodates numerous QF departments and offices. The headquarters was designed with energy efficiency and sustainability in mind earning it a gold rating Operations and Maintenance Global Sustainability Assessment System (GSAS) certification from the Gulf Organization for Research and Development (GORD) (Qatar Foundation, 2021c).

The Education City Stadium, one of the venues for the upcoming FIFA World Cup Qatar 2022™, is another example of sustainability for the world. The stadium was designed with eco-friendly, passive design features, that enable it to maintain an appropriate climate within the structure for players and fans. The stadium was built using 85% regionally sourced, sustainable materials, minimizing the carbon footprint during building, and over 20% of materials were generated from recycled content. Other features, including low-water fixtures, efficient irrigation systems, LED lighting, and the use of native flora in the landscaping, add to its sustainable function. Its features have earned it three GSAS certifications, including a five-star rating in design and build, construction management and energy efficiency (Qatar Foundation, 2020a). The stadium takes advantage of urban connectivity, with its own metro station on the Doha Green Line just 500 m from the stadium, enabling spectators and visitors to reduce their carbon footprint by commuting. It is also connected to Education City's tram (FIFA World Cup Qatar, 2020) (see Chap. 4). Quite significantly, the legacy plans for the stadium post World Cup include using it as a hub for community engagement, while also housing two QF schools, making it a visual educational tool, showcasing how sustainable building practices can be applied to the community.

More broadly, QF has developed carbon reduction strategies to support national climate change mitigation actions. Following a greenhouse gas (GHG) inventory baseline in 2019, QF initiated a series of carbon audits for different building structures within Education City. The audits led to actionable detailed strategies to reduce CO2 emissions for each of the buildings. Showcasing its commitment to sustainability, QF joined the UN Framework Convention on Climate Change (UNFCCC) initiative "climate neutral now" in 2020, which aims to promote voluntary action to achieve carbon neutrality. Through annual GHG reporting to the UNFCCC, QF has joined a handful of non-state actors in Qatar who are part of the portal for non-state actors on climate action (NAZCA) (UNFCCC n.d.; CDP n.d.). QF obtained a Climate Data Portal (CDP) score of B in 2020, making it the best performing among its peers nationally (CDP n.d.b).

5.2.2 *Best Practices for Sustainable Behaviors: Operations and Waste Management*

Beyond the physical and built environment, QF has taken many steps toward encouraging environmental stewardship and has implemented a set of policies to reduce waste and energy consumption and promote a culture of sustainability.

5.2.2.1 QF as Recycling Innovator

QF has a long history of encouraging innovative approaches to recycling in Qatar, such as introducing designated bins that segregate waste at the source (at a time when this was not common in municipal waste management, see Chap. 16).

Since the institutional system for waste management beyond Education City did not collect segregated waste, systems and partnerships had to be created, showcasing how private–public partnerships can be leveraged to promote practices that advance sustainability beyond Education City. QF currently recycles paper, glass, metals, and plastics.

5.2.2.2 QF Helping Qatar Win the Battle Against Plastic

QF has put in place measures to reduce its waste, starting with policies and regulations to limit single-use plastics due to their severe impact on the environment. QF banned single-use plastic bags and requires all vendors to adopt alternatives. Plastic bags were replaced with paper for all classroom-related deliveries and takeaways in QF cafeterias. The use and sale of plastic bottles was stopped in QF-catered events, and at all QF school cafeterias. Food service contracts for events and catering services prohibit the use of plastic and encourage the use of biodegradable and compostable food packaging. QF pantries were populated with biodegradable disposable items, replacing plastic plates, cups, and cutlery. Further, reusable containers were distributed for meals delivered to grooms in the AlShaqab (QF's equine center committed to preserving Qatar's heritage by promoting the Arabian horse and setting the highest standards in horse welfare, breeding, equine education, and research). More detailed information on QF's journey on banning single-use plastics can be found in the case study published in 2020, which gathered the attention of national authorities and positioned QF as a key partner in the state's ambition to deliver a plastic-neutral World Cup tournament (Qatar Foundation, 2020b).

Another initiative focused on reducing single-use plastics is "Precious Plastics", a make-shift workspace, run and supported by QF students, where community members can drop their plastic waste. The workshop is itself a converted shipping container that was donated by the Supreme Committee for Delivery and Legacy. It contains machines and tools which grind, melt, and inject recycled plastic, and can be used to create new products, like furniture, small functional objects, or colorful sheets/bricks that can be used to make other products. People can use their plastic waste to create new products, therefore demonstrating the "true" value of plastics. The aim of this project is to nudge behaviors and increase recycling of plastic waste. The workshop was launched during the FIFA Arab Cup, in November of 2021. The space will be fully activated for the upcoming FIFA World Cup and will have a permanent home as part of QF's upcoming community recycling hub "Green Island" (explained later in this chapter).

5.2.2.3 Embedding Sustainability in Everything We Do

Sustainable and environment-friendly practices are consistently being improved and introduced across the spectrum of QF's operations. Examples include the use of environmentally friendly cleaning chemicals, e-waste recycling, composting, linen-free cafeterias, events, and restaurants, among others.

In 2019, QF launched an e-waste campaign, as part of the Qatar Sustainability Week (QSW), bringing awareness to the threat to health of electronic waste. Electronic waste must not be disposed of in landfills, as they can filter toxins which lead to groundwater pollution resulting in contamination (Qatar Foundation, 2020c; Qatar Foundation, 2021b). Through the positioning of strategic bins in Education City, over 4.5 tons of recycled e-waste were collected in just seven days. The most recent e-waste drive took place from October 2020 to January 31, 2021. The QF operations team partnered with a facility in Singapore, as there was no facility in Qatar to recycle e-waste. Policies, like having employees use their own mobile phones, and contractual stipulations with vendors to ensure all redundant and old equipment are recycled, were also introduced to reduce the QF workforce e-waste footprint.

QF uses produce from its nursery for school meals, reducing the carbon footprint of the food produced. Menu engineering and food waste monitoring are in place to reduce wastage, and recent stipulations were introduced to mandate food service contracts to localize their supply chains, minimizing carbon miles of the food served within Education City. QF kitchens are gas free and use energy-efficient electrical machines. Closing the loop, beyond policies to encourage reduced food waste, QF mandates that all food waste is composted, and encourages landscape contractors to use the compost generated in their landscaping, which also includes the requirement of using existing plants and propagating trees and plants in the QF nursery. Maintaining native trees is key to reduce the need for water and additional soil. The use of organic fertilizers (including fertilizers created from garden waste) is prioritized in nurturing the plants around Education City (see Chap. 12 & 16).

With these actions, QF supports the national zero-waste initiative as it continues to innovate and pilot waste management and reduction programs and circular economy principles that can serve as examples for national-scale policies.

5.2.3 *Advocating for Well-Being Through the Interface with Nature*

QF has long advocated for the importance of well-being, which it considers a key component to sustainability. The Education City master plan was achieved by establishing knowledge centers and connecting them through open green spaces, which are recognized as having an important role in supporting and improving individuals' health, well-being, and lifestyle.

Situated in a hot and dry climate, Education City includes several green spaces incorporating sport and recreation areas for basketball, volleyball, and football, a golf course, along with multiple areas for relaxation, fitness, and family enjoyment. Additionally, Education City is home to Oxygen Park, a 130,000 sqm green space, with running tracks extending over 800 m, wrapping around the three main "ground bowls", two of which are used as sports pitches. Its dynamic rolling landscape with alternating sunken and elevated play areas enable a variety of activity spaces and integrated leisure facilities. The Park invites visitors to enjoy the outdoors and be immersed in the refreshing scents of nature while increasing their oxygen intake in the Sensory Gardens. Promoting social sustainability linked to local heritage, traditions, and values, the park has dedicated times every Monday, Tuesday, and Wednesday, where several female-focused sports and community initiatives are run.

Exemplifying how Education City can serve as a testbed for innovation is the "Ability Friendly Program" which supports social well-being through inclusion. What started as the provision of recreational venues to families with children with special abilities, including those with autism, learning difficulties, and hearing, visual, and physical challenges, has now grown into a full-fledged sports program for children with special abilities. The positive impact the program has had on the lives of the children and their families has been widely documented and the program is now collaborating with the Supreme Committee to deliver wider impact by partnering with its Generation Amazing program. From a mobility perspective, QF's Education City is connected to three metro stations, and has its own internal tram, a groundbreaking system that sets new standards for sustainable transport, with an electrically powered technology based on modular on-board energy storage units (Qatar Foundation, 2019b). With a total of 24 stations, the Education City tram enables park-and-ride, car-less connections between the different university campuses, schools, facilities, and open spaces. Showcasing the role of QF as a testbed for sustainable solutions and innovations that can be scaled nationally, the Ministry of Municipalities included several of the landscape practices along with lessons learned and specific recommendations related to accessibility and inclusion provided by the QF team into the national landscape guidelines.

More recently, QF introduced car-free days to encourage alternative mobility from and to, as well as within Education City. The initiative aims to promote more walking and use of public transport, therefore enhancing air quality and improving health through increased physical activity.

5.2.4 Promoting Sustainability Through Community Engagement, Education, Research, Policy Advocacy, and Global Dialogue

5.2.4.1 Engaging the Community in Promoting Sustainable Practices

QF believes in the power of local communities and the role they play in tackling global challenges, including climate change. QF launched a "park and plant" initiative aimed at providing shaded pathways around Education City, building on the action of local communities to develop Qatar's first urban farm (Qatar Foundation n.d.). Starting with 2,500 trees planted around Education City, the park and plant initiative has grown with programs like "adopt a tree", for which individuals can register online; "green lungs" that provides an avenue for corporations or NGOs to support mass planting; and the Education City communal gardens, which are being developed in partnership with Qatar Academy schools and designed by HBKU. Well-being gardens were recently launched within QF's HQ building to promote employees' interaction with nature, providing a therapeutic space within the workplace to grow vegetables and plants (Qatar Foundation n.d.; MENA FM, 2022).

The Qur'anic Botanic Garden (QBG) is another community initiative of QF, which exhibits and conserves the plants mentioned in the Holy Qur'an and the native flora in Qatar. It serves multiple goals, from the preservation of Qatar's botanic diversity and research promotion to the revival of cultural traditions. Further, the QBG promotes education through sustainable interactions with the environment, building stewardship, encouraging respect for the environment we live in, and promoting the planting of trees through Islamic values.

To promote and enhance sustainable practices in urban development beyond Education City, QF launched the Qatar Green Building Council (QGBC). Set up during the peak of Qatar's construction boom and, through its affiliation with the World Green Building Councils, the QGBC started the green building movement and introduced and promoted certifications based on recognized global standards, which consider factors like energy efficiency, environment-friendly materials, and waste reduction (Earthna n.d.b). To achieve the promotion of standards and best practices, QGBC ran several professional training programs aimed at enhancing sustainability knowledge in Qatar, with over 35 training topics on more than 70 courses, reaching over 7, 000 trainees during a ten-year period. As the work of QGBC evolved, other aspects of sustainability were introduced, extending beyond the built environment, to focus on overall health and well-being (Earthna n.d.b).

Beyond the construction sector, tourism is important for the development of Qatar's economy. Recognizing, this, QGBC introduced the Green Key Certification, a leading standard in the field of environmental responsibility and sustainable operations within the tourism industry. By early 2022, the Green Key Eco-Label had been awarded to seven 5-star hotels in Qatar, with another 15 hotels under review. The QGBC has engaged national authorities to promote and encourage this standard for all hotels in Qatar. Within tourism, a key sector relates to the meetings, incentives, conferences, and exhibitions (MICE), where Qatar has become a strong hub. The high global carbon footprint of the MICE industry is well understood and to tackle this, the QGBC introduced the Eco-Event certification, to encourage more environmentally sustainable low-carbon events, including indoor and outdoor events. QF has ensured that recent World Cup related events are Eco-Event accredited and is looking for ways to institutionalize this across all its events.

Institutional and certification-based shifts are important, however a key component of the "green movement" must include community awareness and outreach. With the aim to further engage the community in a wide range of sustainability-oriented activities, QGBC launched and held several editions of the Qatar Sustainability Week. This initiative showcases the progress Qatar has made in the areas of sustainability and green buildings to the wider community and serves as a platform to promote the nation's sustainability vision. In 2019, around 250 events and activities were conducted (including awareness activities, fairs and exhibitions, sports and fitness, nutrition and wellness sessions, school activities, seminars, and conferences).

To recognize the efforts, commitment, and contributions of individuals, institutions, and organizations furthering sustainable development and environmental protection in Qatar and beyond, Qatar Sustainability Awards are awarded annually. Award categories include green buildings, green hospitality, green service providers, building products and technologies, green research, and sustainability initiatives, with hundreds of submissions every year. The awards allow for the development of a repository of information related to national and international activities promoting powerful connections between industry in the private sector, civil society, research and education, and the public sector.

Supporting sustained behavioral change, QF runs a weekly market in partnership with Torba Farmer's Market, a Qatar-based initiative that aims to promote organic and locally grown food and homemade products. The market is held every Saturday at Education City's ceremonial court, with on-site waste segregation bins, composting facilities, reusables, and compostable food packaging, to further promote sustainable practices and bring awareness to their interlinkages with sustainable economic growth and sustainable societies.

5.2.5 *Education for Sustainability*

5.2.5.1 Pre-university Education

QF schools have long incorporated sustainability education within their curriculums at all levels of pre-university education (PreK-12) and focuses on solutions and the role students can play in solving local and global issues.

The curriculum provides a contextual understanding of climate change as well as environmental concerns. However, the curriculum alone is not enough to nurture environmentally cautious future generations. Sustainability is integrated into the school culture and extended to the larger community through action. QF schools have greenhouses and planting spaces to allow students to interact with nature and support sustainable food supply at schools. This has led to a school urban farming project, which will promote small-scale pilots for urban farming within Education City. Schools also participate actively in extracurricular activities like community clean-ups, recycling initiatives, and in promoting awareness and behavior change.

This promotion of active citizenship and sustainable behaviors led to a group of passionate students at a Qatar Foundation school. The "Activists in Action" a group of students from grades 4 and 5 launched a petition to ban single-use plastic bags in Qatar. By April 2022, the petition "Student activists at Qatar Academy Doha want to ban single-use plastic bags in Qatar!", which is available on change.org, had received over 8,340 signatures, nearly reaching its target of 10,000. Since launching the petition, the students have been vocal about sustainability participating in several clean-up drives, taking part in Qatar Sustainability Week and other events, including appearing as guests in the PROGRESS | مُستدام podcast launched by HBKU.

Enforcing the role of action, the "Eco-Schools" program led by the Foundation for Environmental Education (FEE) is one of the largest global sustainability education programs for schools and another example of how sustainability education can be introduced beyond the curriculum (Eco-Schools n.d.). Eco-Schools start in the classroom and expand to the community by engaging the next generation in action-based learning. During a two-year application process, schools are required to complete seven steps that include forming and eco-committee, carrying out a sustainability audit, setting an action plan to address an identified gap or area of improvement, working on the curriculum, promoting awareness about the issue, and developing a sustainability code for the school. Efforts are carried out by students, teachers, staff, and parents, and once awarded, schools become part of the global Eco-Schools network. A "green flag" has been awarded to several schools in Qatar and three universities. At QF, the Qatar Leadership Academy and Qatar Academy Doha, were awarded Eco-School status in April 2021 (Qatar Foundation, 2021a). By early 2022, over 30 schools in Qatar had registered in the program.

Extracurricular programs like Doha Debates provide a portal for youth to debate and exchange ideas via live shows, podcasts, and events. Based in QF, along with Qatar Debate, the national debating organization and THIMUN Qatar (the Model UN program for high school students aged 15–18), these programs have long focused on critical global issues. Sustainability frequently takes center stage, with multiple youth activities focused on climate change. The groups participate in critical dialogues like UN General Assembly, COP meetings, and global youth forums, allowing students to access to participate and observe the real negotiations and multilateral discussions. THIMUN students were active participants during the pre-COP meeting that took place in Milan September 2021, which led to a student declaration that was submitted to the COP26 presidency.

Other innovative methods of integrating sustainability are constantly being explored by QF schools. The Qatar Academy of Science and Technology (QAST), has been piloting an online tool that is meant to connect students with peers from other schools around the world, promoting sustainable behaviors and allowing them to share their work and enabling the provision of credits and certifications to support university applications. This platform goes beyond the school to integrate the students' extracurricular activities and promote a visual showcase of their contributions to global challenges. Through the QF network, students also have the opportunity to engage with research facilities, enabling them to experience real-life learning as well as providing them with a glimpse into the world of research and academia.

Continuously promoting progressive education, the World Innovation Summit for Education (WISE) holds an international summit aimed at promoting forward-thinking policies and best practices from education systems to promote policy change in the education sector (WISE n.d.). The most recent WISE summit held in December 2021 included climate change education as one of its themes, with several sessions and discussions focused on the different issues related to sustainability and education.

5.2.5.2 Higher Education

QF hosts eight universities, including HBKU, which houses a College of Science and Engineering. Within the college, a division of Sustainable Development provides students with a comprehensive training fostering the search for balance between development needs, economic growth, environmental protection, human and social progress, and environmental preservation. Topics covered include, sustainable economy and financing, renewable energy technology and policies, water security, waste management, and food security. The college has had a significant contribution via research projects with national and global impact. Examples of these include research on sustainable cooling technologies for greenhouses and buildings; green and grid independent electric vehicle charging stations, including the design of a stand-alone hybrid and renewable energy-based (solar PV panels and wind turbines) station; ongoing research exploring hydrogen and freshwater production using solar energy; and waste management techniques, development of wastewater treatment systems, and identifying new ways to use waste like the use of polymers in 3D printing.

Beyond the academic and applied research, innovative solutions are being explored at HBKU that leverage technology in the form of mobile applications developed to promote behavioral change. The most recent example of this, QKONs is a digital waste exchange platform that links people to companies, providing an incentive recycle waste, building a circular economy by allowing users to earn economic benefits from their transactions (Al-Ansari et al., 2020). The application is a solution to issues in "sorting and collecting" waste, providing waste generators with the closest location for recycling bins and creates a value for waste through point rewards for their transactions (Qatar Foundation, 2019a).

Partner universities at Education City have also had tangible contributions to sustainability initiatives at Education City. Georgetown University's Center for International Regional Studies (CIRS) department includes environmental studies that address questions related to climate change, the politics of natural resources, food security, water conflicts, and other issues of environmental concern. Within these thematic areas, CIRS compares the Middle East to other regions, investigates cross-national understandings of environmental issues, and aims to translate knowledge into practice. Georgetown University has hosted a sustainability club since 2014, which holds awareness events for students, and interacts with the university's facilities management team to ensure sustainability considerations are top of mind.

"Sahtak Awalan", Your Health First, launched in 2012 by Weil Cornel Medical College-Qatar (WCMCQ), is another example. This multi-year initiative was designed to curtail unhealthy habits and is intended to encourage people to lead healthy lives through educating about sustainability, exercise, and nutrition, to allow them to make informed choices about the foods they buy, and the way they live their lives. The initiative is inclusive aimed at everyone: young and old, nationals and expatriates. It is mostly focused on the nation's youth, advocating preventative measures that will help the next generation stay healthy and rise to the challenges of Qatar National Vision 2030 (QNV 2030). Sahtak Awalan initiative led to other programs like "Khayr Qatarna" which aims to supply fresh fruit and vegetables to local supermarkets and help to improve food security, increase sustainability, and encourage healthy eating. The Greenhouse initiative under Khayr Qatarna set up greenhouses at more than 130 schools, which are being used to grow fruits and vegetables, teaching children about healthy eating, and cultivating their own food.

With the rise of energy and water security needs, Qatar needs to equip industries with a strong scientific evidence-based background. Texas A&M University runs a sustainable water and energy initiative, which aims to enhance working relationships between energy and industrial sectors and the research and educational institutions. Texas A&M is equipped with state-of-the art water and environmental facilities needed to provide the necessary research for these sectors (Texas A&M University at Qatar n.d.).

The unique Education City multiversity construct allows for students to take classes from different universities as well as participate in internships with other Qatar Foundation institutions and centers. Research grants from the Qatar National Research Fund allow for the use of students as interns or fellows, which enable applied research experience for students. Students can also attend, and access workshops and trainings organized by different QF entities outside their own universities. Finally, student clubs provide an opportunity for students to engage and interact on sustainability consideration, providing a platform that can reinforce learning through real-life applications and exchange of ideas. A cross-university, Education City-wide student-led sustainability club is under consideration and would promote wider collaboration.

5.2.6 *Applied Research for Sustainable Development*

5.2.6.1 Qatar Energy and Environment Research Institute

HBKU hosts the Qatar Environment and Energy Research Institute (QEERI), launched in 2011, as a national research center focused on understanding the environmental needs of Qatar to support the environmental ambitions in Qatar's National Vision 2030. Qatar's arid environment poses unique challenges and concerns when it comes to climate change adaptation and mitigation, environmental protection and restoration, and resource management. QEERI houses several centers with a specific focus to support Qatar in addressing its challenges related to energy, water, and environment (QEERI n.d.a).

The Environment and Sustainability Center at QEERI works on understanding the various factors that affect air quality and together with local authorities, established an air quality research program to "enable new discoveries, build capacity, and contribute to national policy directives in air quality and particulate air pollution in Qatar" (QEERI, 2021). To complement this effort, QEERI established two key indicators to assess urban air quality: Air Quality Index (AQI) and Inhalable Particulates Index (IPI) (QEERI, 2021). These indicators are tracked and measured by Air Quality Monitoring Stations (AQMS). Multiple stations have been placed across key areas, such as Education City, and at some of the stadiums, to better monitor air quality around World Cup venues in collaboration with Supreme Committee for Delivery and Legacy.

The QEERI water center tackles water security concerns, crucial for a nation that relies on desalination as the sole source of clean water. QEERI research aims to optimize water resources, treatment materials, processes and technologies that are adapted to Qatar's environment (QEERI n.d.b). The center has partnered with local stakeholders such as Ashghal, the Qatar Electricity and Water Company (QEWC), and Qatargas, to develop more efficient desalination technology, with the Multi-effect Desalination (MED) pilot plant demonstrating a 40% energy efficiency improvement and reduction in greenhouse gas emissions compared to existing technologies (QEERI, 2019). In addition to water security, Qatar is exploring ways to provide alternative energy supply, to adopt a cleaner mix of energy supplies. As such, QEERI launched the Solar Consortium to promote and encourage solar energy adoption and installation in Qatar, with some partners including Kahramaa (the national energy provider). Furthermore, QEERI is working on the development of an Energy Systems Model that aims to support and inform the decision-making process through scenario planning and evidence-based policy recommendations (see Chap. 8).

One critical area of focus for QEERI is understanding the impact climate change has on desert environments. To do so, it partnered with NASA Oasis to build Qatar's first scientific satellite focused on water and climate change challenges in desert areas (NASA, 2020). Orbiting Arid Subsurface and Ice Sheet Sounder (OASISS) is the first mission dedicated to measuring the impact of climate change beneath the surface of hyper-arid deserts and polar ice sheets. The aim is to monitor the impact of rising temperatures on rising sea levels in Qatar, a peninsula that has one of the highest submersion rates, as sea levels pose threats to coastal aquifers.

5.2.6.2 Qatar Science and Technology Park

The QSTP, launched by QF in March of 2009, is a free zone and technology park aimed at conceiving and growing tech enterprises in Qatar (QSTP n.d.). QSTP focuses on facilitating the development of new high-tech services and products, driving applied research and development, and supporting the commercialization of market-ready solutions. QSTP houses over 40 members, with several entities focusing on sustainability-related technologies and solutions, a few of these are outlined below.

The ConocoPhillips Global Water Sustainability Center focuses on the development of advanced wastewater treatment and cost-efficient solutions to reduce water waste from industrial activities and works closely with Qatargas to develop scalable solutions to reduce global water consumption and waste. The Qatar Shell Research and Technology Center works with universities like Texas A&M to enhance the production and efficiency of the hydrocarbons industry, advancing understanding of carbonate reservoirs and addressing challenges like carbon capture and storage and enhanced oil recovery processes to support the production of clean fossil fuels. Infrastructure and Research Development focuses on market solutions to reduce carbon emissions through waste management and efficient recycling technologies and processes. The Total Research Center runs programs on solar energy, restoration of marine habitats including coral reefs rehabilitation, biodiversity, and gas conversation. The Mitsubishi Water Technology Center aims to introduce and develop sea water reverse osmosis technology in Qatar, considering the Gulf's unique sea water conditions. ExxonMobil research focuses on four areas: environment management, water reuse, LNG technology and surface geology. The Qatar Mobility Innovations Center (QMIC) focuses on developing and deploying smart mobility systems and solutions in the areas of intelligent transport, logistics and telematics, road safety and environment. The Gulf Organization for Research and Development (GORD) focuses on research and development in the areas of green building certifications, accreditations, standards setting, voluntary carbon markets, and advisory services on sustainability and climate change. GORD led to the development of GSAS, the first MENA-region specific performance-based system designed to assess buildings and infrastructure on their sustainability impacts. The GSAS rating has become widely used in Qatar including in all FIFA World Cup 2022 construction efforts.

These centers and others within the QSTP are uniquely positioned to collaborate with Education City entities and universities and work closely with national stakeholders to promote new market-ready solutions and enable the private sector with technologies that are contributing to making Qatar more sustainable.

5.2.6.3 Education City as a Testbed for National Pilots

Education City provides a critical value to national sustainability efforts through its ability to bring in together research, academia, policy, industry, and the private sector and serves as a unique space where research findings, new solutions, and innovations can be piloted. Leveraging the ecosystem of research, academia, physical infrastructure, and operations, Education City has lent itself as a testbed and enabled the piloting of several sustainability-related initiatives.

From behavior-changing interventions like banning plastics, providing healthy lunches, giving out reusable bottles and lunch packs to school students and workers, to procurement changes that stipulate contractors act more sustainably to apps like QKONs aimed at enhancing recycling, several solutions have been tested to enable evidence-based policy advice for national stakeholders.

Air quality solutions via initiatives like park and plant, urban farms, planting a green belt around Education City and more recently measuring the impact of car-free days on the air particle composition are being observed and documented. Indoor air quality smart home solutions are being piloted in Education City's community housing compound, through sensors tied into HVAC systems in collaboration with Ministry of Public Health, Ooredoo and Kahramaa. The learnings will enable solutions that can be scaled up nationally for better outdoor and indoor air quality, a concern given the amount of indoor time spent during the hotter months of the year.

Pilots related to asphalt composition, paint colors, and materials for roads to reduce urban heat island effects have been run and results provided in terms of actionable standards and solutions to national entities. Solar Photovoltaic (PV) panels' rooftops have been installed around Education City which have enabled QF to save around 2,590 tons of CO_2 emissions annually. Charging stations for electric vehicles were developed in partnership with Ministry of Transport and Communications and Kahramaa.

Educational models are continuously being disrupted and progressive education systems tested in the area of sustainability which will support national educational development. Community action and behavior change are constantly being encouraged and different policies, incentives, apps, and tools are trialed and tested. Through its vibrant and unique ecosystem, QF is a critical innovative space where new solutions and products can be piloted and tested to provide evidence-based advice to national authorities.

5.2.7 Global Reach: QF's Role in Advancing and Promoting Dialogue

QF has a global reach and promotes dialogue on key topics via multiple platforms. QF had an active presence in the latest COY and COP meetings that took place in 2021. A group of students from the QF schools network participated in the pre-cop youth summit in Milan, where they were able to express their concerns about their future and demanded that government leaders and policymakers make serious strides and commitments to reach the target of 1.5° Celsius. The students were also present at the COP26 meetings in Glasgow, where they presented a declaration on climate change to the meeting of world leaders.

Other QF entities took part in the meeting, including the World Innovation Summit for Health (WISH) who, along with the World Health Organization (WHO), discussed the findings and policy recommendations from their work on climate change and health, and policies to maintain healthy dry cities. During COP26, QF announced a partnership with Rolls-Royce to invest, develop, and scale-up climate-tech business, creating a center of innovation for global climate technology, with two planned campuses in the UK and Qatar, positioning Qatar as among the top five global countries to invest in clean energy R&D and as a pioneer in small, and advanced economies.

5.3 Impact and Challenges

QF has no doubt had a positive impact. Through its multiple initiatives, QF has avoided the use of over 100,000 plastic bottles (in cafeterias, schools, and offices) and over 120,000 single-use plastic bags per month. Furthermore, 20% of plastic packaging was avoided in construction projects around Education City (Qatar Foundation, 2020b). Green building standards have reduced emissions of QF structures ensuring it B rating in 2020, the best performing among peers reporting on GHG emissions nationally to the UN (CDP n.d.b).

Colored bins around campus promote segregation at the source and partnerships with private companies support proper waste management and high rates of recycling. Innovative apps (like QKons) and policies that aim to nudge behaviors (like park and plant, community and school gardens, and car-free days) are routinely being piloted to promote environmentally friendly practices.

QF schools and universities have embedded sustainability education, giving rise to multiple youth activist groups. A key example is the Activists in Action, who launched a petition to advocate for a nation-wide ban of single-use plastic bags and may have contributed to the nation's recent announcement to ban these in May 2022 (The Peninsula, 2022a, 2022b). The Eco-Schools program has educated hundreds of students on sustainability-related issues, with more in the pipeline. The annual Qatar Sustainability Week reaches thousands of participants promoting awareness about sustainability. The weekly Farmer's Market promotes eco-friendly businesses and circular economy practices among those visiting. The Qur'anic botanical gardens provide values-based learning to instill a moral drive to protect the environment.

To fully determine the impact of QF's initiatives and interventions will require some rigorous case studies and assessments, given that many of these are intangible social impact and awareness raising. Nonetheless, it is clear that QF is on the path toward a more sustainable future as it continues to innovate and pilot new thinking and solutions and inspires future generations to change their behaviors.

Despite the visible impact, all success is not without challenges. Mobilizing change at multiple fronts can be a complex and time-consuming exercise. Further, implementing innovative policies can sometimes be difficult. QF must coordinate with multiple stakeholders to ensure and promote systems-wide change. For example, the COVID pandemic led to government imposing measures which forced the use of plastic bottles, bags, and cutlery for food delivery. Despite QF's desire to bans these items, they had to be temporarily reinstated at schools and offices to adhere with health safety measures.

Qatar's arid desert environment, scarce water resources, and low rainfall present a persistent challenge to a sustainable future, making it difficult to enhance walkability at such a large premise like Education City. Research in the sustainability considerations of such environments remains sparse. Nonetheless, QF's investment in education and research are leading to innovative bottom-up, community-based solutions that are solving Qatar's specific challenges and moving the needle toward a more sustainable outlook.

Promoting success stories can be challenging and ensuring messages reach the relevant audience involves a coordinated effort of convening and policy advocacy. To amplify the impact of QF's practices, QF must coordinate action and align with national stakeholders, the private sector and industry, and communities. This will ensure that relevant regulations are set in place (building on lessons learned from QF experience) to promote wider-scale behavioral change.

5.4 Toward a Sustainable Future

Moving forward, Qatar is uniquely positioned to have a leading local and global voice on sustainability due to it being on the receiving end of climate change, facing risks of rising temperatures and sea levels, as well as given its influence on the global energy sector. The World Cup 2022 acts as a visible international event, placing Qatar in the global spotlight, providing an opportunity to showcase its successes and achievements in the sustainability field.

Sustainability continues to be a national and global priority and QF has embarked on a series of important initiatives to support the State of Qatar. As part of its recently launched refreshed strategy, QF featured the selection of sustainability as one of five key themes that the foundation will focus on. Within sustainability, focus areas include: (1) sustainable energy, (2) resource security and management, (3) environmental protection and restoration, (4) sustainable and circular economy, and (5) human and society well-being. The QF strategy recognizes that to achieve impact in these areas, multiple interventions must be employed and aligns a set of implementation tracks, or components of the Foundation's ecosystem to achieve the necessary multisectoral systems shifts. These tracks include education and lifelong learning, research, policy analysis and advocacy, innovation and development, and community outreach and engagement.

QF will continue to lead in education and is currently undertaking a project to review its educational pedagogy in relation to sustainability, with educators from the pre-university education team exploring different methods and tools to educate youth within the QF schools' network and documenting best practices to recommend to national educational institutions. In addition, a team of researchers from HBKU were awarded a research grant aimed at understanding how SDGs are currently taught in Qatar and providing recommendations on how to further empower students via SDG education and Global Citizenship (Tok, 2021; see Chap. 17).

Within community outreach and engagement, through the Education City Stadium, the World Cup tournament taking place in November–December 2022 will serve as a catalyst to promote sustainable practices. With waste segregation and informational campaigns, as well as linking to the "Precious Plastics" initiative outlined above, QF will engage the community to actively take part in and learn about sustainable practices. Other plans include the launch of Green Island, a community-centered one-stop-shop for all recyclables within Education City (The Peninsula, 2022a, 2022b).

QF recognizes the critical role that it can play in shaping and promoting policy both within Qatar and globally. QF is advancing this commitment with the launch of Earthna—Center for a Sustainable Future, designed to provide a holistic view of environmental, social, and economic prosperity as part of Qatar's ambition to enhance sustainability at a national and global level (Earthna n.d.a). By bringing together and disseminating Qatar's policy research and perspectives on topics that have global relevance, the center will drive dialogue, knowledge exchange, and the identification of solutions, supporting nations with similarly harsh climates in addressing sustainability challenges. Earthna will capitalize on Qatar's position to play a regional and global role as well as bridge the North–South gap on climate change. Earthna will focus on defining and contextualizing sustainability needs of dry and arid climates; leverage Qatar's unique position in global energy markets to explore opportunities; and draw upon QF's flagship Education City and its unique ecosystem of research, academia, and innovation as a national testbed to pilot new ideas. With this policy center, the role of QF in continuing to promote sustainability will be solidified as it becomes a key enabling entity, supporting the state and non-state actors to work together toward a comprehensive multi-disciplinary approach to understanding and mitigating climate change and promoting sustainability.

References

Al-Ansari, T. et al., (2020). https://www.research.hbku.edu.qa/projects/33

BBC. (2021, May 26). Shell: Netherlands court orders oil giant to cut emissions. https://www.bbc.com/news/world-europe-57257982

Climate Data Portal. (n.d. a). NAZCA. https://www.cdp.net/en/campaigns/nazca

Climate Data Portal. (n.d. b). Qatar. https://www.cdp.net/en/responses?page=1&per_page=20&queries%5Bname%5D=qatar+&sort_by=project_year&sort_dir=desc

Earthna. (n.d. a). About Earthna, Center for a Sustainable Future. https://www.earthna.qa/about-us

Earthna. (n.d. b). Welcome to Earthna: Center for a Sustainable Future. https://www.earthna.qa/

Eco-Schools. (n.d.). About FEE EcoCampus. https://www.ecoschools.global/ecocampus-about#:~:text=FEE%20EcoCampus%20is%20an%20award,and%20lead%20the%20way%20towards

FIFA World Cup Qatar 2022. (2020). Ten sustainability facts about Education City Stadium. https://www.qatar2022.qa/en/news/ten-sustainability-facts-about-education-city-stadium

Greenpeace. (n.d.). Global. https://www.greenpeace.org/global/

MENA FM. (2022, January 3). Her Excellency Sheikha Hind Visits New Green Lab at QF School. https://menafn.com/1103780652/Her-Excellency-Sheikha-Hind-Visits-New-Green-Lab-at-QF-School&source=139

NASA. (2020, September 23). US-Qatar Partnership Aims to Find Buried Water in Earth's Deserts. https://www.jpl.nasa.gov/news/us-qatar-partnership-aims-to-find-buried-water-in-earths-deserts

Qatar Foundation. (n.d.). Park and Plant. https://qf.org.qa/park-and-plant

Qatar Foundation. (2019a, December 17). Rewarding Waste. https://qf.org.qa/stories/rewarding-waste

Qatar Foundation. (2019b, December 25). Qatar Foundation unveils the pioneering Education City tram. https://qf.org.qa/stories/qatar-foundation-unveils-the-pioneering-education-city-tram

Qatar Foundation. (2020a, August 16). Education City Stadium: A Lesson in Sustainability. https://qf.org.qa/stories/education-city-stadium-a-lesson-in-sustainability

Qatar Foundation. (2020b). Reducing Single-Use Plastic Waste in Education City. https://downloads.ctfassets.net/2h1qowfuxkq7/34z4VioKDaPw0rIaG2DdBB/893611237ec79e7ed258f5e2653224ce/QF_Plastic_Reduction_Brochure_ENG_DIGITAL__1_.pdf

Qatar Foundation. (2020c, November 3). Qatar Foundation takes action against e-waste. https://qf.org.qa/stories/qatar-foundation-takes-action-against-e-waste

Qatar Foundation. (2021a, April 7). QAD Primary and QLA hoist green flags as they become Eco-Schools. https://qf.org.qa/stories/qad-primary-and-qla-hoist-green-flags-as-they-become-eco-schools.

Qatar Foundation. (2021b, October 31). QF's e-waste recycling journey from Doha to Singapore. https://qf.org.qa/stories/qfs-e-waste-recycling-journey-from-doha-in-singapore
Qatar Foundation. (2021c, February 27). Why QF's HQ is setting a sustainable standard. https://qf.org.qa/stories/why-qfs-hq-is-setting-a-sustainable-standard
Qatar Environment and Energy Research Institute. (n.d. a). About QEERI. https://www.hbku.edu.qa/en/qeeri/about
Qatar Environment and Energy Research Institute. (n.d. b). Water Center. https://www.hbku.edu.qa/en/qeeri/water-center
Qatar Environment and Energy Research Institute. (2019, June 16). QEWC and QEERI sign pact for desalination pilot plant. https://www.hbku.edu.qa/en/news/qewc-qeeri-desalination
Qatar Environment and Energy Research Institute. (2021). Scientific Research Raises Level of Urban Air Quality Management: A Decade of Qatar Environment and Energy Research Institute. https://www.hbku.edu.qa/en/news/QEERI-SUST-SRRLUAQ
Qatar Science and Technology Park. (n.d.). Home to some of the best minds and talents in the region and beyond! https://qstp.org.qa/free-zone/members/
Texas A&M University at Qatar. (n.d.). About the Qatar Sustainable Water & Energy Utilization Initiative. https://qwe.qatar.tamu.edu/about-qwe/
The Peninsula. (2022a, March 17). Green Island: QF to build Qatar's first recycling hub. https://thepeninsulaqatar.com/article/17/03/2022a/green-island-qf-to-build-qatars-first-recycling-hub
The Peninsula, (2022b May 25). Qatar cabinet approves draft resolution prohibiting single-use plastic bags. https://thepeninsulaqatar.com/article/25/05/2022b/qatar-cabinet-approves-draft-resolution-prohibiting-single-use-plastic-bags
Tok, E. (2021). SDG Education and Global Citizenship in Qatar: Enhancing Qatar's Nested Power in the Global Arena. https://mis.qgrants.org/Public/AwardDetails.aspx?ParamPid=fhgjelhhnp
United Nations. (2021). Youth4Climate: Driving Ambition. https://www.un.org/youthenvoy/youth4climate-driving-ambition/
United Nations Educational Scientific and Cultural Organization. (n.d.). Education for Sustainable Development. https://en.unesco.org/themes/education-sustainable-development
United Nations Framework Convention on Climate Change. (n.d.). Climate Neutral Now. https://unfccc.int/climate-action/climate-neutral-now
World Innovation Summit for Education. (n.d.). Home. https://www.wise-qatar.org/
World Wildlife Fund. (n.d.). Home. https://www.worldwildlife.org/

Fahad Al-Musalmani joined Qatar Foundation (QF)'s Policy Hub at the CEO office in 2019 as a Policy Development Associate. Prior to joining QF, Fahad worked as Economics Researcher in the Technical Office Governor's Office of the Qatar Central Bank. He graduated from Georgetown University School of Foreign Service in Qatar, majoring in International Economics, and was part of the International Economics Honor Society (Omicron, Delta, Epsilon).

Sylvie Maalouf is Programs Manager of the Policy Hub team at the CEO Office of QF. Sylvie brings experience in strategic planning, analysis and consulting in areas including economics, public and non-profit management, and sustainability, providing policy advice to multinational institutions and public sector entities. As Business Development Leader for the Government and Public Sector practice at Ernst & Young (EY) she developed go-to-market strategies and oversaw the accounts and offerings for the MENA region. As economic expert for the Ministry of Development Planning and Statistics in Qatar, she contributed to the development of the economic component of the National Development Strategy 2018-2022, as well as several Qatar Economic Outlook publications. Before moving to Qatar, Sylvie was a consultant for the World Bank and American Chamber of Commerce in Washington, DC, and the World Trade Organization in Geneva, Switzerland. Sylvie holds a master's degree in Intentional Trade and Investment policy from the George Washington University and a Bachelor of Science in Computer Science with a minor in Business from the American University of Beirut.

Chapter 6
Qatar's Energy Policy and the Transition Towards a Renewable and Carbon–Neutral Future

Steven Wright

Abstract Qatar's successful exploitation of its natural gas resources has been a cornerstone of its rise as a global energy power. Despite progress in diversification, the energy sector remains a primary driver of Qatar's economy, which makes it integral in any assessment of future sustainability. Qatar's future sustainability and the role of its energy sector must be evaluated in light of global changes, such as the shift to renewables, carbon neutrality, and changing geopolitics. This chapter seeks to engage with global trends and provide a sustainability perspective for Qatar. Reflections are made on changes occurring within the European Union and East Asia, and how Qatar can position itself to capitalize on these trends.

Keywords Energy policy · Energy transition · Renewable energy · Carbon–neutral · Qatar

6.1 Introduction

Qatar presents a fascinating case study on how a small state has developed through its energy sector and has navigated the dynamics that this type of political economy entails. With this in mind, the question of sustainability and adaptation to the changing nature of the global energy landscape becomes an issue of central importance. Qatar has exhibited the characteristics of a 'late-stage petro-developmental state', given the role of political strategy and statecraft that have underpinned its rapid development and diversification (Wright, 2021). Amongst the economic development achievements, Qatar had the highest GDP per capita in the Middle East and North African (MENA) region by 2020 (World Bank, 2020). Although the economy remains energy-driven, as most government revenue is derived from this sector, progress towards a more sustainable political economy through diversification has been achieved. In practical terms, more than two-thirds of Qatar's GDP comes from the non-energy

---

S. Wright (✉)
College of Humanities and Social Sciences, Hamad Bin Khalifa University, Doha, Qatar
e-mail: stwright@hbku.edu.qa

L. Cochrane and R. Al-Hababi (eds.), *Sustainable Qatar*, Gulf Studies 9,
https://doi.org/10.1007/978-981-19-7398-7_6

sector (International Monetary Fund, 2022). Central to this is how a low-tax environment, coupled with high levels of fiscal spending, has enabled broader diversification and private sector growth. Nevertheless, what has underpinned this is the economic capacity that Qatar's energy sector has provided. This makes the question of sustainability within Qatar's energy sector a matter of pressing importance as it has broader implications in all sectors and industries covered in this important volume.

In terms of the question of change within the energy sector, it was the former Saudi Arabian Minister of Oil, Sheikh Ahmed Zaki Yamani, who famously stated, "the Stone Age did not end for lack of stone, and the Oil Age will end long before the world runs out of oil" (Fagan, 2000). Yamani's adage rings true today in that the accepted wisdom within the energy sector is that a 'global energy transition' is underway. This transition amounts to not just a change within the energy mix in favour of renewables, but a more systemic and fundamental change within the global system on political, cultural, economic, and social levels. Changing energy consumption patterns does not necessarily mean that there is no future for fossil fuels, but rather that demand will shift to alternative renewable fuels. In addition, research and innovation in this emerging sector can be a driver of economic growth and development. With this in mind, the central issue relating to sustainability within the energy sector is not just a simple question of how existing resources can be monetized, but how can the existing capacities (Obeidat et al., 2020), technical expertise, and comparative advantage be repositioned. For Qatar, this amounts to an assessment on how its energy sector can be used to venture and diversify into emerging sectors that can cater for sustainability.

It is with the above in mind that this chapter seeks to contextualize the global energy transition and how it relates to Qatar in three substantive sections. Firstly, it aims to engage in how global energy consumption has historically been dynamic and how major shifts in the energy mix have led to upheaval in geopolitical terms. Moreover, it seeks to draw attention to historical lessons with regard to how this has resulted in fundamental shifts in the global economy, and discernible changes in politics, culture, and society. Secondly, this chapter will engage in the nature of the energy transition and how it is redrawing the global energy landscape. By recognizing that a new 'energy map' has emerged (Yergin, 2020), it is possible to conclude that achieving sustainability requires adapting to this new dynamic to maximize relevancy, innovation, and economic returns. In the final section of this chapter, engagement is given specifically to how Qatar has begun to adapt to this changing reality and offers some observations on potential areas for future engagement. The central argument of this chapter is that achieving sustainability in the energy sector requires strategic foresight on how the energy sector is changing and how innovation into new sectors can help diversify revenue streams and position the sector for the future. Overall, a significant opportunity for Qatar is identified for achieving sustainability through building on its technical expertise, capacity, and market presence, to venture into new areas of economic activity related to the renewables sector.

6.2 The Geopolitics of Energy: A Globally Transformative History

Energy and geopolitics are inextricably linked and have played a central role in shaping contemporary world history. Energy commodities have a special significance as they have been catalysts for national economic growth and decline, geopolitical conflict and integration, and a driver of both new technological innovations and economic transformations. This volume examines various themes and dimensions concerning Qatar's potential for a sustainable future. Given the centrality of Qatar's energy sector, it is critical to appreciate the broader global context of energy and how it is poised to have a direct impact on the future character of Qatar's political economy. Therefore, it is appropriate to begin with some historical reflections to remind ourselves how energy dynamics have been transformative and impacted global geopolitics.

Within energy studies, important research on historical energy transitions has gravitated towards shifts in the energy mix: such as from wood to coal, from coal to oil, from oil to gas, and how electrification of rural areas has taken place along with its implications (Bridge et al., 2013). The broader conclusion is that energy transitions have been a catalyst behind significant political, social, and economic changes. As shifting patterns of energy demand and the rise of new commodities have historically had a transformative effect, this dynamic is instructive for any evaluation of Qatar's potential for a sustainable future. Such changes in the global energy market have impacted the relationships between energy exporting and importing nations. Moreover, it has also had a far-reaching global impact that fundamentally alters the international system in terms of its political, economic, and social relationships. While the move towards greater use of renewables and a carbon–neutral future is poised to have a disruptive and transformational impact globally, the nature of this change has clear relevance for Qatar, both in navigating the transition and strategically positioning itself from a sustainability perspective.

The quest for energy resources has proven to be a consequential backdrop of modern world history, with coal being the first global energy commodity. Coal proved to be an enabler for the British Empire to expand its influence through industrialization overseas. It also allowed for a projection of military power as the adoption of the steam engine proved revolutionary for the British Royal Navy to secure and expand the trade routes and territorial possessions of the Empire on a global basis (Gray, 2017). On the other hand, the lack of energy resources has also proved to be a driver of war and territorial expansionism. For example, coal proved to be intertwined with geopolitics, economic growth, and stability in Europe and was an important factor in the origins and outcome of the two World Wars. After all, it was only through the establishment in 1951 of the European Coal and Steel Community, which was founded by France, Germany, Italy, Belgium, Luxembourg, and the Netherlands, that an equitable arrangement with steel and coal resources could be achieved within the continent. This allowed the inherent competition over natural resources between

Germany and France to be finally resolved, thereby bringing a new era of peace and stability to the European continent that laid the foundations of the European Union.

In his capacity as First Lord of the Admiralty, Winston Churchill was to take one of the more consequential decisions in world history by re-purposing the British Royal Navy as one based on oil rather than coal. The strategic advantages of this transition were clear as it allowed for a more advanced naval force that had greater speed and efficiency, yet it also marked an era where oil became a strategic commodity that shaped future foreign policy and national interests. Moreover, it allowed the British Navy to be a faster fleet than its German rival, yet it also marked a new era whereby naval military superiority depended on access to oil. Daniel Yergin reminds us that, "…Churchill, on the eve of World War I, had captured a fundamental truth, and one applicable not only to the conflagration that followed, but to the many decades ahead. For oil has meant mastery throughout the twentieth century" (Yergin, 1992).

Throughout the twentieth century, control over oil resources has been central to understanding the origins and outcomes of several conflicts. During World War II, the Japanese attack on Pearl Harbor was motivated by the strategic desire to protect Japan's oil resources in the East Indies. The decision by Nazi Germany to launch a surprise invasion of the Soviet Union, which proved consequential in the outcome of the war, was motivated by a need for Germany to capture the oil fields in the caucuses (Yergin, 1992). Beyond this, oil resources have played a central role in several conflicts in the latter half of the last century. Key examples include the Nigerian Civil War 1967–70, the Iran-Iraq War 1980–88, Iraq's invasion of Kuwait in 1990, and the 2003 Iraq War. Therefore, there is justification to recognize that energy resources should be considered distinct from other commodities, given their central role in security calculations.

The weaponization of energy resources first took place with key Arab states responding to the Yom Kippur War in 1973 by implementing an oil embargo that triggered an 'oil shock' to the global economy. Rapid rises in oil prices are highly consequential for the global economy, given they have led to significant periods of inflation, which resulted in a decline in economic growth and prosperity. Following the 1973 oil shock, the world economy was subjected to a second oil shock due to the Iranian revolution in 1979. The subsequent Iran-Iraq War (1980–88) compounded the geopolitical risk from regional instability and added to the turbulence in the oil market. The sensitivity of the oil market to geopolitical risk was a trend that was to follow in subsequent decades, most notably with Iraq's invasion of Kuwait in 1990, along with the broader global challenges brought on by the War on Terror, which saw a joint US and British invasion of Afghanistan in late 2001, and subsequent invasion of Iraq in 2003. Nevertheless, the pattern of energy commodities being impacted through geopolitical instability without subsequently having an impact on the global economy is a recurrent pattern in world history. More recently, Russia's invasion of Ukraine in 2022 exposed how coal, oil, and gas resources could become weaponized commodities with particular regard to the European Union, given its inherent dependency on Russian energy imports. Periods of geopolitical instability have fed into concerns about supply disruptions of oil to the global market, so sharp increases in oil prices that accompanied key periods, such as Iraq's invasion of Kuwait in 1990, resulted in far-reaching global consequences given the economic impact on a global basis. For example, the Gulf conflict of 1990–91 contributed to a recession in the United States which then saw a shift from a Republican administration to a Democratic one headed by Bill Clinton in 1993. Indeed, the interlinked nature of the global energy market to economics and politics is what makes energy commodities distinct and so consequential.

While the weaponization of oil exports has underlined the distinctive character of energy resources as a commodity, it has also proved transformational in the global political economy. A key example of this was how Japan responded to the oil price shock of 1973. Japan's reliance on imported energy put it at risk through supply security and price volatility. As a result, the oil shock left Japan with its first trade deficit since 1964 and sparked public fear, as prices skyrocketed due to a 20% inflation rate and the economy contracted for the first time in postwar history. Additionally, it triggered a strategic change in Japan's industrial base away from high-energy sectors and towards electronics and semiconductors. Indeed, Japan's energy instability and sensitivity to global oil geopolitics and geoeconomics were the catalysts for an economic and industrial revolution (Yergin, 1992). It was also a clear example of how national energy security calculations shift as a response to price volatility. The impact of the 1973 oil shock was a catalyst behind Japan's expansion of nuclear power and its progressive adoption of natural gas as an alternative to coal-fired power stations.

On a broader level, access to low-cost energy has been an enabler of rapid advances in economic growth, particularly within the OECD economies (Nye, 1999). Accessible, low-cost oil has been a central enabler of globalization, given it has enabled low-cost transportation and trade (Dicken, 2007). On a broader level, however, it is also observable how economic development is correlated with energy consumption patterns. Fluctuations in energy prices, which limit their affordability and availability, are undoubtedly key inhibitors to economic growth, which can perpetuate stark inequity within the international system.

When taken together, this brief sketch on the relationship between energy, political economy, and broader geopolitics underscores the far-reaching and transformative nature energy commodities have had. It is also clear that it has direct implications for economic development and the stability of the international order. On this basis, we can recognize that as Qatar is a leading exporter of liquefied natural gas (LNG) and gas-to-liquid fuels (GTL), the transformative nature of the global energy transition will undoubtedly be a critical factor facing Qatar's energy sector and its sustainability in the future. As highlighted above, the impact of energy pricing and geopolitics can have a far-reaching impact globally. The nature of the global transition towards a renewable energy future has the potential to recalibrate Qatar's international linkages based on demand variables but also the manner in which it strategically positions its energy sector. In this regard, the above historical observations underscore the volatility and far-reaching impact of the global energy market on the international system. While Qatar's political economy has shown clear potential in terms of diversification from energy export dependency and has the character of being a petro-developmental state, its economic capacity and composition remain driven by energy export revenue. It is with this in mind that the next section of this chapter will consider the nature of the global energy transition and what characteristics it is expected to have. This will provide the context for examining Qatar's energy sector and the considerations needed from a sustainability perspective.

6.3 Conceptualizing the New Energy Landscape

It is uniformly recognized within energy studies that an 'energy transition' has commenced on a global basis. The transition is driven by government policy and investments to comply with carbon–neutral targets to offset the risks of climate change. Research and innovation are proving to be a key enabler of a progressive adoption of renewable energy sources, given that they are becoming more competitive with traditional fuel sources. Although the motivations behind the energy transition are established, there is no clear consensus on how this transition will manifest. Indeed, significant variance can be observed globally in terms of how nation-states are engaging and adapting to the energy transition (Hafner & Tagliapietra, 2020a). However, what is clear is that it has the potential to have a transformational impact on social, technical, cultural, economic, and political systems globally (Davidson & Gross, 2018). At the very least, it is expected to be as significant as in previous transitions, given the multidimensional impact such shifts have historically had. While the interrelationship between geopolitics and energy is well-established, scholarship remains in its initial stages in assessing the broader political, economic, and geopolitical implications of the new energy transition towards renewables and the decline of fossil fuel consumption (Hafner & Tagliapietra, 2020b).

According to the International Renewable Energy Agency (IRENA), the new energy landscape is projected to have profound implications:

> This ongoing transition to renewables is not just a shift from one set of fuels to another. It involves a much deeper transformation of the world's energy systems that will have major social, economic and political implications which go well beyond the energy sector. The term 'energy transformation' captures these broader implications. The global energy transformation will have a particularly pronounced impact on geopolitics. It is one of the undercurrents of change that will help to redraw the geopolitical map of the 21st century. The new geopolitical reality that is taking shape will be fundamentally different from the conventional map of energy geopolitics that has been dominant for more than one hundred years. (IRENA, 2019)

A key observation from IRENA's study on the global energy transformation is that the dynamics of change are far deeper than a more simplistic transition from a move towards increased use of cleaner alternative fuels would imply. The broader observations are that the move towards renewables and the decline of fossil fuels would have much deeper repercussions within societies with the resources to adopt renewable fuel sources progressively. Innovation, digitalization, and electrification of key sectors of the economy have the potential to be transformative within advanced economies. This aligns with work that has been done on the 'Green Deal' scenario, which argues that the seismic nature of the energy transition will lead not just to economic growth through innovation and new industrial commitments made towards carbon neutrality, but also broader means of spurring global economic development basis on a more equitable basis (Leonard et al., 2021). In this regard, BP observed that the role of LNG as a 'green' transitional fuel could be expected to increase at least to 2030 (Dale, 2022). BP also noted that depending on the speed of the energy transition, demand for gas would increase potentially beyond 2050.

In advance of the 2021 United Nations Climate Change Conference (COP26) in Glasgow, the International Energy Agency (IEA), published a roadmap for the global energy sector for it to achieve net zero by 2050 (Bouckaert et al., 2021). The IEA's roadmap sought to provide a pathway for how states would achieve the legally binding disclosures on targets set by the Paris Agreement (COP21) in 2015. The COP21 was noteworthy in that it was a legally binding agreement adopted by 196 parties that limited global warming to below 2 °C, but preferably a limit of 1.5 °C was to be targeted. From this, states were to provide declarations on their emissions, despite there being no legal consequence for not meeting the targets set. This report by the IEA was noteworthy in that it advocated no additional investment in fossil fuels beyond those already committed to as of 2021. It is worth noting here that while this is a progressive goal, it is questionable whether it is realistic in practical terms. Moreover, it outlined the ambitious target of upwards of 90% of electricity generation being from renewables and a goal of over 60% of total energy supply by 2050 coming from renewable energy. The position taken by the IEA is significant in that it underlined the transformational nature of the energy sector, which would amount to fossil fuels "declining from almost four-fifths of total energy supply today to slightly over one-fifth by 2050" (Bouckaert et al., 2021). By 2021, more than 120 countries had made pledges towards significantly reducing emissions by 2030 and achieving a net zero by around 2050. Furthermore, more than 100 countries have committed to making a 30% cut in methane emissions by 2030. While it is recognized that this is a future target rather than an assessment of what has been accomplished, it is instructive in that it provides an early indication of the scale of the transformation needed for governments to achieve their pledges. Despite such pledges being made, it has been noted that there is a credibility, action, and commitment gap, particularly with the major emitters (Stockwell et al., 2021).

For energy-exporting countries, the energy transformation is expected to create systemic risks concerning their fiscal ability to sustain their respective economies. This was an argument advanced by the IMF in an influential study conducted in 2020 (Pescatori et al., 2020). It was argued that the energy-exporting states of the Gulf face clear risks to their political economy due to projected shifts in energy demand as part of the energy transition. They argued that the broader Middle East is projected to confront various risk factors that would potentially destabilize their economy in the 2030s due to growing populations and stagnant revenue streams. IMF econometric modelling indicated that given the rising demand for fossil fuels, maintaining fiscal expenditures, currency pegs, and social welfare programmes by Gulf countries would become unsustainable. This prognosis has clear implications for any assessment of future sustainability and underlines the necessity for Qatar and other energy-exporting countries to navigate the energy transformation strategically.

It is noteworthy that key oil-producing states face the challenge of how they will adapt their energy sector to the changing energy landscape. Central to this is how they can reduce emissions of their energy sector, monetize their reserves, and reposition their energy exports as a cleaner 'blue-energy' export commodity. The case of Saudi Arabia is particularly striking and serves as a useful comparison for Qatar. Saudi Arabia's state-owned oil company Aramco pledged to be carbon neutral by 2050, and Saudi Arabia as a country by 2060. Its 2060 carbon–neutral goal focuses on what Saudi Arabia will do internally, and it does not have a bearing on oil exports as emissions associated with oil exports—dubbed 'Scope 3' emissions—will be omitted from the count.

Saudi Arabia remains committed to exploiting and monetizing its resources to a maximum as an energy exporter (Krane & Young, 2021). It is worth recognizing that the length of time necessary for Saudi to monetize its reserves requires a climate strategy distinct from that of its smaller rivals, particularly shareholder-owned international oil corporations (IOC). Saudi Arabia's proven reserves are over 260 billion barrels, more than five times the combined reserves of the five largest international oil companies, ExxonMobil, Shell, Chevron, Total, and BP. Saudi Aramco has almost 50 years of production remaining from domestic reserves at present rates, which might increase further if more discoveries are made (Krane, 2021). Jim Krane notes:

> Given the strong likelihood of additions to its proven reserves, the Saudi government has made allowances for Aramco to maintain its monopoly over the Saudi oil concession for as long as 100 years – until the year 2117. By contrast, IOCs' proven reserves of just over 200 billion barrels would be collectively depleted in nine to 15 years based on current output rates. (Krane, 2021)

Therefore, adopting a carbon–neutral strategy can be interpreted as a means for Saudi Arabia to prolong and monetize the use of its oil sector, while progressively focusing on blue energy exports as a transitional fuel. This differentiates from Qatar's LNG, which currently is not processed into and exported as blue ammonia. 'Sustainable' blue hydrogen produced from hydrocarbons in Saudi Arabia might serve as transitional 'premium fuel' until sufficient green renewable energy is available. Nevertheless, a key issue for Saudi Arabia and other major energy exporters will be how to achieve a circular carbon economy which rests on the achievement of 'four Rs': "*Reduce*: energy efficiency, renewable energy and other low carbon energy such as nuclear; *Reuse*: carbon capture and utilization (CCU) and emissions to value (E2V); *Recycle*: natural sinks such as forests and oceans, bio-energy and hydrogen; and *Remove*: carbon capture and storage (CCS) and direct air capture" (Belaid & Al Sarihi, 2022).

The trend towards monetizing reserves while also seeking to integrate new technologies to cater for a reduction in emissions and the development of blue energy exports can also be observed in the United Arab Emirates. The UAE unveiled its Hydrogen Leadership Roadmap at the COP26, targeting a 25% market share of low-carbon hydrogen by 2030. This was outlined as central to the UAE's net-zero objective by 2050. With the decline of the UAE's oil and gas industry, gaining an early mover advantage in alternative energy sources underlines how the UAE is strategically seeking to position itself as part of the energy transition. Nevertheless, it is also recognized that the 'green hydrogen' sector remains in its infancy for all countries engaged in it. As a sector, green hydrogen faces key issues relating to the cost implications of large-scale hydrogen production, infrastructure investments, high volume storage, transport and logistics, safety considerations, and the uncertainty associated with matching supply and demand (Eljack & Kazi, 2021).

In addition to the above variables, it is important to recognize that the geopolitical impact of the 2022 Russian invasion of Ukraine is a significant factor that is fundamentally redrawing the global energy map. The overarching effect on energy geopolitics of the invasion amounts to a fundamental revision in the role of Russia as an energy supplier. As of 2019, the EU-27's reliance on Russian energy amounted to 27.1% of total oil imports and 45.7% of natural gas. While debates over the dependence of the European Union on Russian energy have been long-standing, the response to the Russian invasion has seen the EU strategically commit to ending oil, gas, and coal imports from Russia as soon as feasible. Achieving this goal will give impetus to the European Union's 2050 goal of achieving the European Green Deal that commits to carbon neutrality and renewable energy sources taking more significance.

Beyond this, however, ending European imports of Russian fossil fuels also has global implications in terms of the pricing of energy commodities and geopolitical supply and demand dynamics. Indeed, the prospect of sanctions, litigation, or risk assessments on making payments to Russia also has the impact of deterring countries from importing Russian energy, which in practical terms means a narrowing of potential importers. From a macro perspective, both China and India can be expected to be the main alternative target markets for Russia's exports. While there are practical issues facing the supply of greater volumes of energy commodities to China, the repositioning of Russia's energy sector also then means that competition for the East Asian market will increase. Prior to the 2022 invasion, projected increase in demand for oil and gas was largely from the non-OECD states, and in particular from China (Wright, 2017). The dynamic of China being able to source increased amounts of its energy needs from Russia places it in a more competitive position regarding energy exports from the Middle East and Australia. The expansion of the Russia-China 'Power of Siberia' pipeline to 38 billion cubic meters (bcm) by 2025 (versus flows of only 5 billion cubic meters in 2020) and a push to boost domestic output mean that further strong growth in China's LNG import volumes beyond the early 2020s is far from guaranteed as an export market for both Qatar and Australia (Wright, 2020). It is with this change dynamic in mind that the broader factor of the future need for Qatari LNG needs to be considered as thus far, the projection has been that future demand will come from East Asia and, in particular from China, but the repositioning of Russia's energy sector which may include the supply of discounted oil and gas, necessitates a more competitive market for Gulf region supplies to gain market share.

In the final analysis, the central conclusion on the changing international energy landscape is that a comprehensive energy transition is taking place, which will have widespread implications. Furthermore, it has been observed that the impact can be expected to span consumption patterns of fossil fuels, opportunities to offset the emissions from the energy sector, in addition to the broader socioeconomic implications on energy-exporting states. The above observations have clear implications for Qatar and necessitate adaptation in order to ensure sustainability. Given this observation, the subsequent sections of this chapter will engage on the specifics of Qatar's energy sector and how it will need to respond to these broader trends.

6.4 A Sustainable Future for Qatar's Energy Sector?

Qatar's energy sector has been central to providing the country with the capacity to achieve development. Although oil exports initially drove Qatar's development, it was the strategic decision to focus on the gas sector that emerged as the defining feature of its political economy. Following the first shipment of LNG, which took place in 1996, Qatar has progressively emerged as the leading global exporter of LNG. This has come to define and drive the capacity to further economic growth. In terms of conceptualizing Qatar's political economy, it is appropriate to consider it as a 'late-stage petro-developmental state' (Wright, 2021). This is on the basis that Qatar's political economy should be understood in political terms, with the role of the energy sector providing the capacity for achieving development. Rentier-based conceptualizations have proven useful in understanding state behaviour and challenges. It lends itself to the phenomena known as the Dutch Disease, which essentially is a causal relationship between an economic boom triggered by a particular sector such as oil and gas, and a decline in growth, productivity, and diversification in other sectors. In this sense, it emphasizes structural challenges to the prospects for sustainable development. While such observations are valuable, they do not satisfactorily account for the political character of Qatar's economic development and how it has been achieved, hence the justification for looking at Qatar's prospects through the lens of the developmental state approach. While Qatar's economy remains energy-driven, it is observable that the majority of government revenue (85% in 2020) is sourced from this sector. Despite this, more than two-thirds of GDP comes from the non-energy sector, which underlines the character of diversification and growth that has been achieved in the private sector. Furthermore, it illustrates that income from natural resources has strategically been used to cover government costs, which has supported the growth of state capitalist activities and the private sector through a low-tax environment. Nevertheless, it remains clear that the energy sector is central to Qatar's political economy. Hence, an assessment of how this can be related to the question of sustainability is of central importance (Meza et al., 2022).

In terms of the export markets for Qatari LNG, it is observable in Fig. 6.1 that as of 2020, 60% of Qatar's LNG exports are directed towards the Asian market, of which India, South Korea, Japan, Taiwan, and Pakistan are the principal importing markets. This compared to 26% of supplies going to Europe. Although China is not the largest export market for Qatar's LNG at around 10% of total exports, its strategic importance as a destination market is longer-term. As of 2022, Qatar's LNG capacity stood at 77 million tons per year, but it is expected to increase capacity to 126 million

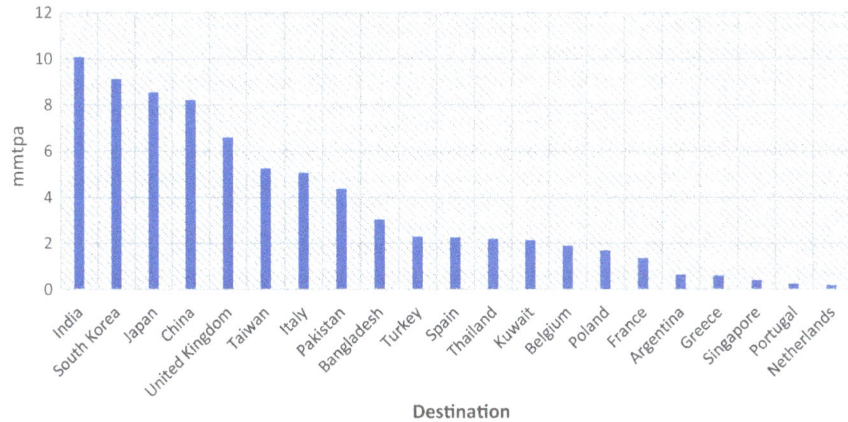

Fig. 6.1 Qatar LNG exports by Destination (2020) (*Source* Woodmac, MEES, authors own calculations)

tons per year by 2027. This raises the question of what the target destination markets will be once Qatar has increased its export capacity by around 50 million tons per year. As highlighted above, the projected future demand for LNG was primarily expected to be in the non-OECD, mainly from China (Piet & Wright, 2016). Given the repositioning of the Russian energy sector to increasingly focus on China as an export market, this brings it into direct competition with Qatar over future gas export capacity and Australia, which became the world's largest exporter of LNG in 2020. A strategic shift can be observed whereby the European Union will be increasingly seeking to secure gas significant volumes of gas to offset what Russia had been supplying it. What is important here is that the 2022 Russia-Ukraine war can be anticipated to have brought into question Qatar's need to focus on China as an energy export market and instead opened an opportunity to reposition itself as a strategic supplier to Europe.

As highlighted above, Qatar is developing its LNG export capacity with the target of 126 million tons per year by 2027. In practice, however, this export capacity is likely to become available from 2026 to 2029 gradually. Nevertheless, the potential for Qatar to increase the share of its gas exports to Europe has the potential to allow it to have a more balanced distribution of supply between the European and East Asian markets. The current reliance on the East Asian market has become more competitive owing to the emergence of Australia as the leading global supplier of LNG. However, it is noteworthy that the demand for LNG from Japan is projected to decline progressively. Japan has traditionally been the main export market for Qatar's LNG but has declined due to an ageing population and energy efficiency savings. Compounding this is Japan's 2030 Energy Plan, which was published in 2021, as it set a target of reducing the consumption of LNG from 37 to 20% of the energy mix. With its LNG demand projected to be reduced, the share of renewables is expected to increase, which underlines the scale of the energy transition that has been discussed above. Nevertheless, with Japan's LNG consumption in decline and China being able to source more of its natural gas needs from Russia, there is reason to conclude that Qatar's uncommitted LNG will increasingly be directed towards meeting European demand. Given this, what is important here is what will an increased integration with the European market mean for Qatar? How should it position itself as an energy supplier within the context of the European Green Deal and its targets for a carbon–neutral Europe by 2050?

For Qatar to align with the global trend towards carbon–neutral targets in order to mitigate climate change, in addition to the recognition that growth in demand will primarily be coming from the OECD rather than China as has previously been the case, it is necessary to consider approaches for adapting to this changed reality for sustainability purposes. In March 2022, Qatar Energy announced a revised sustainability strategy that targeted 11 million tons a year of carbon capture and storage (CCS) along with progressive growth in the role of solar energy (Martín-Pomares et al., 2017). One of the major projects, according to QatarEnergy, is the deployment of carbon capture and storage (CCS) technology to collect over 11 million tons of CO2 per year in Qatar by 2035. These projects have the potential to cut the carbon intensity of Qatar's LNG facilities by 35% and its upstream facilities by at least 25%, demonstrating Qatar's commitment to sustainably delivering cleaner LNG at scale to assist the energy transition (Anonymous, 2016).

The focus on CO2 capture and sequestration is part of the broader changing nature of the energy sector and within Qatar in particular (Al-Moftah et al., 2021). In line with the trend towards carbon neutrality, it is increasingly important for energy suppliers to demonstrate that their energy production offsets emissions for it to support targets for growth in renewables and the achievement of a carbon–neutral economy. Although the energy transition will necessitate a growth in renewables and a decline in fossil fuels, the role of natural gas and oil in an energy mix is expected to remain. Therefore, key exporting regions such as the European Union can be expected to be increasingly observant of the carbon emissions of any fossil fuel imports that they are receiving. This has a direct impact on the interests of Qatar to reposition its energy sector as one which is environmentally sensitive to safeguard demand and thereby support Qatar's future sustainability.

The importance of reducing the carbon footprint of Qatar's energy sector is a central aspect of its future growth and development. It was noted by Qatar's Minister of Energy, Saad Al-Kaabi, that:

> One of the most important environmental elements of the NFE North Field East project is its CCS that will integrate our wider CCS scheme at Ras Laffan, which once fully operational will be the largest of its kind in terms of capacity in the LNG industry, and will be one of the largest ever deployed or developed anywhere in the world. (Ingram, 2021)

As a sector, the liquification of gas for export is an energy-intensive process that comes with a clear carbon impact. A central aspect of Qatar's domestic electricity mix and the extent to which it has a carbon impact. Therefore, offsetting carbon emissions within the electricity sector can be an important means of reducing the national carbon footprint of as a fossil fuel exporter. The electricity sector within Qatar is predominantly gas powered, but the expansion into the solar sector is an important development that allows Qatar to position its LNG exports as more environmentally sensitive. The Al Kharsaah Solar Power Plant is a photovoltaic plant currently which will produce 800 MW of energy, making it the world's biggest solar power plant utilizing high-efficiency, half-cut bifacial solar modules (Byrne, 2022). A key benefit of Qatar investing in solar energy is that it allows it to free up gas for LNG export purposes rather than using it for its own domestic electricity needs.

The benefit of Qatar increasing its renewable energy capacity is that it allows it to offer a less carbon-intensive LNG product and the ability to venture into the 'blue energy' sector. As highlighted above, a central part of the energy transition concerns the move into the hydrogen economy. Increased use of hydrogen as a fuel source has enormous environmental potential, but it faces practical issues with regard to its cost. Essentially, the difference between blue hydrogen and green hydrogen depends on the method of production (Yu et al., 2021). The first is low-carbon blue hydrogen, which is made from natural gas using standard carbon-intensive processes but with CO2 collected and stored using CCS technology. The second is green hydrogen, which is created by electrolysis, which uses renewable energy to divide water into hydrogen and oxygen. Given the processes that are needed for the production of green hydrogen, its overall cost may be upwards of five times more expensive than conventional LNG, while blue hydrogen is double the price of LNG (Ingram, 2020). Although LNG is more cost-effective than renewables, the long-term trend indicates that renewables will continue to become more cost-effective. The steady expansion of the hydrogen sector is a longer-term issue for the LNG industry as the world's first cargo of blue ammonia arrived in Japan from Saudi Arabia in 2020, which underlined its potential to serve as a fuel source for existing gas-fired power stations. While LNG can be converted to blue ammonia, making it a cleaner fuel source than it is presently, the main challenge will be how to do it at a cost that is competitive. Nonetheless, the move to a hydrogen-based economy and the recognition that blue ammonia from LNG may one day be a viable fuel source opens up opportunities for research and development and technological innovation to make that option more cost-effective. For Qatar, blue ammonia is likely to have the most significant potential, given that it can be processed from existing resources (Okonkwo et al., 2021).

What can be observed here is that as Qatar has a comparative international advantage in its energy sector through the expertise it is accumulated, it has the potential to take a leading position in the energy transition by investing in both the blue/green hydrogen and ammonia sectors. When this is complemented with CCS, the carbon impact of Qatar's LNG can be reduced significantly, positioning itself for the market as part of the energy transition. By ensuring market relevancy, it caters for sustainability, growth, and profitability. A key observation here is that enhancing the role of renewables in the energy production cycle will allow Qatar to position itself as a more attractive energy supplier to countries seeking to achieve carbon neutrality. In essence, adapting to the energy transition both allows for more innovation and growth into new sectors of the economy for Qatar and enhances its attractiveness as an energy supplier to countries seeking to achieve carbon neutrality strategically.

6.5 Concluding Observations

It is worth recalling that 'sustainability' as a concept was defined in 1987 by the United Nations Brundtland Commission as, "meeting the needs of the present without compromising the ability of future generations to meet their own needs" (Redclift, 2005). Therefore, when considering the question of sustainability in Qatar's energy sector, it is important to recognize that achieving this goal has implications on how it adapts to the energy transition and advances greater diversification of the economy. The stakes concerning Qatar's energy sector are arguably one of the more important issues within this book. The economy is energy-driven, and an inability to adapt to changing realities would have broader implications in Qatar's political economy.

In conjunction with the energy transition towards renewables and carbon neutrality, the Russian invasion of Ukraine in 2022 has served as a systemic shock to the global energy market. This has manifested itself with an upending of Russia's position as a reliable energy exporter. The strategic decision by the European Union to cease imports of Russian energy amounts to a fundamental shift in market dynamics. Given the scale of Russia's energy exports and the size of the European market, it has profound implications on the flow of trade in oil and natural gas. It has been identified in the above text that a consequence of this geopolitical instability is that Qatar is likely to see strong growth in its trade with the European market based on market fundamentals. Furthermore, given the leading manner in which the European Union is moving towards a green economy, an opportunity is identified for Qatar to integrate with this market shift to achieve further diversification within its energy sector, which enhances its relevancy and sustainability.

In the context of this, this chapter has made observations on the historical significance of the energy transition and how the energy landscape is being redrawn in a way that has clear implications for Qatar's trading relationships and foreign relations and offers potential for research and innovation in new sectors. The energy transition presents Qatar with an opportunity to build on its existing capacity, technical expertise, and market presence to venture into the renewable sector. As a result, Qatar's principal exports could progressively factor in renewable energy to make it a more attractive and cleaner export product. Also, investing resources in research and innovation in the renewable sector and engaging in joint projects globally in the renewables sector opens up new opportunities for economic growth and diversification within the energy sector. In the final analysis, it is through adapting to this change that Qatar's energy sector will be able to be more sustainable by diversifying into emerging areas that have the potential to maximize relevancy and profitability.[1]

[1] Author's note: This article was made possible by a National Priorities Research Program Standard (NPRP-S) 12th Cycle grant no. NPRP12S-0210–190,067 from the Qatar National Research Fund (a member of Qatar Foundation). The findings herein reflect the work, and are solely the responsibility, of the author.

References

Al-Moftah, A. M. S., Marsh, R., & Steer, J. (2021). Life cycle assessment of solid recovered fuel gasification in the state of Qatar. *ChemEngineering, 5*, 81.

Anonymous. (2016). Qatar finally agrees to Boost Dolphin gas deliveries to UAE. *MEES.*

Belaid, F., & Al Sarihi, A. (2022). Energy transition in Saudi Arabia: Key initiatives and challenges.

Bouckaert, S., Pales, A. F., McGlade, C., Remme, U., Wanner, B., Varro, L., D'Ambrosio, D., & Spencer, T. (2021). Net zero by 2050: A roadmap for the global energy sector.

Bridge, G., Bouzarovski, S., Bradshaw, M., & Eyre, N. (2013). Geographies of energy transition: Space, place and the low-carbon economy. *Energy Policy, 53*, 331–340.

Byrne, M. (2022). Qatar's Al Kharsaah Solar PV plant poised for startup. *MEES.*

Dale, S. (2022). Energy outlook: 2022 Edition. BP Global Group.

Davidson, D. J., & Gross, M. (2018). A time of change, a time for change. *Oxford Handbook of Energy and Society.*

Dicken, P. (2007). *Global shift: Mapping the changing contours of the world economy.* SAGE Publications Ltd.

Eljack, F., & Kazi, M.-K. (2021). Prospects and challenges of green hydrogen economy via multi-sector global symbiosis in Qatar. *Frontiers in Sustainability, 1*, 612762.

Fagan, M. (2000). Sheikh Yamani predicts price crash as age of oil ends. *The Telegraph.*

Gray, S. (2017). *Steam power and sea power: Coal, the Royal Navy, and the British Empire, C. 1870–1914*, Springer.

Hafner, M., & Tagliapietra, S. (2020a). *The geopolitics of the global energy transition.* Springer Nature.

Hafner, M., & Tagliapietra, S. (2020b). The global energy transition: A review of the existing literature. *The Geopolitics of the Global Energy Transition* (pp. 1–24).

Ingram, J. (2020). Qatar: Massive LNG Tanker Order. *MEES.*

Ingram, J. (2021). Full steam ahead for Qatar's $29bn LNG Expansion. *MEES.*

International Monetary Fund. (2022). Non-oil revenue for general government for Qatar [QATGGRXOGDPXOPT]. FRED, Federal Reserve Bank of St. Louis.

International Renewable Energy Agency. (2019). *A new world: The geopolitics of the energy transformation.* International Renewable Energy Agency.

Krane, J. (2021). The bottom of the barrel: Saudi Aramco and global climate action. *Working Paper.*

Krane, J., & Young, K. E. (2021). A net-zero Saudi Arabia? Not so fast. *Al-Monitor.*

Leonard, M., Pisani-Ferry, J., Shapiro, J., Tagliapietra, S., & Wolff, G. B. (2021). The geopolitics of the European Green Deal. European Council on Foreign Relations: Bruegel Policy Contribution.

Martín-Pomares, L., Martínez, D., Polo, J., Perez-astudillo, D., Bachour, D., & Sanfilippo, A. (2017). Analysis of the long-term solar potential for electricity generation in Qatar. *Renewable and Sustainable Energy Reviews, 73*, 1231–1246.

Meza, A., Koç, M., & Al-Sada, M. S. (2022). Perspectives and strategies for LNG expansion in Qatar: A SWOT analysis. *Resources Policy, 76*, 102633.

Nye, D. E. (1999). *Consuming power: A social history of American energies.* MIT Press.

Obeidat, S. M., Al Bakri, A. A., & Elbanna, S. (2020). Leveraging "Green" human resource practices to enable environmental and organizational performance: Evidence from the Qatari oil and gas industry. *Journal of Business Ethics, 164*, 371–388.

Okonkwo, E. C., Al-Breiki, M., Bicer, Y., & Al-Ansari, T. (2021). Sustainable hydrogen roadmap: A holistic review and decision-making methodology for production, utilisation and exportation using Qatar as a case study. *International Journal of Hydrogen Energy, 46*, 35525–35549.

Pescatori, M. A., Matsumoto, M. A., Mirzoev, M. T. N., Zhu, L., Roos, M. E., Yang, Y., & Zhang, M. T. (2020). The future of oil and fiscal sustainability in the GCC region. International Monetary Fund.

Piet, R., & Wright, S. (2016). The dynamics of energy geopolitics in the Gulf and Qatar's foreign relations with East Asia. *Energy Relations and Policy Making in Asia.* Springer.

Redclift, M. (2005). Sustainable development (1987–2005): An oxymoron comes of age. *Sustainable Development, 13*, 212–227.
Stockwell, C., Geiges, A., Ramalope, D., Gidden, M., & Hare, B. (2021). Climate Action Tracker Global update–Glasgow's 2030 Credibility Gap: Net Zero's Lip Service to Climate Action. *Warming Projections Global Update.* Climate Action Tracker.
World Bank. (2020). *World Development Indicators* [Online]. Available: https://data.worldbank.org/indicator/NY.GDP.PCAP.PP.CD [Accessed].
Wright, S. (2017). Qatar's LNG: Impact of the changing East-Asian market. *Middle East Policy, 24*, 154–165.
Wright, S. (2020). *Advancement of environmental sustainability through LNG: The case of Qatar-China relations.* Routledge.
Wright, S. (2021). *Conceptualising Qatar's political economy as a developmental state.* Springer.
Yergin, D. (1992). *The Prize: The Epic Quest for Oil, Money, and Power*. Simon & Schuster.
Yergin, D. (2020). *The new map: Energy, climate, and the clash of nations.* Penguin Uk.
Yu, M., Wang, K., & Vredenburg, H. (2021). Insights into low-carbon hydrogen production methods: Green, blue and aqua hydrogen. *International Journal of Hydrogen Energy, 46*, 21261–21273.

Steven Wright is an Associate Professor of International Relations and an Associate Dean in the College of Humanities and Social Sciences, Hamad bin Khalifa University. He previously held positions as an Associate Dean and the Head of the Department of International Affairs at Qatar University. His area of research expertise is on the international relations and political economy of the Arab Gulf states and energy geopolitics. He obtained his Ph.D. in International Relations from the University of Durham in the UK. He has been an invited keynote speaker at a wide range of events hosted by leading organizations such as the World Bank, NATO, various governments, and several multinational corporations.

Chapter 7
Qatar in the Energy Transition: Low Carbon Economy Challenges and Opportunities

Sara A. Al-Mohannadi and Dhabia M. Al-Mohannadi

Abstract Climate change is accepted today as an indisputable fact resulting from human activity, predominantly the burning of fossil fuels. In order to prevent the catastrophic impacts of climate change, the world agreed in Conference of the Parties (COP) 21 to limit the global temperature rise above pre-industrial levels to 2 °C. To meet the Paris Agreement, a profound transformation of global energy systems is required from fossil fuel-based to low or zero carbon sources. This creates a risk for hydrocarbon-producing countries such as Qatar, which depend on fuel exports for income. In the short term, Qatar is well suited for the energy transition due to its abundance of natural gas, a widely acceptable transition fuel. Technological improvement in renewable energy as well as fast moving competition in the hydrogen market, showcase Qatar's hydrocarbon export vulnerabilities. This chapter will provide a background on energy transition elements, Qatar's efforts locally and internationally in the transition to a low carbon economy and highlights the opportunities and vulnerabilities for Qatar to build resilience.

Keywords Energy transition · Low carbon · Climate change · Sustainability · Qatar

7.1 Introduction

Qatar is a small peninsula with extreme weather conditions, hyper aridity, and water scarcity. The discovery of oil first, and gas later changed the country, which generated the wealth of Qatar today and enabled water desalination critical to the modern water system (see Chapter 11). The hydrocarbon economy aided the population increase

S. A. Al-Mohannadi
School of Oriental and African Studies (SOAS), University of London, London, UK
e-mail: 694165@soas.ac.uk

Qatar University Law, Doha, Qatar

D. M. Al-Mohannadi (✉)
Chemical Engineering Program, Texas A&M University at Qatar, Education City, Doha, Qatar
e-mail: dhabia.al-mohannadi@qatar.tamu.edu

L. Cochrane and R. Al-Hababi (eds.), *Sustainable Qatar*, Gulf Studies 9,
https://doi.org/10.1007/978-981-19-7398-7_7

from 28,000 people in 1939 to 2.8 million people living in Qatar in 2022 (Planning and Statistics Authority, n.d.b). These resources also facilitated investments in education, healthcare, infrastructure, culture, and sports, making Qatar a city-state of the future.

As the world moves toward adopting cleaner fuels and lessening dependency on fossil fuels, like many hydrocarbon-producing countries Qatar is faced with the challenges of the energy transition. The challenges are not in terms of local energy mix per se, but primarily economic. Qatar's major income—about 34% of the Gross Domestic Product (GDP)—is obtained by selling hydrocarbons (Planning & Statistics Authority, 2020), largely Liquefied Natural Gas (LNG). Fortunately for Qatar, natural gas is the cleanest form of fossil fuels and is classified as an energy transition fuel, used to shift away from coal and other forms of carbon intensive fuels. Moreover, natural gas is often used to backup renewable energy systems to overcome intermittent production and seasonality issues. This puts Qatar in a unique position, as it supports the world to decarbonize and is a seller of this in-demand commodity. While the current demand for natural gas is high, it puts Qatar at a risk, not due to lack or ability of production but a probability of demand decline in the future when the world demands more cleaner low carbon fuel. A similar effect was observed following the drop in oil prices at the end of 2014, which caused a sharp reversal in Qatar's economic growth dynamics. The nominal GDP growth rate declined by 20.2% in 2015 due to the decline in oil and gas export revenues and resultant restricted public spending (Planning & Statistics Authority, 2018). Qatar will have to adapt to the changing markets and the growing need for a renewable energy or carbon neutral source. The country's leadership anticipated this trend in 2008 and put forth a Qatar National Vision 2030 (QNV 2030). The vision is a master plan aimed at economic diversification, human development, and environmental protection to sustain the high living standards of its people while maintaining growth (see Chapter 2). The recent change of Qatar's state-owned hydrocarbon company from Qatar Petroleum to Qatar Energy and the introduction of the new Ministry of Environment and Climate Change (MOECC) indicate the shift of the country away from hydrocarbons to fit in the world's movement toward the energy transition.

7.1.1 From Hydrocarbons to Renewable Energy Systems

The energy transition is a complex and integrated problem; the current fossil fuel-based system allows the generation of both forms of energy, heat, and power. It is reliable, and able to generate power and heat consistently and throughout the year, unlike seasonality dependent renewable energy. Liquid fuels supply chains are established and well connected. Fuels can be stored for a long time and purchased for strategic supplies. They can be easily embedded in central utility generation, or decentral production. These attributes and more make fossil fuels the preferred power and heat source for most countries (Thiel & Stark, 2021). Nonetheless, fossil fuel prices might fluctuate due to geopolitics or demand thus creating price instability.

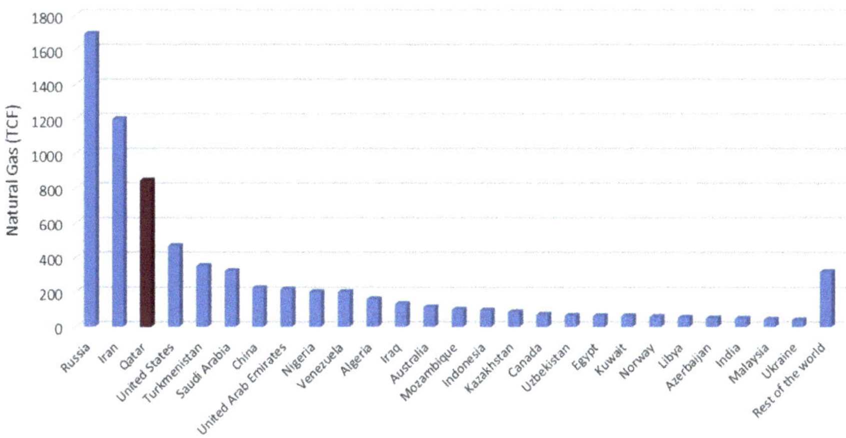

Fig. 7.1 World proven natural gas reserves in 2020 (U.S. EIA, 2020)

While the power generation via renewable energy has advanced significantly in terms of efficiency, deployment, and reduction in cost, it still suffers from intermittency issues and relatively large energy storage costs. On the other hand, renewable heat generation feasibility and scalability are still highly geographical based where renewable energy such as geothermal energy is at the forefront. This makes it harder for other countries to produce heat from renewable energy with low emissions (Thiel & Stark, 2021). Exceptions include the use of nuclear energy or bioenergy for heat generation. All these factors contribute to natural gas being recognized as the transition fuel to a low carbon economy. Natural gas demand has steadily increased over the last decade. The current production of natural gas is led by Qatar, followed by Russia, Australia, and the United States. Figure 7.1 shows natural gas proven reserves and production by country.

The supply and demand of natural gas became prominent after the Russian-Ukraine conflict, where the price shot up, and Europe had to face the reality of low carbon energy readiness as well as the source of its energy supplies. Qatar currently sells the majority of its LNG to Asia (Planning and Statistics Authority, n.d.a). Japan, a major Qatari LNG client, has adopted an energy transition plan—to convert to hydrogen by 2050 (Japan's Ministry of Economy Trade and Industry, n.d.). Qatar and Singapore recently signed the first LNG long-term emission detailed contract that requires shipments imported from Qatar to detail carbon emissions associated with LNG production and transport (Shiryaevskaya, 2020),[1] meaning that Qatar will have to reduce its own emission locally to lower its total supply chain carbon emission (see Chapter 6). The small population and heavy industrial activity, which includes oil and gas extraction, natural gas liquification and petrochemical production, in addition to

[1] According to the Intergovernmental Panel on Climate Change (IPCC) carbon accounting should provide transparent, consistent, comparable, complete, accurate, verifiable, and efficient recording and reporting of changes in carbon stocks and/or changes in greenhouse gas emissions by sources and removals by sinks.

the use of natural gas for all water, power, and food production, make Qatar one of the highest emitters of carbon dioxide (CO_2) per capita—although contributing less than 0.3% of the total world CO_2 emissions (Ritchie et al., 2020).

7.1.2 Challenges and Vulnerabilities

The ongoing transformation of the energy system toward a low carbon one will have profound challenges (Sim, 2020) in terms of geopolitical considerations and domestic arrangements. The energy transition will be associated with revenue and job volatility, especially for hydrocarbon-producing countries that depend on fossil fuel exports as the main income source. Hydrocarbon fluctuating demand and price volatility will have a negative impact on economic growth for hydrocarbon exporting nations, as observed in the oil drop in 2014 (Planning & Statistics Authority, 2018). In Qatar, hydrocarbon wealth has been used to fund jobs for citizens, more than half of whom work in the public sector. Any reduction in income will correlate with an increase in unemployment (Sim, 2020). This will mean that there will be more pressure for growth needed in the non-hydrocarbon private sector and entrepreneurship. Together with the shifting markets, the challenges posed to Qatar include: (1) how Qatar can reduce its carbon footprint locally, and (2) what would Qatar do to sustain its energy market, if energy is the best solution to go with. This requires a systematic assessment of Qatar's potential challenges and opportunities from a technical and policy perspective.

7.2 Qatar's Energy History

Qatar historically has been an energy innovator. Qatar's oil discovery started in 1939 with Dukhan field, and the first crude export was in 1949. Since then, Qatar has become an oil-rich country, but is a relatively small player in the oil market in comparison with Saudi Arabia, Iraq, Kuwait, and United Arab Emirates (UAE). In the 1970s, Qatar's first gas discovery was marked, a major gas reserve in the north of Qatar that it shares with Iran. The field, which covers about 6,000 square kilometers, lies offshore to the northeast of Qatar. It is considered the largest single concentration of non-associated natural gas in the world with total proven reserves of more than 900 trillion cubic feet, representing 20% of world total, and making Qatar the third country in the world with the largest reserves of natural gas after Russia and Iran (Qatar Ministry of Foreign Affairs, n.d.b). The field remained untapped, especially during the turmoil period of the Iranian revolution and then the Iraqi-Iranian war. It was in 1984 when Qatar began developing the field. The first operation was in 1991. Qatar partnered with then Exxon Mobil and took a leap of faith, based on the vision of the Father Emir, His Highness Sheikh Hamad Bin Khalifa Al Thani, to monetize the gas and invest in LNG. LNG technology was developed in the 1940s and was deployed commercially at a small scale. Qatar's innovation came in managing the supply chain from production to transportation. This included getting the product to customers with investments in regasification terminals and new large LNG ships. The first LNG facility started producing in 1991, and in 1997 QatarGas sold its first shipment of LNG to Japan. Qatar's strategic geographical location and the merging of extraction and downstream operations make it one of the lowest cost producers of natural gas. The partnership with ship producers to innovate on LNG transport, in creating large-scale LNG ships, put Qatar on the map as an energy leader.

7.3 Qatar's Local Emissions and Solutions

Qatar is among the countries that will be affected by sea level rise. The rise could damage coastlines and marine life, and climate scenarios envision more weather extremes that could mean heavy local flooding and more frequent sandstorms (Planning & Statistics Authority, 2011; see Chapter 9). Qatar's greenhouse gas emissions stem from the industrial and power production sectors (Planning & Statistics Authority, 2011; Fig. 7.2). The major production and oil and gas activities lie in the industrial cities, namely Ras Laffan Industrial City, Mesaieed Industrial City, and Dukhan Industrial City. Ras Laffan industrial city includes LNG plants, gas-to-liquid conversion (GTL), multiple industrial facilities and an industrial port. Mesaieed industrial city has a crude oil refinery as well as hydrocarbon, petrochemicals, iron, and steel production. Dukhan is the first site in which oil exploration and production started (Qatar Ministry of Foreign Affairs, n.d.a). The majority of Qatar's emission comes from industrial activities (extraction, manufacturing, and heat production) and power and water production, which is mostly carried out in the forementioned industrial cities.

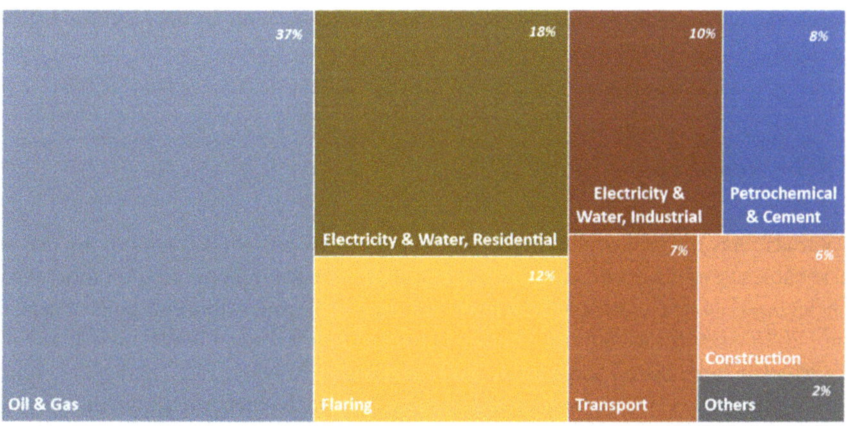

Fig. 7.2 Qatar emission breakdown by sector (Planning & Statistics Authority, 2011)

7.3.1 Qatar's Mitigation Options

This unique emission profile and the close proximity of the emission sources give the opportunity for Qatar to capture large amounts of CO_2 at a lower cost. Capturing CO_2 from industrial sources already exists using well-established technologies, such as physical adsorption and chemical absorption. While these technologies are energy intensive, for carbon capture itself, Qatar has these capture technologies in major facilities as they are used for natural gas desulfurization and in natural gas processing (van Ewijk & McDowall, 2020). It is worth noting that Qatar produces a large amount of CO_2 alongside the extraction of hydrocarbons, which are separated before any processing. This provides already separated CO_2 at a high purity that can be re-used for chemical production or carbon capture and storage (CCS). Qatar is well suited for CCS, a process that puts extracted CO_2 back into the geological formations. Qatar has one CCS project, to capture 5 million ton per annum of CO_2 from the LNG facility and store it underground, with plans to increase the capacity by 2030 (Qatar Energy, n.d.).

7.3.2 Carbon Capture Utilization and Storage (CCUS)

The untapped advantage for Qatar is to apply carbon capture utilization and storage (CCUS). Carbon utilization means to produce value added products, as a utility or in enhanced hydrocarbon production. Each usage varies in sequestration efficiency. Some can sequester CO_2 for a long time, others for shorter periods. Nonetheless, there exist industrial processes that can convert CO_2 on a commercial scale, such as urea production, dry reforming, and methanol production (Al-Mohannadi & Linke, 2016). The facilities already exist in Qatar cities. Intra-plant integration can help save mitigation cost and generates income streams (Al-Mohannadi & Linke, 2016), which can motivate companies to reduce their emissions without the need for policy interface.

7.3.3 Renewable Energy (R.E.)

Qatar also has the potential to avoid producing emissions locally through the use of renewable energy. Solar potential in Qatar is high, with sunlight running for more than 9 h per day. However, Qatar did not have any large-scale renewable energy production until 2021 when Siraj—a Joint Venture between QatarEnergy, QEWC, and TotalEnergies—was announced to generate 800 megawatts (MW) in 2022 and plans more than 3 gigawatts (GW) in the future. The more renewable energy deployment locally will lessen the amount of natural gas consumed and thus avoids the production of combustion related emissions (Mac Kinnon et al., 2018). It is worth noting that the planned solar energy power plants are backed up with fossil-based generators, to supply power during the night. The same company is working with Qatar Foundation to deploy R.E. systems applications on buildings and WOQOD to investigate the installation of Photovoltaic (PV) systems in WOQOD stations (QatarEnergy, 2021b). The latter will help reduce emissions from the transport sector, especially as the market moves toward electric vehicles (EVs) (Al-Buenain et al., 2021). Currently, the number of EVs in Qatar is limited, but the country has put forth the Green Car Initiative, to target 100% e-powered public transport and to have EVs at 10% of total cars by 2030 (Al-Buenain et al., 2021). These initiatives will help Qatar become less carbon intensive locally, but still a long way to reduce emissions from the industrial sector.

7.4 Qatar's International Efforts in Climate Mitigation

Qatar's international commitments reflect its genuine interest in combating climate change (see Chapter 3). While adapting to climate change halt Qatar's economic development path, Qatar's role in climate change international arena is undisputed. Qatar hosted COP back in 2012 which laid the grounds for the 2015 Paris Climate Agreement. Qatar is also a signatory to the Kyoto Protocol and ratified the United Nations Framework Convention on Climate Change.

Preventing environmental degradation is a part of QNV 2030 where it puts Environmental Development as the fourth pillar, which includes preventing air pollution and damages to natural habitats among several things. In addition, Qatar issued its National Environment and Climate Change Strategy before the Glasgow COP26 which provides the foundational policy framework for adapting and mitigating climate change. The highlight of the action plan states a reduction of greenhouse gas (GHG) emissions by 25% by 2030 and a 55% shift to sustainable technology in water desalination. The action plan also includes an impressive 100% account for all wastes and the closing and rehabilitation of 100% of unsanitary landfills. However, the remaining targets of the action plan seem to be redundant. For example, reducing 50% of per capita in food waste at retail and consumer levels is insufficient. Through proper management, 100% of food waste can be allocated to the working class or the poorer communities. The construction workers alone amount to half a million amidst a two million population (de Bel-Air, 2014). There is room for improvement in the action plan; however, it does provide a good starting point.

An indication of Qatar's good intentions in responding to climate change is its devotion to providing the voluntary Intended Nationally Determined Contributions (INDC). Qatar has communicated two INDCs in response to the UNFCC's decisions 1/CP.19, 1/CP.20, and 24/CP.18. The 2021 INDCs reflect Qatar's focus on reducing its overall emissions in comparison with the business-as-usual scenario. The two INDCs show Qatar's long commitment to climate change and the Paris Agreement and its determination to upholding climate change commitments according to its national circumstances and capability.

Qatar response to its international obligations is also illustrated in the initiation of its first UNFCCC Clean Development Mechanism (CDM) project. The Al-Shaheen Oil Field Gas Recovery and Utilization Project reduced its flaring to a significant 90% reduction in addition to reducing GHG emissions by capturing flared gas and allocating it into clean electricity. The 'zero flaring' should be a standard applied to all industrial production in Qatar. There are some commitments from Qatar Energy to reduce methane emissions. Methane itself is a larger contributor to the greenhouse effect that could leak from the production process, thus increasing net emissions of the LNG process. Moreso, startup operations and shutdown will result in flaring the excess gas, increasing emissions.

Internationally, Qatar's investment arm, Qatar Investment Authority (QIA), is investing in the energy transition and funding technologies that enable low carbon electricity generation. According to QIA Director, this is aligned with QIA's mandate to deliver long-term value for future generations through responsible sustainable investments.[2] QIA is a founding member of the One Planet Sovereign Wealth Fund Working Group and has helped to produce a framework to integrate climate change analysis into investment decisions. Following Glasgow (COP 26), QIA announced that it will invest GBP 85 million in Rolls-Royce to fund low carbon technologies. In addition, recently, QIA invested in clean energy generation in sub-Saharan Africa (Qatar Investment Authority n.d.) alongside its prior solar energy investments and energy storage solutions. In terms of international collaborations, His Highness Sheikh Tamim bin Hamad Al Thani committed $100 million in 2019 to support small island developing and least developed countries to address climate change challenges. This gives an opportunity, a small nation itself, to showcase an example of efficient carbon management that could be replicated by other nations.

7.5 Embracing the Low Carbon Energy Transition

The new shift from Qatar's national oil company, Qatar Petroleum to QatarEnergy,[3] indicates the understanding of the changing market perception. It should be followed by an overhauling shift into strategic investments, transparent accounting, deployment of renewables, research, and development and to look for win–win partnerships that are unique to Qatar's roles and opportunities. The shift is timely, Qatar indeed is no longer a petroleum country but also should not be an only gas exporter. Qatar focused on gas production and among the declining of oil production. The recent expansion project aims to boost the country's LNG annual production from 77 to 126 million tons and will confirm Qatar's dominance as an LNG exporter. It will increase

[2] QIA fund was established in 2015 to enhance Qatar's economy through diversifying investments, creating new revenue streams, and managing the surplus hydrocarbon income to reduce the risk of Qatar's depending on energy prices. The fund predominantly invests in global markets in addition to investing in Qatar beyond the energy sector.

[3] Qatar Energy is a state-owned public corporation responsible for all phases of the oil and gas industry in the State of Qatar.

the country's LNG exports by 64%, and as a result, Qatar will have to manage both the methane and CO_2 emissions that will be produced. Recently, QatarEnergy put forth a plan to curb emissions from operations through applying energy efficiency, avoiding flaring and methane emission reduction. In addition to growing its renewable energy capacity by 2–4 GW by 2030 and implementing carbon capture and sequestration technology to capture its own CO_2 from 7 to 9 million tons per annum by 2030. It is estimated that these steps would be able to achieve a net carbon intensity reduction of 15% from upstream and about 25% from the LNG facilities by 2030, while reducing 0.2 wt.% methane intensity target by 2025 and zero routine flaring by 2030. QatarEnergy aims to have a mixed grid of 90% gas-based and 10% renewables by 2030 (QatarEnergy, n.d.).

7.5.1 *Near-Term Gains*

It is easier and cheaper for Qatar to produce LNG than other places in the world (King, 2021). Qatar has aided in the energy transition of several countries leveraging the low-cost production locally of natural gas, strategic georgical location, integrated supply chain, and favorable long-term gas contracts. Japan and Qatar's energy relationship can be taken as an example. Japan was the first Qatari LNG importer, and after Fukushima nuclear accident, Japan switched off nuclear power and increased imports of Qatari gas. Similarly, LNG can help mitigate coal emissions, and emerging markets, such as China, India, and Brazil, would need natural gas as part of their energy mix. This seems to be the near-term direction of natural gas, QatarEnergy has signed a 20 years' LNG agreement with the Republic of Korea (QatarEnergy, 2021c) and starting from 2022, Qatar will be supplying China with 3.5 million tons of LNG per year for over 15 years. While Europe is a massive investor and deployer of renewable energy, the Russian-Ukraine war has exposed the reliance of the European Union (EU) on Russian gas. Talks are starting with Qatar to become a supplier for the EU with similar moves from the United States (Kozhanov, 2022). Countries with major energy demands, such as Germany, have approached Qatar (Concha, 2022). The energy transition plans have put natural gas as a transition fuel, which extends Qatar's LNG life in the near term.

7.5.2 *Upcoming Threats*

Geopolitics are unstable, and energy security is a priority for hydrocarbon importing countries. Prices and demand of any fuel may rise and fall according to weather, geopolitics, and supply. The price of LNG moves with the price of oil, however, renewable energy does not. Together with the motivation to lower the CO_2 footprint, massive investments were allocated for renewable energy technologies to improve their production, efficiency, and storage. The cost of renewable energy has significantly dropped. PV modules cost has fallen by around 80% since 2010, similarly, wind energy by 38% on average since 2009 with improved efficiencies (International Renewable Energy Agency, 2017a). Energy storage facilities at large scales are being piloted with significant effort in research to lower costs (International Renewable Energy Agency, 2017a). This includes shifting lithium-ion batteries to other cheaper formats using abundant resources (Kebede et al., 2022). Coupled with the increased deployment of green hydrogen production and electrification of the transport sector, the threat of demand reduction of fuel is a reality that hydrocarbon producers must address. Qatar's existing infrastructure and history provide opportunity to be part of that energy transition. Technically, Qatar produces natural gas at a low cost and can produce cheap hydrogen using commercial technologies. To deal with the emissions, CCUS options and renewable energy can be integrated to lower the emissions. Early movers will be the winners in the next phase of the energy transition.

7.5.3 *The Next Low Carbon Fuel*

Hydrogen is a new frontier. The issue of hydrogen today lies in the complex supply chain as well as the energy intensive cost and transportation (IEA, 2019). Leveraging lessons learned from LNG, of integrating production, and working with ship manufacturers and end-users, Qatar—if moves swiftly—can take the leadership in this emerging market. Gas abundance and emission reduction technologies allow the production of efficient blue hydrogen. Indications of moving toward that direction can be seen in QatarEnergy's recent agreement with Korea's Hydrogen Convergence Alliance (H2Korea) for the development of the hydrogen sector in both countries, encouraging the growth of the hydrogen industry and expansion of the hydrogen supply (QatarEnergy, 2021a).

Liquid hydrogen transportation is complex and energy intensive (Ibrahim et al., 2021). This is due to the compression of hydrogen from gas to liquid. It produces major amounts of CO_2, as fossil fuels are used to power the compression. At the same time, issues of safety in storage and transportation especially in the form of pipelines, such as hydrogen imperilment, are still an active research area (Eljack & Kazi, 2021). Nonetheless, hydrogen can be transported in other formats such as ammonia, methanol, or in liquid organic carriers. Qatar already produces Ammonia at a large scale via Qatar Fertilizer Company (QAFCO) in the largest single combined ammonia-urea facility in the world (QAFCO, n.d.). The process is produced by converting methane (natural gas) to hydrogen and CO_2. The process itself produces a large amount of CO_2. However, this CO_2 can be combined with ammonia to produce urea, it can be also sequestered or used in CO_2 utilization, thus, lowering the carbon intensity of the end product. Other countries in the region have begun working on this, such as Saudi Arabia which invested in blue ammonia and sent the first shipment to Japan (Aramco, 2020).

On the other hand, transporting hydrogen in liquid organic hydrogen carriers (LOHC) could be another opportunity to lower hydrogen transportation costs. LOHC are petroleum products in which hydrogen could be dissolved and could be transported at atmospheric pressure. This means that hydrogen will not need to be compressed and thus lowering the energy intensity, and associated emission from the transportation process. The UAE, which has larger petroleum reserves than Qatar, has invested in liquid organic hydrogen carriers (CNBC, 2022), thus creating a secondary use for their main export in the low carbon economy. Similarly, Australia, with its own natural gas, has invested heavily in hydrogen production (Australian Renewable Energy Agency, n.d.). Australia's advantage is its proximity to Qatar's largest clients. For Qatar to be on the next wave, and remain an energy leader, hydrogen should be a part of the future mix. This can be done through investing in research and development (R&D) to improve production and transportation processes to enable wide hydrogen adaptation. Qatar has two advantages: the reliability and the relationships it has built alongside the trust of clients. Qatar did not break contracts during the blockade with blockading countries (such as UAE) (Alkhalisi, 2017) or during COVID-19 crisis (QatarGas, 2020). The hydrogen market will likely be introduced in the 'hard to abate sectors' with application of hydrogen fuels in industries such as heavy vehicles, shipping, industry, and in decarbonizing the heating sector. Knowing this, Qatar should invest in hydrogen production targeted toward those technologies (IEA, 2019).

7.6 The Path Forward

The energy transition, albeit working for Qatar in the short term, exposes it to vulnerabilities as a hydrocarbon producer in the long term. There will be a strong need to think of a holistic solution that does not just take into account managing local emissions or the next fuel, but a new market beyond responding to climate mitigation needs. There will probably be new markets emerging, and thereby opportunities and threats. Qatar will have to be adaptable and resilient, relying on innovation and evidence-based policies rather than reacting to climate policy changes and fluctuating demands. To achieve this, Qatar should invest in system analysis methods development and create knowledge for its own future.

7.6.1 Threat: Hydrocarbon Asset Desertification

Priorities of current natural gas clients are in achieving energy security, with it comes diversification of suppliers to lower costs and elevate supply chain instabilities. In addition, the need for climate action is of utmost importance, coupled with technological advancements of renewable energy which could result in a quicker transition pace. The risk is, then, a slow adaptation that could lead to severe impacts on the economy. Thus, economic diversification should be embedded in any future energy or climate policy (QNV, 2008). Qatar has already invested heavily in the oil and gas sectors, in the LNG market, and in petrochemicals (KPMG, 2021). Climate targets would lead to asset desertification, where billions of dollars' worth of assets will be unneeded which would lead to loss of value and companies' losses (International Renewable Energy Agency, 2017b). This was observed in coal companies. The failure to recognize market trends, slow response to shifting policies, and refusal to acknowledge newer technologies led to a downfall of the industry. Likewise, joint-venture partners, with pressures from shareholders, could decide to sell off assets to lower the company emissions by dropping the most pollutant assets in their portfolios.

7.6.2 Deployment of Systems Analysis: Evidence-Based Policies

Having the wrong policy without systems analyses could lead to disastrous effects (OECD, 2020). Policies should be studied holistically with the energy mix and carbon mitigation. Early or premature carbon policies could fail in achieving their economic and environmental objectives. Economically, tough carbon policies could lead companies to leave to other countries or regions with no climate policies in place. Thus, reducing profit and jobs in the country with the positive climate policy. Environmentally, the total CO_2 will not change as the company will continue emitting CO_2 into the shared atmosphere. This was observed in the European experiment with the Emission Trading Scheme (ETS) (European Commission, n.d.). While ETS initially reduced CO_2, it has been seen that some coal production picked up in Germany, where they were able to purchase carbon credits (Marin et al., 2017) and elongate the life of a carbon polluting technology. Thus, there is a need to conduct the analyses using diverse inputs and modeling methods.

Energy systems modeling is a quantitative method that can be combined with policy analysis, treatment, and carbon capture, and incorporate the uncertainty of economics and environmental parameters to guide policy design. Reducing the cost of climate action requires a comprehensive screening through the available emission reduction pathways to select the best ones. Optimization-based decision support methods have been developed to guide cost-optimal energy transitions (Chang et al., 2021) and CO_2 integration networks (Tapia et al., 2018). The different tools provide minimum-cost integrated systems as solutions for achieving a defined level of CO_2 reduction (Al-Mohannadi et al., 2020). Energy transition optimization models can also consider variations in renewable energy supply and demand to determine the optimal design of energy systems (Limpens et al., 2019). Minimum marginal abatement cost (mini-MAC) curve has been proposed to analyze the cost of integrated CO_2 source and reduction pathways and including renewable energy and CCUS pathways (Lameh et al., 2021). Using the Mini-MAC method, and accounting for about 80% of Qatar's total emissions, the below curve was created (Fig. 7.3).

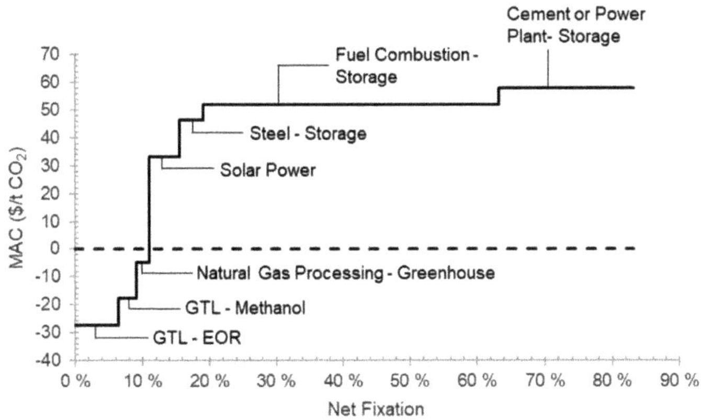

Fig. 7.3 Qatar's abatement cost profile (Lameh et al., 2020)

The mini-MAC method was used to analyze emissions from gas-to-liquid facilities, steel production, power production from natural gas, cement, and natural gas processing. The reduction options considered were enhancing oil recovery (EOR), hydrogenation of CO_2 to methanol, georgical storage, CO_2 in greenhouses, and avoiding emissions by deploying solar power. It can be seen that Qatar's emissions from the GTL facility can be used to generate profit if used in enhanced oil recovery and methanol production. The emissions from the natural gas processing facility can be implemented in greenhouses to enhance the yield of food production. Only after this utilization would Qatar apply solar energy, here it was capped at the 800 MW announced capacity. Larger segments could be implemented. The carbon sequestration option would only emerge as the last emission reduction technique.

Methods, such as the aforementioned, could guide policy and technological design. Qatar should invest in developing multiscale analysis that investigates the effect of disturbances such as geopolitical instability, market changes, or climate change policies on the cross-border supply chains. By applying scientific and engineering holistic thinking to developing strategies, it will enable stakeholders to assess different scenarios for various objectives such as environmental, economic, or security and stability. The approaches should be designed with a Qatari perspective to aid strategic planning and help guide future investments and partnerships.

7.6.3 Support Material, Energy and Resources Exchange, and Integration

In terms of near-term technological solutions, there are possibilities to reduce CO_2 in the industrial sector through having a common infrastructure that allows the exchange of power, heat, and other materials. On an industrial cluster level, Qatar should have a common utility sector, which allows industrial feed-in of excess electricity and heat.

On an operational level, connect utilities, governments, and private sector partners to encourage innovation in areas that suit both government goals and private sector interests.

Taking the example of LNG supply, the liquification process of the natural gas requires a large amount of energy and is produced by burning some of the gas feedstock to supply power. Electrifying the compression using renewable energy or the use of mix grid (Lee et al., 2012) would reduce emissions and at the same time save natural gas feedstock that can be sold (Chiu, 2008). Coupled with detailed accounting, Qatar would be leading the carbon neutral LNG and would be in a better position to incorporate the next low carbon fuel.

Regionally, Qatar has an opportunity to export electricity to the rest of the Gulf Cooperation Council (GCC) via a common grid (Al-Maskati & Al-Asaad, 2007). The grid project will allow Qatar to feed into the grid, thus supplying electricity instead of transporting gas and thus reducing emissions. To be carbon effective, any electrical generation from hydrocarbons needs to be supplemented by CCUS. Electricity export is possible and has been deployed in Norway-EU (Buil, 2021). Norway generates a mix of hydropower, wind, and natural gas that is connected with the EU. At the same time, Norway is monetizing its natural gas wealth to feed into its sovereign wealth fund and to create local subsidies for low carbon technologies such as electric vehicles.

7.6.4 Investments in Research and Development (R&D)

Climate change even though is a global problem, solutions will have to be customized to the environment. Qatar has huge potential for CCUS deployment but needs a holistic plan for climate change from local R&D to support existing research initiatives. Knowledge is emerging from the R&D investments, such as those made via the Qatar National Research Fund and at research institutions, such as Qatar University, Hamad Bin Khalifa University, and Texas A&M University at Qatar. Use existing institutions and resources to develop region-specific technological research in niche areas to provide a comparative advantage (Meltzer et al., 2014). There is a need to create multidisciplinary groups compromised from academia, public, and private stakeholders to anticipate trends and empower these actors to innovate.

7.6.5 Leaning in Energy Diplomacy and Taking Climate Leadership

Qatar is well positioned to lead the climate transition, having access to the transition fuel with potential cost-effective local mitigation strategies. If it takes the opportunities available, it can be an example for others to follow. Qatar also represents both a hydrocarbon exporter and a developing nation that will be severely impacted by climate change. The small population and large industrial activity have given Qatar the reputation of being the highest emitter per capita, whereas, in reality, Qatar's gas exports helped reduce global emissions by replacing more carbon intensive fuels. According to an analysis outlined in the Qatar Development Strategy, "Qatar would be ranked much lower if only emissions stemming from consumption were measured"(Planning & Statistics Authority, 2011). By communicating these challenges, threats, and perspectives, Qatar's emission reduction efforts would be more effective at the international level.

In conclusion, methods would aid the government decision-making and regulation, and should be done quickly. Qatar must decide on a route beyond natural gas and should not waste time. It is well positioned, given its access to expertise and know-how that exist from years of being an energy leader. The reputation of being a reliable, stable, and trustworthy supplier should be maintained with data transparency of emission sources and accounting, and to provide the next product of the future. Qatar will have to diversify in the hydrocarbon sector itself to build resilience and continue monetizing the valuable resource.

References

Al-Buenain, A., Al-Muhannadi, S., Falamarzi, M., Kutty, A. A., Kucukvar, M., & Onat, N. C. (2021). The adoption of electric vehicles in Qatar can contribute to net carbon emission reduction but requires strong government incentives. *Vehicles, 3*(3), 618–635. https://doi.org/10.3390/vehicles3030037

Alkhalisi, Z. (2017). Qatar keeps gas flowing to UAE despite blockade. CNN Money. https://money.cnn.com/2017/06/07/news/qatar-uae-gas-pipeline/index.html

Al-Maskati, N. B., & Al-Asaad, H. K. (2007). The GCC interconnection grid: ME power & water conference. GCC Interconnection Authority. https://www.gccia.com.sa/Data/PressRelease/Press_9.pdf

Al-Mohannadi, D. M., Kwak, G., & Linke, P. (2020). Identification of optimal transitions towards climate footprint reduction targets using a linear multi-period carbon integration approach. *Computers and Chemical Engineering*, 140, 106907–. https://doi.org/10.1016/j.compchemeng.2020.106907

Al-Mohannadi, D. M., & Linke, P. (2016). On the systematic carbon integration of industrial parks for climate footprint reduction. *Journal of Cleaner Production, 112*(5), 4053–4064. https://doi.org/10.1016/j.jclepro.2015.05.094

Aramco. (2020). World's first blue ammonia shipment opens new route to a sustainable future. https://www.aramco.com/en/news-media/news/2020/first-blue-ammonia-shipment

Australian Renewable Energy Agency. (n.d.). Hydrogen energy. https://arena.gov.au/renewable-energy/hydrogen/

Buil, N. (2021). Norway and Germany open power link boosting EU's green energy supply. Reuters. https://www.reuters.com/business/energy/norway-germany-open-power-link-boosting-eus-green-energy-supply-2021-05-27/

Chang, M., Thellufsen, J. Z., Zakeri, B., Pickering, B., Pfenninger, S., Lund, H., & Østergaard, P. A. (2021). Trends in tools and approaches for modelling the energy transition. *Applied Energy*, 290, 116731–. https://doi.org/10.1016/j.apenergy.2021.116731

Chiu, C. H. (2008). History of the development of LNG Technology. AICHE. https://folk.ntnu.no/skoge/prost/proceedings/aiche-2008/data/papers/P139095.pdf

CNBC. (2022). German firms sign deals in UAE as Berlin seeks alternative energy supplies. https://www.cnbc.com/2022/03/21/german-firms-sign-deals-in-uae.html

Concha, J. (2022). Germany in talks with Qatar for LNG supplies. *Energy Intelligence Group*. https://www.energyintel.com/0000017f-ad7e-dc0d-a17f-fffebeaf0000

de Bel-Air, F. (2014). Demography, migration, and labour market in Qatar. http://hdl.handle.net/1814/32431

Eljack, F., & Kazi, M. K. (2021). Prospects and challenges of green hydrogen economy via multi-sector global symbiosis in Qatar. *Frontiers in Sustainability, 1*. https://doi.org/10.3389/frsus.2020.612762

European Commission. (n.d.). EU Emissions Trading System (EU ETS). https://ec.europa.eu/clima/eu-action/eu-emissions-trading-system-eu-ets_en

Ibrahim, Y., Al-Mohannadi, D. M., & Linke, P. (2021). Modelling and optimization of hydrogen production in an industrial cluster accounting for economic cost and environmental impact. *Chemical Engineering Transactions, 88*, 439–444.

International Energy Agency. (2019). The future of hydrogen: Seizing today's opportunities. https://iea.blob.core.windows.net/assets/9e3a3493-b9a6-4b7d-b499-7ca48e357561/The_Future_of_Hydrogen.pdf

International Renewable Energy Agency. (2017a). Renewable power: Sharply falling generation costs. https://www.irena.org/-/media/Files/IRENA/Agency/Publication/2017/Nov/%20IRENA_Sharply_falling_costs_2017.pdf

International Renewable Energy Agency. (2017b). Stranded assets and renewables: How the energy transition affects the value of energy reserves, buildings and capital stock. https://www.irena.org/-/media/Files/IRENA/Agency/Publication/2017/Jul/IRENA_REmap_Stranded_assets_and_renewables_2017.pdf

Japan's Ministry of Economy Trade and Industry. (n.d.). Basic hydrogen strategy (key points). https://www.meti.go.jp/english/press/2017/pdf/1226_003a.pdf

Kebede, A. A, Kalogiannis, T., Van Mierlo, J., & Berecibar, M. (2022). A comprehensive review of stationary energy storage devices for large scale renewable energy sources grid integration. *Renewable & Sustainable Energy Reviews, 159*. https://doi.org/10.1016/j.rser.2022.112213

King, L. (2021). Qatar sees green role for LNG as world gasps for more energy. Forbes. https://www.forbes.com/sites/llewellynking/2021/12/04/qatar-sees-green-role-for-lng-as-world-gasps-for-more-energy/?sh=c1634e166601

Kozhanov, N. (2022). Qatar is no short-term savior, but it may still play a role in strengthening EU energy security. Middle East Institute. https://www.mei.edu/publications/qatar-no-short-term-savior-it-may-still-play-role-strengthening-eu-energy-security

KPMG. (2021). Qatar industrial landscape 2.0: Resilient and stronger. https://assets.kpmg/content/dam/kpmg/qa/pdf/2021/04/tl-qatar-industrial-landscape-2.0-resilient-and-stronger-.pdf

Lameh, M., Al-Mohannadi, D. M., & Linke, P. (2020). Developing minimum cost targets for carbon reduction in different geographical regions. *Chemical Engineering Transactions, 81*, 97–102.

Lameh, M., Al-Mohannadi, D. M., & Linke, P. (2021). Minimum marginal abatement cost curves (Mini-MAC) for CO_2 emissions reduction planning. *Clean Technologies and Environmental Policy, 24*(1), 143–159. https://doi.org/10.1007/s10098-021-02095-y

Lee, A., Zinamen, O., & Logan, J. (2012). Opportunities for synergy between natural gas and renewable energy in the electric power and transportation sectors. National Renewable Energy Laboratory. https://www.nrel.gov/docs/fy13osti/56324.pdf

Limpens, G., Moret, S., Jeanmart, H., & Maréchal, F. (2019). EnergyScope TD: A novel open-source model for regional energy systems. *Applied Energy*, 255, 113729–. https://doi.org/10.1016/j.apenergy.2019.113729

Mac Kinnon, M. A., Brouwer, J., & Samuelsen, S. (2018). The role of natural gas and its infrastructure in mitigating greenhouse gas emissions, improving regional air quality, and renewable resource integration. *Progress in Energy and Combustion Science, 64*, 62–92. https://doi.org/10.1016/j.pecs.2017.10.002

Marin, G., Marino, M., & Pellegrin, C. (2017). The impact of the European emission trading scheme on multiple measures of economic performance. *Environmental and Resource Economics, 71*(2), 551–582. https://doi.org/10.1007/s10640-017-0173-0

Organisation for Economic Co-operation and Development. (2020). Systemic thinking for policy making: The potential of systems analysis for addressing global policy challenges in the 21st century. https://www.oecd.org/development/systemic-thinking-for-policy-making-879c4f7a-en.htm

Planning and Statistics Authority. (n.d.a). Foreign trade. https://www.psa.gov.qa/en/statistics1/ft/Pages/default.aspx

Planning and Statistics Authority. (n.d.b). Monthly figures on total population, 2022. https://www.psa.gov.qa/en/statistics1/StatisticsSite/pages/population.aspx

Planning and Statistics Authority. (2011). National Development Strategy (NDS), Qatar First National Development Strategy 2011–2018.

Planning and Statistics Authority. (2018). National Development Strategy (NDS), Qatar Second National Development Strategy 2018–2022.

Planning and Statistics Authority. (2020). Qatar economic outlook 2020–2022. https://www.psa.gov.qa/en/knowledge/Doc/QEO/Qatar-Economic-Outlook-2020-2022-En.pdf

QAFCO. (n.d.). Home. https://qafco.qa/

QatarEnergy. (n.d.). Home. https://www.qatarenergy.qa/en/Pages/Home.aspx

QatarEnergy. (2021a). QatarEnergy and H2Korea Sign Hydrogen Cooperation Agreement. https://www.qatarenergy.qa/en/MediaCenter/Pages/newsdetails.aspx?ItemId=3687

QatarEnergy. (2021b). QatarEnergy and Qatar Foundation Sign Solar Energy Cooperation Mous. https://www.qatarenergy.qa/en/MediaCenter/Pages/newsdetails.aspx?ItemId=3688

QatarEnergy. (2021c). Qatar Petroleum and Korea Gas Corporation (KOGAS) signs a 20-year SPA to supply 2 million tons of LNG annually. https://www.qatarenergy.qa/en/MediaCenter/Pages/newsdetails.aspx?ItemId=3677

QatarGas. (2020). 2020 Sustainability report: Navigating Unprecedented challenges. https://www.qatargas.com/english/sustainability/Sustainability%20Reports/Sustainability%20Report%202020_English.pdf

Qatar Investment Authority. (n.d.). Home. https://www.qia.qa/en/Pages/default.aspx

Qatar Ministry of Foreign Affairs. (n.d.a). Industry.

Qatar Ministry of Foreign Affairs. (n.d.b). The North Gas Field.

Ritchie, H., Roser, M., & Rosado, P. (2020). CO_2 and Greenhouse Gas Emissions. Our World in Data. https://ourworldindata.org/co2-and-other-greenhouse-gas-emissions

Shiryaevskaya. (2020). Qatar, Singapore natural gas deal will detail carbon emissions. Aljazeera. https://www.aljazeera.com/economy/2020/11/9/bbnew-qatar-singapore-natural-gas-deal-seeks-to-reduce-emissions

Sim, L. C. (2020). Low-carbon energy in the Gulf: Upending the rentier state? *Energy Research & Social Science, 70*, 101752–101752. https://doi.org/10.1016/j.erss.2020.101752

Tapia, J. F. D., Lee, J. Y., Ooi, R. E., Foo, D. C., & Tan, R. R. (2018). A review of optimization and decision-making models for the planning of CO_2 capture, utilization and storage (CCUS) systems. *Sustainable Production and Consumption, 13*, 1–15. https://doi.org/10.1016/j.spc.2017.10.001

Thiel, G. P., & Stark, A. K. (2021). To decarbonize industry, we must decarbonize heat. *Joule, 5*(3), 531–550. https://doi.org/10.1016/j.joule.2020.12.007

van Ewijk, S., & McDowall, W. (2020). Diffusion of flue gas desulfurization reveals barriers and opportunities for carbon capture and storage. *Nature Communications, 11*(1), 1–10. https://doi.org/10.1038/s41467-020-18107-2

U.S. Energy Information Administration (EIA). (2020). https://www.eia.gov/international/data/world/natural-gas/dry-natural-gas-reserves

Sara A. Al-Mohannadi is a lecturer at College of Law, Qatar University. Sara obtained her BSc in Law from Qatar University in 2014 and her LLM from Duke University in 2016. Her work specializes in commercial and environmental law. Al-Mohannadi worked as a legal advisor with the Qatar National Compensation Council where she managed human rights cases. She is currently a Ph.D. candidate at SOAS, University of London focusing on distributional justice in climate transition. Her research interests include International Environmental Law, Environmental procedural rights, and climate change legal reform.

Dhabia M. Al-Mohannadi is an engineering assistant professor of Chemical Engineering at Texas A&M University at Qatar (TAMUQ). Her research area is process system engineering where her work focuses on process integration, techno-economic modeling and optimization, sustainable design and operation processes, carbon dioxide mitigation and natural resource management. This work often develops inter-disciplinary approaches that help decision-making and strategic planning. Dr. Al-Mohannadi is the recipient of HH the Amir of Qatar Education Excellence Award (2019), winner of Oryx GTL Qatar postgraduate PhD student award future research leaders in Qatar in 2016 and 2018, the Best Paper Poster Award at the 2015 Modeling and Optimization for Energy Saving and Pollution Reduction (PRES) Conference, and the Merit Award in 2020 and 2021 as an advisor.

Chapter 8
A Systems Perspective on the Sustainable Development of Qatar

Marcello Contestabile

Abstract Sustainable development is a critical component of the Qatar National Vision 2030 and of the national strategy implementing it. The State of Qatar is developing policies aimed at protecting the environment, sustaining economic prosperity, and developing a knowledge-based society. The three pillars of sustainability—environment, economy, and society—are closely interlinked and need to be treated as such by policymakers. In this chapter, we first discuss the systemic nature of the sustainability challenges of Qatar, particularly focusing on the complex transition the country is facing as it moves away from the current rentier economic model, with its social and environmental implications, towards a new model where new, knowledge-based industries develop in areas such as climate-tech and services. We then discuss systems analysis frameworks and tools at the disposal of governments to support sustainability policy design, and we illustrate one tool in particular that we are developing at QEERI: an energy systems model. We conclude with recommendations on equipping Qatar with an analytical toolkit to support sustainability policymaking.

Keywords Sustainable development · Systems analysis · Systems modelling · Energy transition · Climate change mitigation

8.1 Introduction

All main definitions of sustainability explicitly recognize its multiple dimensions: environmental, social, and economic (Brundtland, 1987; United Nations, 2015; United Nations General Assembly, 2005). Sustainable development is one where environmental protection goes hand in hand with social development and economic prosperity. The Qatar National Vision 2030 (General Secretariat for Development Planning, 2008) clearly reflects this, identifying the need for action to address these dimensions simultaneously. In particular, Qatar aims to protect its environment,

M. Contestabile (✉)
Qatar Environment & Energy Research Institute (QEERI), Hamad Bin Khalifa University, Doha, Qatar
e-mail: mcontestabile@hbku.edu.qa

L. Cochrane and R. Al-Hababi (eds.), *Sustainable Qatar*, Gulf Studies 9,
https://doi.org/10.1007/978-981-19-7398-7_8

promote the development of a knowledge-based society and diversify its economy away from the fossil fuel sector. The QNV 2030 is operationalized through a national strategy (General Secretariat for Development Planning, 2011, 2018) and implemented through a range of sector-specific policies. Addressing the different dimensions of sustainability simultaneously, however, is necessary but not sufficient: policymakers need to also understand and account for the complex interconnections among the different dimensions and sectors—in other words, they need to take a whole-systems approach to sustainability.

Before discussing the systemic nature of sustainability problems though, it is worth briefly touching on the notions of sustainability and sustainable development. In this chapter, we have so far used the two terms without distinguishing between them—which is not uncommon—however in the literature they are typically associated with different meanings. The notion of sustainable development has its roots in research work from the 1970s, such as the report "The Limits to Growth" (Meadows et al., 1972), showing how unconstrained economic growth based on a business-as-usual scenario would eventually lead to mankind overshooting the planet's carrying capacity and facing a subsequent collapse in population and industrial capacity. The first formal definition of sustainable development can be found in the Report "Our Common Future" (Brundtland, 1987), which articulates it as "development that meets the needs of the present without compromising the ability of future generations to meet their own needs". Since then, the concept of sustainable development has become pervasive; however, it has also been widely criticized for being vaguely defined and, by some, for implying that economic growth could be pursued indefinitely.

More recently, the notion of sustainability has emerged as distinct from sustainable development and decoupled from the idea of sustained economic growth, possibly also encompassing the perspective of pursuing economic "degrowth" in favour of social and environmental gains. In other words, sustainability can be pursued through transitioning away from development models where economic capital increases at the expenses of environmental and social capital, and overall capital degrades, to one where such total capital is preserved or increased in a durable manner. While conceptually sound, such definitions are obviously difficult to operationalize due to the uncertain and possibly subjective nature of any measurement of social and environmental capital. Moreover, different schools of thought still exist and the debate around the different notions of sustainability is still very open. For a more detailed discussion, we refer interested readers to a recent critical review of sustainability concepts (Ruggerio, 2021).

The problem of which sustainability notions and frameworks are more appropriate for Qatar is very interesting—particularly considering the country's economic dependence on the exploitation of its fossil hydrocarbon endowment—however, we will not be addressing it in this chapter. Instead, we will focus on the systemic nature of the sustainability problem—which remains critical irrespective of the particular framework one may want to adopt—and how it can best be addressed by policymakers. Scholars and practitioners have long recognized that the different dimensions of the sustainability problem are interconnected through a complex set of relationships, and that balancing them cannot be reduced to seeking tradeoffs. Therefore, from a policy perspective, these dimensions should certainly not be treated in isolation and a whole system approach should be adopted instead (Senge, 1990). Systems thinking in sustainability, while not new, is drawing increasing attention from academics and practitioners (Future Earth, 2019; Gallopin, 2003) and we believe that such as approach can be particularly valuable for Qatar.

Other chapters in this book focus on specific environmental, social, and economic challenges that affect the State of Qatar today. In this chapter, we want to reflect on how these dimensions are intimately interconnected in the particular case of Qatar, and how policies introduced to address challenges in each of them can best be designed with a system perspective in mind. Failing to do this risks substantially limiting the efficiency and effectiveness of such policies, and possibly leads to unintended consequences. A set of tools are available to support systems analysis that are increasingly being used by policymakers and other stakeholders worldwide. We recommend that these are applied to the sustainability challenges of Qatar as well.

In the remainder of this chapter, we provide a qualitative discussion of the main systemic problems that the sustainability transition in Qatar is faced with (Sect. 8.2), then we provide an overview of the main tools available to support systems analysis (Sect. 8.3), and we focus on one tool in particular—i.e. an energy systems model—that we are developing at QEERI to support policymaking in the area of energy policy and climate change (Sect. 8.4). We conclude with recommendations on how systems thinking and analysis can further be built into sustainability policymaking in Qatar (Sect. 8.5).

8.2 The Systemic Nature of the Sustainability Challenges of Qatar

In this chapter, we do not aim to provide a comprehensive treatment of all systemic problems associated with the sustainability transition of Qatar. Instead, we will limit ourselves to outlining the main ones and illustrating how the different dimensions are closely interconnected. A more comprehensive analysis is in fact a step we would recommend for future research aimed at better informing sustainability policy in Qatar, as discussed in Sect. 8.5. Let us first begin by briefly discussing how the different dimensions of environmental, economic and social sustainability are interconnected in the case of Qatar and, more broadly, the countries of the Gulf Cooperation Council (GCC).

Since the second half of the twentieth century, the State of Qatar has experienced dramatic economic and social development, driven by the extraction and commercial exploitation of its vast hydrocarbon resources, and has achieved great prosperity—its GDP per capita ranking 4th in the world in 2020 (World Bank, 2022). However, the rentier state[1] economic model of Qatar and other GCC countries—where most of the country's revenues originate from international sales of fossil fuels—is coming under increasing pressure due to environmental reasons, particularly global climate change. In a post-Paris Agreement world that seeks to limit global average temperature rise to 1.5 degrees above pre-industrial times, and where use of unabated fossil fuels will need to rapidly reduce (IEA, 2021), the rents associated with them are bound to decrease.

This means that Qatar and other GCC countries will need to rapidly transition away from the current economic model and create new, more sustainable industries that can at least in part make up for the revenue gap from fossil fuel exports. Indeed, such transformation is already evident in the GCC, where countries such as Saudi Arabia and the United Arab Emirates in particular are taking substantial steps towards the development of a domestic climate-tech sector, focusing on technologies such as renewable energy and carbon capture, utilization and storage that both complement and support the fossil fuel sector. Developing new industries, however, requires substantial investments, clear strategies, and careful planning, especially in view of a future where fossil fuel rents may decrease while the scale of investment needed to replace existing infrastructure with a more sustainable one increases. Such strategies and plans need to be informed by system-level analysis, or else they risk being ineffective or inefficient, possibly jeopardizing the future economic prospects of the country.

The current economic model of Qatar and the GCC has also led to the development of an informal social contract based on the redistribution of fossil fuel rents to citizens through various mechanisms. One such mechanism is the provision of free or highly subsidized energy services, such as lighting, cooling, transport, and water, the latter mostly coming from energy-intensive seawater desalination due the lack of renewable freshwater resources. This, combined with the harsh climate of the region, has resulted in the GCC countries having among the highest energy consumption per capita in the world. Such high consumption, satisfied primarily through the use of fossil fuels, has added pressure on the already fragile natural environment and ecosystems of the Arabian Peninsula. To compound the problem, increasing energy demand tends to bring additional water use, and vice versa. This situation has been described as the energy-water nexus or, taking a broader perspective, the energy-water-food nexus. The systemic nature of the nexus and the importance for policymakers of considering the multiple dimensions it spans are being increasingly stressed by both scholars and practitioners (Borgomeo et al., 2018).

[1] The rentier state concept has emerged from the political science literature to describe the political, social and economic systems of oil-exporting countries, such as those of the Gulf (Beblawi, 1987). In economic terms, a "rent" is a profit that comes from selling a natural resource at a price that is substantially higher than its production cost including normal profit margins, as is the case of oil and gas exports. Gulf rentier state economies are typically dominated by the oil and gas industry and the public sector, at the expense of other industries and the private sector. Government rents from oil and gas exports are redistributed to citizens through an extensive welfare system and generous subsidies. Rentier state economies have generally proven resistant to reforms.

The strong interdependence between the economic and environmental dimensions of sustainability in Qatar and the GCC also deeply intersects with the social sustainability dimension. The rentier state economic model of the past has allowed GCC citizens to benefit from well-remunerated and secure public sector jobs, and generally enjoy high levels of social prosperity and stability on a scale that is not known in other countries of the Middle East and North Africa (MENA). The environmental pressures that are forcing the transition away from the current model will also entail substantial social changes. Developing new sectors of the economy is a highly complex process that requires, inter alia, a transformation of the national innovation system, duly supported by an education system that produces graduates with the necessary skills to match the new sectors that the country seeks to develop. Economic diversification will also require the private sector to grow in size and to improve its productivity, so as to be able to compete on international markets. As Qatar and other GCC countries move away from a planned economy to a market-based economy, labour productivity will, therefore, need to increase and wages reduce, and a complete social transformation will be required to maintain prosperity and stability.

GCC countries have long been aware of the risks of relying on the revenues generated from fossil fuel exports. Economic diversification and social transformation policies have therefore been on the agenda of GCC countries for some time, and yet results achieved so far are generally modest. Over the last few years, however, the rapidly accelerating pace of international climate action has made economic diversification an ever more pressing issue. As the GCC countries commit to the rapid and drastic changes required, the complexity of decisions that policymakers are facing grows exponentially, and with it the importance of tackling economic, social, and environmental sustainability challenges from a systems perspective.

The following section provides an overview of the main systems analysis tools that are at the disposal of scholars, policymakers, and other stakeholders and that can support the generation of evidence necessary to inform effective and efficient policy decisions.

8.3 Systems Analysis Tools to Support Sustainability Policy

Due to the complex nature of sustainability challenges, no single framework exists that is used to tackle all sustainability policy problems—in Qatar or elsewhere—across all dimensions (environmental, social, and economic) and all sectors of the economy. However, a wide range of tools exist that have been applied to policy analysis in several countries worldwide and that can be beneficial to Qatar as well. Different categorizations are possible, however at a high level we can distinguish between essentially two types of tools:

a. Theoretical frameworks that can be operationalized through qualitative and quantitative indicators
b. Computer-based models that cover different domains and use different approaches such as optimization and simulation.

Within these two categories, very many different tools exist and a systematic discussion of them is outside the scope of this chapter. Instead, in order to provide the reader with a basic understanding of sustainability policy research and practice, we will focus on selected tools that are particularly well-known globally and relevant to Qatar in particular.

Each of these tools is designed to address specific sets of questions and therefore needs to be used for the intended purpose, taking their relative strengths and weaknesses into due consideration. It is also important to keep in mind that none of these tools can be used off the shelf and that substantial work is needed to adapt them to the study of specific problems in a given context, which requires relevant expertise. Lastly, none of these tools should be intended as offering an accurate representation of reality, but rather simplified representations that focus on particular problems and allow us to better understand them for the purpose of deriving policy-relevant insights.

For this reason, a solid approach to studying policies to promote the sustainability transition—in Qatar or elsewhere—is one that is driven by specific questions, recognizes the complex, systemic nature of the challenge, and combines the use of multiple tools, carefully chosen to complement one another and ensure robustness of the insights so derived. Here follows a brief discussion of theoretical frameworks and computer-based models in turn.

8.3.1 Theoretical Frameworks that Can Be Operationalized for Policy Analysis

Sustainability problems are often described as those where current practices are unsustainable and need to be replaced by new ones. This may involve the development and adoption of new technology, and the social, institutional and behavioural changes that goes with it. Transitioning to more sustainable technologies is certainly of critical importance to Qatar, as outlined in Sect. 8.2. A number of frameworks have been developed over the last decades that aim to describe the mechanisms underlying technological and social transitions—also referred to as socio-technical transitions—around sustainability problems. Stemming from the fields of technology studies and evolutionary economics (Nelson, 1985), a theory of socio-technical transitions has started to emerge in the late 1980s (Freeman et al., 1987; Lundvall et al., 1988) which has developed into multiple strands (Savaget et al., 2019). The transitions literature is complex and we will not discuss it in detail. However, we will briefly discuss two particularly notable frameworks that have found many applications in sustainability-related problems: the Multi-Level Perspective (MLP) on Socio-Technical Transitions

(Geels, 2002) and the Functions of Innovation Systems (Hekkert et al., 2007). Both frameworks treat sustainability transitions as systemic problems, characterized by multiple dimensions and actors. Both start from the premise that current paradigms owe their stability to reinforcing feedback loops linking technology, institutions, users, regulation, markets, and infrastructure, and that targeted policy interventions are required to overcome systemic barriers to change and effect the desired transition.

The MLP in particular focuses on the processes required for a transition to occur that involve interactions at different levels, defined as "landscape" (the external context, which in the case of Qatar can be international climate change policy), the "socio-technical regime" (the technology, in a broad sense, that needs to be replaced with a more sustainable one) and the "niche" (the new technology as it emerges from research, development, and early markets). The MLP describes technological transitions as a process where landscape pressures contribute to destabilizing dominant technologies ("regimes") and therefore create a window of opportunity for new technologies to emerge. The three levels and the overall process of technological transition postulated by the MLP are schematically illustrated in Fig. 8.1.

In the MLP, which can be applied to the study of any technological transition, the process of technological substitution does not happen spontaneously but is essentially driven by policy. Therefore, the MLP provides a useful tool for both ex-ante and ex-post policy analysis. The MLP is particularly popular among scholars and practitioners, and has been applied to a vast number of cases across many countries worldwide, including the GCC where one instance of its use is in the study of the transition to renewable energy in Saudi Arabia (Alrashoud, 2020). The MLP has also strongly influenced policy thinking however its direct use in policymaking processes to date remains limited.

The Functions of Innovation Systems framework, on the other hand, draws on a longer tradition of national and sectoral innovation systems analysis that has been directly used to inform government policy in many instances. The strong correlation between rates of technological innovation and economic growth is known since the 1960s, and the early models depicting innovation processes as linear have been replaced by a system approach already in the early 1990s (Lundvall, 1992). There is clear evidence of the use of national innovation systems frameworks for innovation policymaking aimed at creating knowledge-based economies since at least the late 1990s (OECD, 1997). In this context, the functions of innovation systems framework represent a development of the national innovation systems framework: while the latter focuses on studying innovation systems from the point of view of their structure—hence providing a static and institution-centred picture—the former focuses on the processes—or functions—that an innovation systems need to perform (Hekkert et al., 2007)—hence capturing systems dynamics and agency. Figure 8.2 below provides a schematic representation of such functions and how the link with one another.

The national innovation systems and functions of innovation systems frameworks can be seen as complementary approaches to the study of innovation systems, and their distinctive elements can also be combined as appropriate. Compared with

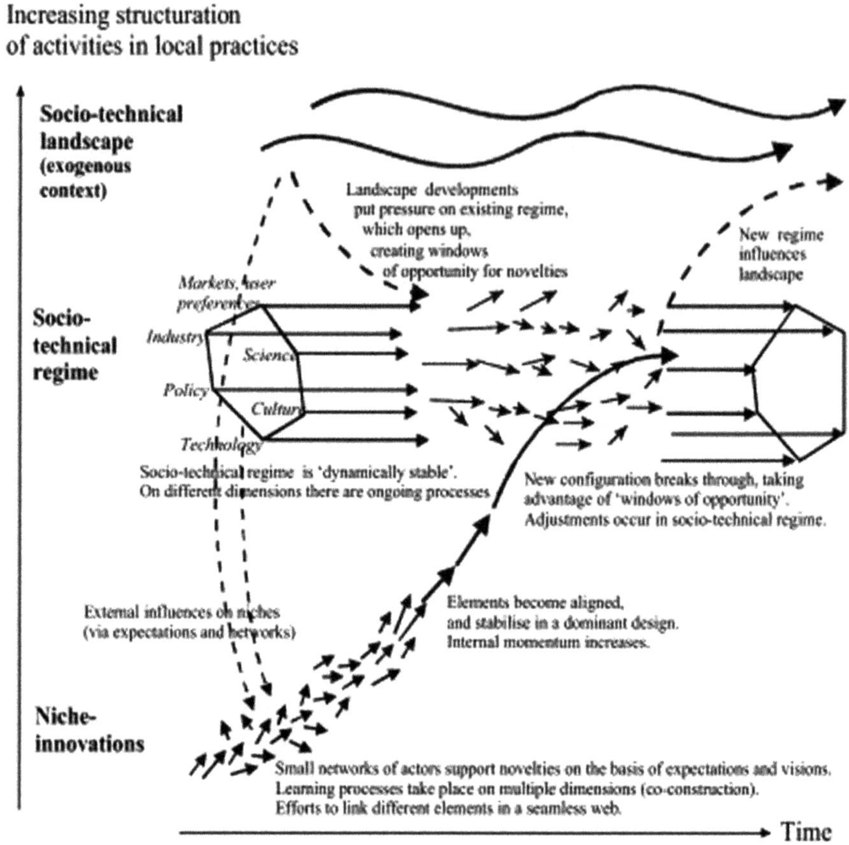

Fig. 8.1 Illustration of the multi-level perspective on socio-technical transitions (Geels, 2011)

the MLP, innovation systems frameworks are also easier to operationalize, which explains their more extensive use in policymaking.

To summarize, both MLP and innovation systems frameworks have been extensively applied by scholars and practitioners and, considering the importance that the State of Qatar places on sustainable technological innovation and the need to transition to a knowledge-based economy, we recommend their use to support policymaking. The frameworks are self-standing and can be operationalized and used on their own. However, they can also be used alongside the formal computer-based models that we will discuss in the next sections, and offer very valuable complements to them.

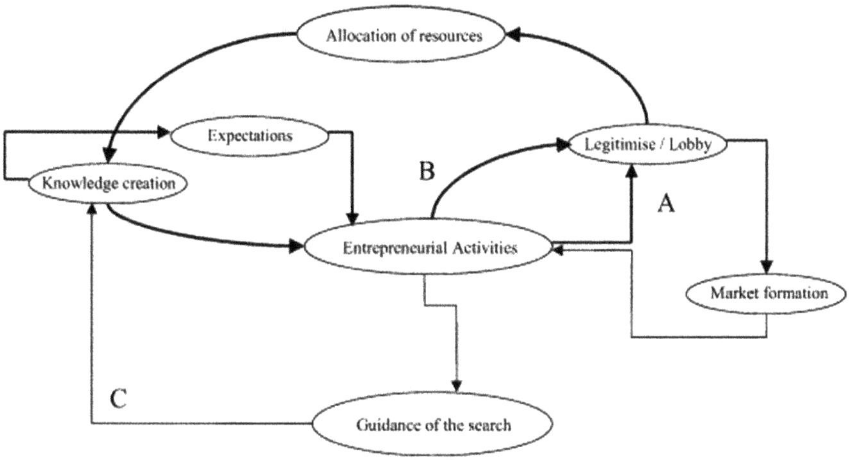

Fig. 8.2 Illustration of the main functions of technological innovation systems (Hekkert et al., 2007)

8.3.2 *Computer-Based Models that Are Used to Explore and Design Different Policies*

The recognition of the complexity of the sustainability problems under study has brought to the realization that computer-based tools are necessary in order to capture the multitude of elements and the dynamics linking them, in a way that the human brain alone is not able to. A range of computer-based models have, therefore, been developed to address different kinds of policy problems, which have found many applications at national, regional, and global scales. Here we will not provide a comprehensive overview of these tools, but rather we will discuss the main categories of models and their intended applications, as well as some of their main limitations, and their relevance to the sustainability challenges of Qatar.

When discussing the main types of models, we can make a first distinction among economic, engineering, and natural systems models. Economic models are built based on mathematical representations of economic theory and use real economic data as inputs. They are best used to study economic policies—such as pricing reforms, subsidies, and taxes, to name a few—and assess their likely impacts on the performance of the economic system under study or of specific sectors within it. Engineering models, on the other hand, are built based on mathematical representations of physical technical systems—such as power generation plants, buildings—and allow to test the effects of policies supporting the deployment of certain technologies on the energy consumption, emissions, and overall costs and performance of the system under study. Lastly, models of natural systems—of which very many exist, operating at different scales and representing physical as well as living elements of the natural environment—allow us to test the impact of anthropogenic activities on environmental quality, ecosystems, and global climate.

It is important to note that each of the model types discussed above can be used independently; however, they can also be used jointly, by either soft-linking or hard-linking them. In addition, there are also models that combine engineering and economic approaches—such as the TIMES energy systems model we are developing for Qatar that we will discuss in Sect. 8.4. Moreover, models that combine engineering and economic approaches can be further coupled with models of the natural environment, resulting in so-called Integrated Assessment Models (IAMs); these models are typically used to study climate change policy at the global level.

Another useful distinction is between models that optimize and those that forecast or simulate. Optimization models are useful to identify and characterize optimum pathways to reaching certain policy targets; the type of analysis these models lend themselves to is often referred to as "back-casting", because it works backwards from the desired end state and identifies all those policies, technologies, and infrastructures that need to be deployed at different points in time in order to achieve it at the least cost to society. Optimization models typically represent the system they study in terms of neoclassical economic theory, i.e. economic actors have perfect foresight, fully rational behaviours, and markets can adjust instantaneously so they are constantly in equilibrium. However, optimization models can also be made more realistic by introducing constraints to the knowledge and rationality of the economic actors, which allows to identify optimum pathways that are also potentially easier to implement practically.

Forecasting or simulation models on the other hand allow testing the real-world impact of different policies based on the behaviour of economic actors such as firms, governments, and consumers, as well as of the man-made and natural environment in which they operate. These models project into the future using different techniques, including econometrics, agent-based, and system dynamics modelling, each of which has its own strengths and limitations. In particular, agent-based techniques allow exploring future dynamics—which could be unexpected and counterintuitive—that directly arise from the complex interaction among the relevant agents, each of which is assumed to behave according to certain rules. Agent-based modelling is a very powerful approach which, however, requires a detailed understanding of the behaviours of all relevant agents, be they institutions, firms, or consumers. Econometric methods, on the other hand, can be applied effectively even when such detailed understanding of the problem is lacking, however their use required extensive datasets, typically time-series, and panel data. Both optimization and forecasting models can be used to study a range of problems occurring over different timescales—typically spanning years to decades.

The purpose of forecasting or simulation models is clearly very different to that of optimization models, and yet the two approaches are complementary: they can be used to address different problems, but they can also be applied to the same problem to gain a better understanding of it. One notable example where an optimization and a simulation model have been used together is the high-profile IEA study "Net-Zero by 2050" (IEA, 2021) where global pathways to net-zero have been identified and assessed from the point of view of social optimum (optimization) as well as based on the expected real-world responses to the relevant policies (simulation). For practical purposes, however, generally only one kind of model is used to address a particular policy problem, the choice being based on both the nature of the problem and the availability of tried and tested tools that fits it. Such considerations have informed our choice of the TIMES modelling platform, as is further explained in Sect. 8.4.

Lastly, it is worth mentioning that in principle more comprehensive tools could be developed that combine both optimization and simulation approaches, as well as economic, engineering, and natural environment components, all in one. However, modelling practice suggests that, as the complexity of a model increases, its transparency and ease of interpretation decreases, and so does its practical value. Therefore, when tackling a complex policy problem, it is often preferable to use a set of self-standing tools—including theoretical frameworks and computer-based models of different kinds—that can complement and corroborate one another, rather than attempting to develop a single tool that combines multiple methods and approaches.

Having provided a high-level categorization of computer-based models for policy analysis, here we will briefly discuss some types that are particularly relevant to sustainability problems such as those that characterize Qatar.

Let us begin from the problem of mitigating environmental damage while sustaining economic prosperity and social development. As discussed earlier, in Qatar and other GCC countries, this problem is inextricably linked with the imperative of diversifying the economy away from fossil fuel export and energy-intensive industries, and towards knowledge-based, high value-added new sectors that can generate the necessary economic value and social development (General Secretariat for Development Planning, 2008). Policymakers in Qatar and other GCC countries have been grappling with this problem for long, and achieved some degree of success, however dependence on the hydrocarbons industry in general remains high. Achieving the desired degree of economic diversification is difficult due to the presence of systemic barriers to change, overcoming which requires a set of distinct yet interlinked policy decisions. Computer models that allow to study the likely response of the country's economy to combinations of different policies are therefore suitable tools to aid decision-making.

A range of economic models exist, both simulation and optimization-based, that can be applied to sustainability problems. Among them, particularly worth mentioning are Computable General Equilibrium (CGE) models, which capture the structure of the current economic system and the behaviour of its main actors (firms, governments, and consumers), and simulate the impacts that policy changes can bring to it, including on employment, expenditure, and income. These models are capable of representing the economic system under study in detail and can reveal complex dynamics leading to indirect or unintended effects of given policies. CGE models have emerged from Input–Output analysis and in their more recent declination—known as Dynamic Stochastic General Equilibrium (DSGE) models—also offer the ability to study not just long-term effects of policy measures but also the adjustment path that the economic system goes through as it evolves from one equilibrium point to another, thus offering further valuable insight to policymakers. A CGE model of the Qatari economy is available, currently owned by the Planning and Statistics Authority (PSA). CGE models have already been applied to sustainability problems in other GCC countries, such as the macroeconomic impacts of renewable energy policies in Saudi Arabia (Blazquez et al., 2017), and can be a critical tool to study sustainability policies in Qatar as well.

Closely linked with the problem of diversifying the economy away from fossil fuels, the other increasingly pressing issue for Qatar and other GCC countries is reducing carbon emissions domestically and the carbon content of the fuels and other products they export. This requires, inter alia, replacing the existing civil and industrial infrastructure—developed over decades around the rentier economic model previously discussed—with one that is more energy efficient, has low emissions, and is compatible with the new economic model the country is aiming for. Such a transformation of all sectors of the energy system of Qatar requires substantial investments and decades to be completed, due to the large scale and long lifetime of the infrastructure. It is also a complex problem of systemic nature, as moving towards a low carbon, efficient energy system means that the different sectors within it (buildings, transport, power and water, and industry) will become increasingly interlinked and that decisions made today will have major consequences on future systems cost and environmental performance. The models that are typically used to study energy transition pathways are called energy systems models.

Different types of energy systems models exist, based on both optimization and simulation approaches. In industrialized countries such as the UK, these models have now been used for around two decades, and are now being increasingly used by developing countries too. In the GCC these tools have so far not been used by policymakers, however the picture is changing rapidly as the oil and gas exporting countries of the region are finally embracing the need for aggressive climate action. As GCC countries set themselves increasingly stringent carbon emissions reduction targets, the complex policy decisions that they are faced with will make energy systems models a critical tool for policy analysis for years to come. Recognizing the importance that such tools will have for Qatar, at the Qatar Environment and Energy Research Institute (QEERI) we are currently developing an energy systems model of Qatar; this is discussed in the following section.

Lastly, we recognize that several research teams in Qatar have developed and used different models of natural systems to conduct research around the environmental issues faced by the country. Due to space constraints, we will not discuss them here. However, we want to emphasize the potential for these models to be further leveraged for sustainability research purposes.

8.4 Developing an Energy Systems Model of Qatar

QEERI's mission is to conduct market-driven research to address the energy, water, and environmental challenges of Qatar and other countries characterized by arid climates. Research teams at the Institute are developing a range of cutting-edge technology solutions that can be deployed in Qatar and exploited commercially in other markets, thus supporting two national imperatives: protect Qatar's environment and promote the development of a knowledge-based economy. At QEERI we are acutely aware of the systemic and multi-dimensional nature of the challenges we are facing and this is why, alongside research and development activities, we also focus on the economics and policy research.

Motivated by the need to better inform both QEERI's R&D agenda and government policy in a number of areas, back in 2019 at QEERI we realized the importance of equipping Qatar with an energy systems model of the same standard as those used by leading governments worldwide. Thanks to funding from the Qatar National Research Fund (QNRF) through the grant NPRP13S-0204-200,250, in partnership with Imperial College London, and with the support of Kahramaa (Qatar's national electricity and water utility company) and the Abdullah Bin Hamad Al-Attiyah International Foundation for Energy & Sustainable Development, we are currently developing an energy systems model of Qatar that we expect to be ready for initial use by early 2023.

Since we first conceived the idea, the policy relevance of the model has further increased. In particular, having set its first carbon emission reduction target in 2021—a reduction of 25% of carbon emissions by 2030 relative to a business-as-usual scenario based on policies implemented up to 2019—Qatar will need to progressively tighten it in future, which will require continuous analysis aided by the necessary tools. At the same time, the Qatari government is introducing policies in a number of other areas (water, food, transport, buildings, etc.), all of which bear consequences in terms of climate change policy and that therefore need to be addressed in a systemic manner.

Based on the nature of the problem at hand—i.e. exploring least-cost technology and policy pathways to decarbonization and environmental protection in Qatar—and taking into consideration the need for a tool that is tried and tested and on which policymakers can rely, our choice has fallen on a particular kind of techno-economic optimization model called TIMES. A brief discussion of the nature of the model, its use to support policy analysis so far, and the particular value we believe add to Qatar follows below.

8.4.1 What Is a TIMES Model and How It Works

The name TIMES is an acronym that stands for "The Integrated MARKAL-EFOM System". The TIMES modelling framework has evolved from the MARKAL ("MARKet Allocation") framework originally developed in 1978 by the Energy Technology Systems Analysis Programme (ETSAP) of the International Energy Agency. The development of the MARKAL model was initially motivated by the need to explore alternatives to oil at a global level, due to supply security considerations. However, as the climate change issue became pressing on the policy agenda, MARKAL, and its successor, TIMES, provided policymakers with a very valuable tool to inform the development of energy-environmental policy strategies at national, regional, and global levels. We will further discuss the policy relevance of TIMES in Sect. 4.2, however first let's briefly discuss its structure and functioning.

The TIMES modelling framework combines the engineering approach—it is a technology-rich, bottom-up model of the energy system—with the economic approach—it balances demand and supply within the energy system through partial equilibria at sector level. It is an optimization modelling framework that allows building least-cost energy systems transition pathways consistent with pre-defined policy targets by making decisions on equipment investment and operation, and on import and export of energy commodities. In doing so, the model also takes into account all technology and resource availability constraints set by the user. The TIMES modelling framework revolves around a specific objective function—an algorithm based on which the optimum solutions to the problem are found—and it can be used as a basis to develop TIMES models for any market or geographic area of choice (Loulou et al., 2005). No two TIMES models are the same; however, they operate according to the same principles.

The TIMES modelling framework is graphically illustrated in Fig. 8.3.

As the figure shows, TIMES models account for all steps of energy conversion, from fuel extraction and refining or renewable power generation—including energy imports and exports—through to transmission, distribution, and end-use in all sectors of the energy economy (buildings, transport, power generation, and industry). In the model, all relevant energy conversion and end-use technologies—both present and future—are included, and their adoption and use strategies are determined by the decisions of the relevant agents in the system—such as power companies, passenger car

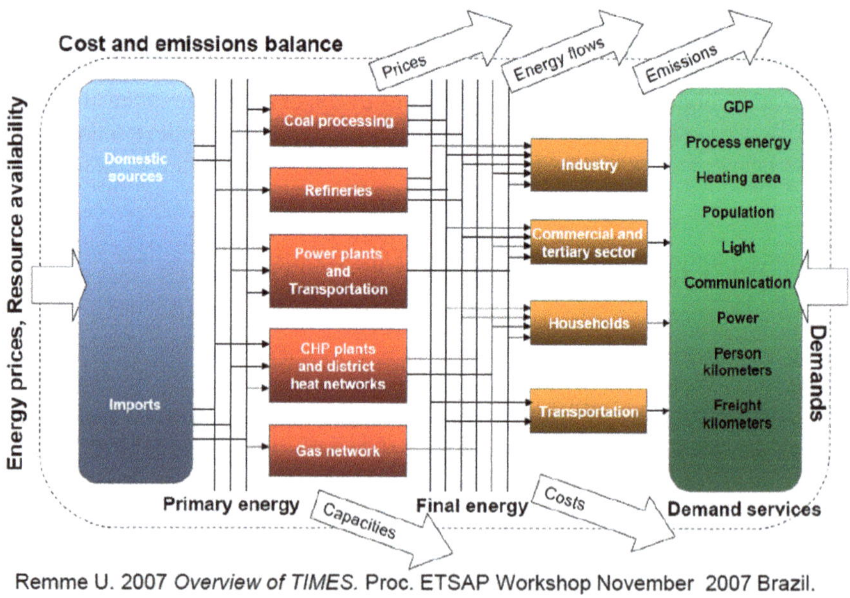

Fig. 8.3 Overview of TIMES modelling framework (Remme, 2007)

owners—who are characterized as rational and having perfect foresight. Constraints to agents' behaviour or technology deployment can be introduced in the model to simulate the likely real-life effect of different policies. Model scenarios are driven by energy service demand projections—e.g. demand for cooling in buildings, transport of passengers and goods, etc.—which constitute an input to the model. Given the policy targets, service demand projections, and other constraints set by the user, the model computes transition scenarios that fulfil all requirements at the least cost to society. Policy analysis with TIMES is conducted by generating a number of scenarios that test different assumptions and policy approaches, and by comparing and contrasting them so as to derive the desired insights.

8.4.2 Use of TIMES in Policy Analysis Internationally

The TIMES model has been extensively used by governments worldwide to conduct energy-environmental policy analysis in support of major pieces of legislation. At the forefront is the UK, that has been using MARKAL and TIMES models to inform its climate change policy for almost two decades. Table 8.1 below provides a non-exhaustive list of instances of relevant use of TIMES by national governments.

Apart from national governments, the TIMES model is also used by non-governmental research organizations to explore global energy transition and decarbonization scenarios, as well as major corporations in the oil and gas and other sectors, who use TIMES for planning purposes.

Table 8.1 List of main countries where TIMES has been used for national energy-environmental policymaking (Gargiulo, 2021)

Country	Application of TIMES
UK	Committee on Climate Change (CCC) carbon budgets
Ireland	Negotiations with the EU on Effort Sharing Decision for 2030 Low Carbon Energy roadmap to 2050
Denmark	Analysis supporting the Danish Energy Agency's target of 100% renewable energy by 2050
Sweden	Climate policy to reach net-zero by 2045
Finland	Energy and Climate Roadmap 2050
Greece	National Energy Planning Roadmap to 2050 National Renewable Energy Action Plan National Energy Efficiency Action Plan
USA	Analysis of effect of policies for the Office of Air Quality Planning and Standards
South Africa	National Climate Change Response Green Paper Integrated Resource Plan
Egypt	National Energy Strategy to 2035

8.4.3 *Policy Relevance of TIMES in Qatar*

The Qatar TIMES model can generate evidence to support policymaking in a number of areas of interest to major national stakeholders in Qatar.

Firstly, the TIMES model is a major tool used worldwide to inform climate change policy, as illustrated in the previous section. This makes it relevant to the Ministry of Environment and Climate Change and complements the methods used so far to inform Qatar's national climate change strategy and related targets. Clearly another key stakeholder in this area is QatarEnergy, because reducing carbon emissions from industry is going to be a critical part of any climate change plan. TIMES can help both entities assess which sectors and sub-sectors should be prioritized, and which should be tackled at a later time, so as to achieve the country's overall carbon emission reduction targets in a cost-effective manner.

Linked to Qatar's climate change policy are policies on energy efficiency (the Tarsheed programme), electric mobility, and solar energy, to name the main ones. These are overseen by Kahramaa who, as previously mentioned, is supporting the development of the Qatar TIMES model. As also illustrated in the previous section, TIMES has been used in several other countries to assess and design energy efficiency, sector-specific and technology-specific policies from a systems perspective. Kahramaa can, therefore, use the model to further inform and refine its policy programmes.

Lastly, TIMES can also be used to assess R&D priorities and innovation gaps at national level, which makes it relevant to the Qatar Research, Development, and Innovation Council (QRDI), the entity overseeing national policy in this area in fulfilment of the QNV 2030. The use of TIMES can, therefore, complement the analysis already undertaken by QRDI.

8.5 Recommendations for Future Work to Support Sustainability Policy in Qatar

In this chapter, we have illustrated the systemic nature of the sustainability policy challenges faced by Qatar, as environmental pressures—both global and local—force the country to rapidly transition away from a rentier economic model, diversify its economy, and create a knowledge-based society to preserve or increase its total economic, social and environmental capital. The complex decisions faced by policymakers today and in future, therefore, demand a systems approach and can be effectively supported through the deployment of a range of systems analysis tools, some of which we have briefly discussed.

Before these tools can be put to use in Qatar, however, we need to start from raising awareness among policymakers in government and industry on the need for a systems approach and the availability of tools to support it. The challenge is to gradually move away from a paradigm where individual entities make policy decisions in their respective domains without taking into full consideration the interlinkages among them, and where such decisions are mainly informed by one-off studies by international consultancy firms that may not be able to deploy the necessary tools, let alone tailor them to the needs of Qatar. A system approach inevitably requires a higher degree of coordination and the development and deployment of specific tools. Developing tools that are tailored to the national circumstances requires technical competencies and timeframes that are not compatible with those of typical consultancy projects. In other words, a systems approach to sustainability policymaking ideally requires government, industry, local experts, and international consultants to all work together in a continuous manner. Raising awareness and building capacity within government and industry is, therefore, going to be of critical importance. Thankfully, several entities exist in Qatar that are well placed to achieving this, especially if they work closely together.

The following step is to map the main systemic challenges that Qatar is faced with on the way to realizing its QNV 2030 and the relevant stakeholders. We have briefly discussed them earlier in the chapter: economic diversification away from the oil and gas sector, reduction of national carbon emissions and of the carbon content of fuel and commodity exports, the development of a national innovation system focusing on high-added value products and services, and the development of a knowledge-based society. These challenges are all interlinked, and a clear understanding of how they connect with one another is essential to addressing them.

Once the picture is clear, Qatar needs to equip itself with a toolkit to support sustainability policymaking. The tools chosen need to be robust and well-accepted internationally, to give policymakers confidence, and selected so as to complement one another and allow for better validation of the insights they generate. Adapting the tools to the specific circumstances of Qatar is a process that requires time and resources, and needs to be seen as an investment for the country, not an ad-hoc, policy-specific source of insight. Once the tools are available, they can be applied to national policymaking for years to come. For the tools to remain relevant over time, obviously they will have to be continuously updated and refined, which also requires resources. However, if correctly managed, the development and maintenance of the tools will not cost more than typical consultancy projects but will deliver substantial value to the country.

Such toolkit can take different forms; however, it will inevitably need to include a model of the country's economy and an energy systems model. It will also need to include tools to conduct innovation systems analysis. Developing such tools for Qatar requires specific economic and technology policy expertise, as well as sector-specific and technology-specific knowledge. Much of the necessary expertise and knowledge is already available in the country, and we call for the contributors to this book to come together and work jointly to help address the sustainability challenges of Qatar from a systems perspective.

Acknowledgements The author gratefully acknowledges the support of the Qatar National Research Fund (QNRF) through the grant NPRP13S-0204-200250 "Developing an energy systems model of Qatar as a tool to support national energy-environmental policy analysis".

References

Alrashoud, K. A. (2020). Multi-level perspective of technological transitions: Analysis of the residential solar photovoltaic systems policies in hydrocarbon-rich Saudi Arabia. In K. A. Alrashoud (Ed.), *Managing technology transition in Saudi Arabia: Residential solar photovoltaic systems development* (pp. 107–162). Springer Singapore.

Beblawi, H. (1987). The rentier state in the Arab World. *Arab Studies Quarterly, 9*(4), 383–398. http://www.jstor.org/stable/41857943

Blazquez, J., Hunt, L. C., & Manzano, B. (2017). Oil subsidies and renewable energy in Saudi Arabia: A general equilibrium approach. *The Energy Journal, 38*(KAPSARC Special Issue).

Borgomeo, E., Jägerskog, A., Talbi, A., Wijnen, M., Hejazi, M., & Miralles-Wilhelm, F. (2018). *The water-energy-food nexus in the Middle East and North Africa—scenarios for a sustainable future.* Retrieved from International Bank for Reconstruction and Development/The World Bank, 1818 H Street NW, Washington, DC 20433. https://documents1.worldbank.org/curated/en/927041530193545554/pdf/127749-WP-28-6-2018-1-16-49-WeBook.pdf

Brundtland, G. (1987). *Report of the World Commission on environment and development: Our common future.* Retrieved from United Nations. https://sustainabledevelopment.un.org/content/documents/5987our-common-future.pdf

Freeman, R., Freeman, C., & Freeman, S. (1987). *Technology, policy, and economic performance: Lessons from Japan.* Burns & Oates.

Future Earth. (2019). *A systems approach: Imperative to achieve the sustainable development goals.* https://futureearth.org/wp-content/uploads/2019/06/SDG-systems_issue-brief.pdf

Gallopin, G. (2003). *A systems approach to sustainability and sustainable development.* Retrieved from Santiago, Chile. https://repositorio.cepal.org/bitstream/handle/11362/5759/S033119.pdf?sequence=1

Gargiulo, M. (2021). *VEDA-TIMES course material.*

Geels, F. W. (2002). Technological transitions as evolutionary reconfiguration processes: A multi-level perspective and a case-study. *Research Policy, 31*(8), 1257–1274. https://doi.org/10.1016/S0048-7333(02)00062-8

Geels, F. W. (2011). The multi-level perspective on sustainability transitions: Responses to seven criticisms. *Environmental Innovation and Societal Transitions, 1*(1), 24–40. https://doi.org/10.1016/j.eist.2011.02.002

General Secretariat for Development Planning. (2008). *Qatar National Vision 2030.* https://www.gco.gov.qa/wp-content/uploads/2016/09/GCO-QNV-English.pdf

General Secretariat for Development Planning. (2011). *Qatar National Development Strategy 2011~2016—Towards Qatar National Vision 2030.* https://www.psa.gov.qa/en/nds1/Documents/Downloads/NDS_EN_0.pdf

General Secretariat for Development Planning. (2018). *Qatar Second National Development Strategy 2018~2022.* https://www.psa.gov.qa/en/knowledge/Documents/NDS2Final.pdf

Hekkert, M. P., Suurs, R. A. A., Negro, S. O., Kuhlmann, S., & Smits, R. E. H. M. (2007). Functions of innovation systems: A new approach for analysing technological change. *Technological Forecasting and Social Change, 74*(4), 413–432. https://doi.org/10.1016/j.techfore.2006.03.002

IEA. (2021). *Net zero by 2050—A roadmap for the global energy sector.* https://iea.blob.core.windows.net/assets/deebef5d-0c34-4539-9d0c-10b13d840027/NetZeroby2050-ARoadmapfortheGlobalEnergySector_CORR.pdf

Loulou, R., Remne, U., Kanudia, A., Lehtila, A., & Goldstein, G. (2005). *Documentation for the TIMES model—Part I.* https://iea-etsap.org/docs/TIMESDoc-Intro.pdf

Lundvall, B.-A. (1992). National systems of innovation: towards a theory of innovation and interactive learning. Pinter Publishers.

Lundvall, B.-A., Dosi, G., & Freeman, C. (1988). Innovation as an interactive process: From user-producer interaction to the national system of innovation, *349*, 369.

Meadows, D., H., Meadows, D. L., Randers, J., & Behrens, W. W. I. (1972). *The limits to growth; a report for the Club of Rome's project on the predicament of mankind.* Universe Books.

Nelson, R. R. (1985). *An evolutionary theory of economic change*. Harvard University Press.
OECD. (1997). *National innovation systems*. https://www.oecd.org/science/inno/2101733.pdf
Remme, U. (2007, November). *Overview of TIMES: Parameters, primal variables & equations*. Paper presented at the Proc. ETSAP Workshop.
Ruggerio, C. A. (2021). Sustainability and sustainable development: A review of principles and definitions. *Science of the Total Environment, 786*, 147481. https://doi.org/10.1016/j.scitotenv.2021.147481
Savaget, P., Geissdoerfer, M., Kharrazi, A., & Evans, S. (2019). The theoretical foundations of sociotechnical systems change for sustainability: A systematic literature review. *Journal of Cleaner Production, 206*, 878–892. https://doi.org/10.1016/j.jclepro.2018.09.208
Senge, P. (1990). The fifth discipline: The art and practice of the learning organization.
United Nations. (2015). *Transforming our world: The 2030 agenda for Sustainable Development (A/RES/70/1)*. https://sdgs.un.org/sites/default/files/publications/21252030%20Agenda%20for%20Sustainable%20Development%20web.pdf
United Nations General Assembly. (2005). *Resolution A/60/L.1 adopting the 2005 World Summit Outcome 48 "Sustainable development: managing and protecting our common environment"*. https://www.un.org/en/development/desa/population/migration/generalassembly/docs/globalcompact/A_RES_60_1.pdf
World Bank. (2022). *GDP per capita, PPP (current international $)*. https://data.worldbank.org/indicator/NY.GDP.PCAP.PP.CD

Marcello Contestabile is Principal Economist at the Qatar Environment and Energy Research Institute, where he leads economic and policy research, and Visiting Fellow at Imperial College London. Originally, an electrochemist with experience developing lithium-ion batteries, over the last 20 years he has conducted extensive techno-economic and policy analysis of low-carbon technologies, particularly with focus on road transport. He has a masters and PhD in Energy Policy and Technology from Imperial College London, where he served as a Research Fellow and held multiple research grants. Prior to moving to Qatar, he also worked as a Research Fellow at KAPSARC in Riyadh, Saudi Arabia, and as a Managing Consultant at specialist firm E4tech in London. In his current role at QEERI, he has been awarded funding to develop the first energy systems model of Qatar for energy-environmental policy analysis and is promoting multi-stakeholder dialogue around clean hydrogen energy in Qatar.

Chapter 9
Sea Level Rise and the National Security Challenge of Sustainable Urban Adaptation in Doha and Other Arab Coastal Cities

Laurent A. Lambert and Cristina D'Alessandro

Abstract The warming of the global ocean and the melting of ice caps have been continuously and increasingly rapidly driving the phenomenon of sea level rise (SLR) over the past century, threatening the safety and standards of living of the world's 800 million inhabitants of coastal cities. Despite renewed commitments to fight the causes of climate change during the COP26 climate negotiations in Glasgow, the current policies of the world's largest polluting countries still put humanity on a dangerous path toward high levels of global warming and SLR for the decades and centuries to come. Based on the latest scientific publications, including the IPCC's Assessment Report 6, this chapter sheds light on how this phenomenon is expected to affect in a multi-dimensional manner the safety and standards of living of coastal city inhabitants across the Arab region, and especially in the Arabian Gulf sub-region, in the decades and centuries to come. Studying the case of Doha, we highlight several policy challenges and opportunities that could influence the hazards as well as the levels of vulnerability and exposure to which individual Arab coastal cities are exposed to. The authors conclude that collectively fighting the causes of climate change, better planning urban and coastal development, as well as innovating for the climate adaptation of Arab coastal cities should be understood by policymakers, the private sector, and populations alike as a national security challenge that requires urgent individual and collective action.

Keywords Arab region · Coastal cities · Climate adaptation · Climate change · Environmental risks · Extreme environmental events · Sea Level Rise (SLR) · Urban governance

---

This chapter is partly based on a short article published on the website of the Middle East Institute in 2019 under the title 'Climate Change, Sea Level Rise, and Sustainable Urban Adaptation in Arab Coastal Cities.' It can be retrieved from: https://www.mei.edu.

L. A. Lambert (✉)
Doha Institute for Graduate Studies, Doha, Qatar
e-mail: laurent.lambert@dohainstitute.edu.qa

C. D'Alessandro
University of Ottawa, Ottawa, Canada

L. Cochrane and R. Al-Hababi (eds.), *Sustainable Qatar*, Gulf Studies 9,
https://doi.org/10.1007/978-981-19-7398-7_9

9.1 Introduction

The literature on climate change issues and risks in the Middle East and North Africa (MENA) region has largely focused on the issues of temperature increase, water stress, and flooding events, as these phenomena have already been affecting the populations in an increasingly disastrous manner (Sieghart & Betre, 2018). This chapter highlights how the lesser-documented issue of Sea Level Rise (SLR) makes Doha, and more generally the coastal cities of the Arab region, increasingly vulnerable to various climate change risks. Indeed, the warming of the global ocean and the melting of the world's glaciers and ice caps have been continuously and increasingly rapidly driving the rise of the world's sea level for over a century (IPCC, 2021a). Based on a recent body of research on ongoing trends and calculated projections of SLR, and via the case study of Doha, we investigate two aspects of the SLR challenge. First, how this phenomenon is expected to directly affect the safety and standards of living of millions of inhabitants of the region's coastal cities in the decades to come. Second, what are the policy challenges and opportunities that could influence both the nature of the risks (which hazards) and the levels of risks (based on a place's specific exposure and vulnerability to these risks) to which Doha and many other Arab coastal cities are increasingly exposed.

9.2 The Literature on SLR, Related Urban Risks, and Economic Impacts

Over the past dozen years of fast accumulating scientific evidence and growing literature on climate change and SLR, the outlooks for our planet have appeared increasingly dystopian for the global environment and human habitat, particularly in coastal areas of the Global South (IPCC, 2007, 2013, 2021a, 2021b). Increasing average surface temperatures and heat waves have caused decimating consequences for populations across the Sahel, the horn of Africa, and the Arab region, while warmer ocean waters have already led to devastating consequences for marine life (IPCC, 2021a, 2021b; Stone, 2007). Since 2016, half of the Great Barrier Reef—a world biodiversity hotspot—has largely bleached due to the rising temperature of the water (Hughes et al., 2018). The ARC Centre of Excellence for Coral Reef Studies has estimated that 93% of tropical reefs have already suffered fatal bleaching, because water temperatures are too warm for corals, leading to their bleaching and death (ARC Centre of Excellence, 2016). The main problem with this trend is that coral reefs constitute hotspots of marine biodiversity and species reproduction sites, and that their partial loss negatively affects the global marine biodiversity and fish stock. In the already warm waters of the Gulf of Aden, the Arabian Gulf, the Red and Mediterranean Seas, which constitute the habitat and traditional source of seafood for the region's coastal cities and villages, bleaching corals are plainly evident in many areas. The overall marine fish stock has been declining almost everywhere in the Arab region, partly due to climate change and the death of corals, but also due to over-fishing and the pollution from offshore oil and gas production (World Resource Institute, 2011). This decline in marine biodiversity represents a significant threat to many communities of fisherman and to coastal tourism, especially along the Red and Mediterranean Seas.

According to the world's most authoritative report on climate change, the United Nations' International Panel on Climate Change's (IPCC) (2022) sixth assessment report, coral reefs are expected to decline by no less than 70–90% at 1.5 °C of increased global mean temperature compared to pre-industrial average temperatures. This would certainly very negatively affect the marine biodiversity, livelihood, and food security of local populations all around the Red and Mediterranean Seas, the Arabian Gulf, Gulf of Aden, and well beyond. However, for many Arab states, the region's greatest risks are arguably on-shore, or more precisely, at the moving barrier of what is off- and what is on-shore.

9.2.1 Ongoing Dynamics of SLR and Related Risks

In its 2014 landmark fifth assessment report, the IPCC had already summarized our planet's climatic crisis in a sober manner: "Warming of the climate system is unequivocal, and since the 1950s, many of the observed changes are unprecedented over decades to millennia. The atmosphere and ocean have warmed, the amounts of snow and ice have diminished, and sea level has risen" (IPCC, 2014, p. 2). In a more recent report of the IPCC, the scientific panel of the United Nations went further: "[s]ince the late nineteenth century many indicators of the global climate system have changed at a rate unprecedented over at least the last two thousand years" (IPCC, 2021a, 2021b, p. 367).

The notion of global warming essentially refers to ocean—and seas—warming, given that oceans play a dominant role in the storage of additional heat, with, between 1971 and 2010, more than 90% of the energy accumulated (IPCC, 2014, p. 4). The Arctic ice sheet was measured as melting six times faster than in the 1980s and, annually, Antarctic ice alone has been producing 0.6 mm of global sea level rise every year (Rignot et al., 2019; Shepherd et al., 2018).

SLR is the result of a combination of two factors: the thermal expansion of the oceans (warmer water takes more volume) and the increase in loss of land ice (glaciers and ice sheets). As the response times for these drivers of change are long yet different, Clark et al. (2016) have established that SLR will continue for centuries, irrespective of (potential) strong climate action by the international community in the coming years and decades. Only the pace of SLR can be altered. But for now, the pace is not good. Under four of the five different scenarios of greenhouse gas emissions for the decades to come, called 'Share Social-Economic Pathways' (or SSPs), the IPCC's (2022) projections consider that the 1.5 °C global warming level will be reached in the near term (2021–2040), and that even under the most optimistic (but unlikely) scenario (SSP1–1.9), this dangerous threshold is nevertheless more likely than not to be reached during that short-term time frame (IPCC, 2022).

These climate risks and others have been increasingly understood by economists and the global business community. On January 15, 2022, the World Economic Forum (WEF) released its annual Global Risks Report on the major threats to the world economy. The respondents of the Global Risk Perception Survey ranked "extreme weather" and "climate action failure" among the top five short-term risks to the world. More importantly, they considered that the five most menacing long-term threats are all environmental and, more specifically, "climate action failure," "extreme weather," and "biodiversity loss" ranked as the three most potentially severe risks for the next decade (WEF, 2022). These environmental concerns predated the COVID-19 pandemic. Two years prior, the same annual global survey had already highlighted that environment-related risks account for three of the top five global risks by likelihood according to global business leaders, and four of the top five risks by economic impact, namely "Failure of climate-change mitigation and adaptation," "Extreme weather events," "Water crises," and "Natural disasters." And even then, this was not the first time, but the third year in a row that such a result was found. In 2019, the survey report had bleakly concluded, "[o]f all risks, it is in relation to the environment that the world is most clearly sleepwalking into catastrophe" (WEF, 2019, p. 15). What is new in 2022, however, is that the increasing concern with climate action failure is interpreted by the WEF as reflective of the survey respondents' lack of trust in the global ability to contain climate change, as economic issues and risks have deepened during the pandemic. Does that mean that the world has a decreased level of resilience toward climate risks?

Out of a growing concern for US national interests and the counter-productive policies of the Trump administration (2017–2021), US President Biden, shortly after his election, launched high-profile climate diplomacy initiatives, including the ambitious 'Leaders Summit on Climate.' In April 2021, the US and the world's largest industrialized countries gathered to express their resolute commitment to fight climate change and announced more ambitious action targets (The White House, 2021). The Summit has had some positive influence on several important countries' renewed commitments to the Paris Agreement, but the updated trajectories of global warming, based on the Summit's new commitments, indicated that the world was still on a trajectory of rapid increase in temperature, of around 2.4 °C above to pre-industrial temperatures by the end of this century (Climate Action Tracker, 2021). These commitments are thus still far from the collective ambition of limiting it "well below 2 °C above pre-industrial levels and pursuing efforts to limit the temperature increase to 1.5 °C above pre-industrial levels," as it was agreed upon and inked in the Paris Agreement, in December 2015 (United Nations, 2015, p. 4). Although 1.5 °C or even 2.4 °C of global mean temperature increase may not appear as a massive temperature increase for many, it is worth resituating it in a geological perspective, with the great glaciations that covered the entire northern European and American continents with more than 3,500 m of ice resulting in a 120 m drop in sea level taking place at only 4 to 5 degrees below current mean global temperatures (Poitou & Braconnot, 2021).

And this kind of temperature difference is currently not far from possible future temperature increases.

The recent analysis made by the Climate Action Tracker (a well-respected coalition of scientists and specialized NGOs) of the actual climate policies by the largest emitting countries of greenhouse gases (GHGs) indicates that the world's leading economies are still far from being in line with their official commitments. If no major policy reforms were to materialize, the Climate Action Tracker assessed that the world is currently on a trajectory of global increase in the order of an estimated 2.9 °C by the end of the century (Climate Action Tracker, 2021). This is relatively in line with a recently released 3 °C estimation from the IPCC's 6th Assessment Report (IPCC, 2021a, 2021b). These are particularly high levels of temperature increase that had previously been assessed as capable of generating over 2 m of SLR by the end of this century, and several meters more during the next (DeConto & Pollard, 2016). Such a high level of SLR would see the disappearing of many coastal areas across the world. Recent studies, with different methodological approaches and a greater understanding of the dynamics of frozen parts of the world, have projected differing estimations of SLR for this century and over the next.

In 2019 the IPCC special report was released on the ocean and the cryosphere (i.e., the frozen parts of the world), which then foresaw a maximum of around one meter of SLR by the end of the century (IPCC, 2019), which would have severe consequences in the MENA region. This projection of a single meter of SLR is now considered optimistic by many. It was arguably based on a methodology that did not fully integrate all the complex dynamics of ice-melting in Antarctica and Greenland and the major issues of our planet's environmental thresholds also called tipping points. In 2020, a major study published in the journal *Nature* provided solid evidence of potentially much higher ranges for SLR to be expected in the forthcoming decades, along the century, and beyond. The authors have demonstrated that the Antarctic ice sheet, which stores more than half the planet's freshwater resources, reacts differently to different temperature increase thresholds, and not in a simple, linear manner. Beyond some thresholds, ice loss and SLR dynamics are irreversible at the human timescale. The new study finds that the ice sheet's temperature sensitivity to melt corresponds to +1.3 m of sea-level equivalent per degree of warming temperature up to 2 degrees C. above pre-industrial levels, but almost doubling to +2.4 m of SLR per degree of warming between 2 and 6 degrees Celsius (Garbe et al., 2020). As the authors concluded, "[o]ur results show that if the Paris Agreement is not met, Antarctica's long-term sea-level contribution will dramatically increase and exceed that of all other sources" (Garbe et al., 2020, p. 538). It also means that if there is no drastic and rapid reduction in our global emissions of GHGs, our current trajectory of global temperature increase could lead to a catastrophic level of SLR of more than 3 m by the end of the century. Based on their coast elevation and topography, some Arab coastal cities like Abu Dhabi, Aden, Alexandria, Algiers, Basra, Casablanca, Cairo, Doha, Dubai, Kuwait City, Manamah, Tripoli, Tunis, among others, would not exist anymore as we know them.[1] Large swathes of coastal land would be submerged under seawater, at the very least several days a year. Long-term submersion would be particularly acute in the low-lying areas of Southern Iraq and the Arabian Gulf as can be seen on Fig. 9.1.

With over 3 m of SLR and as per currently available technology, it is worth mentioning that it would then not be possible to safely protect them by seawalls

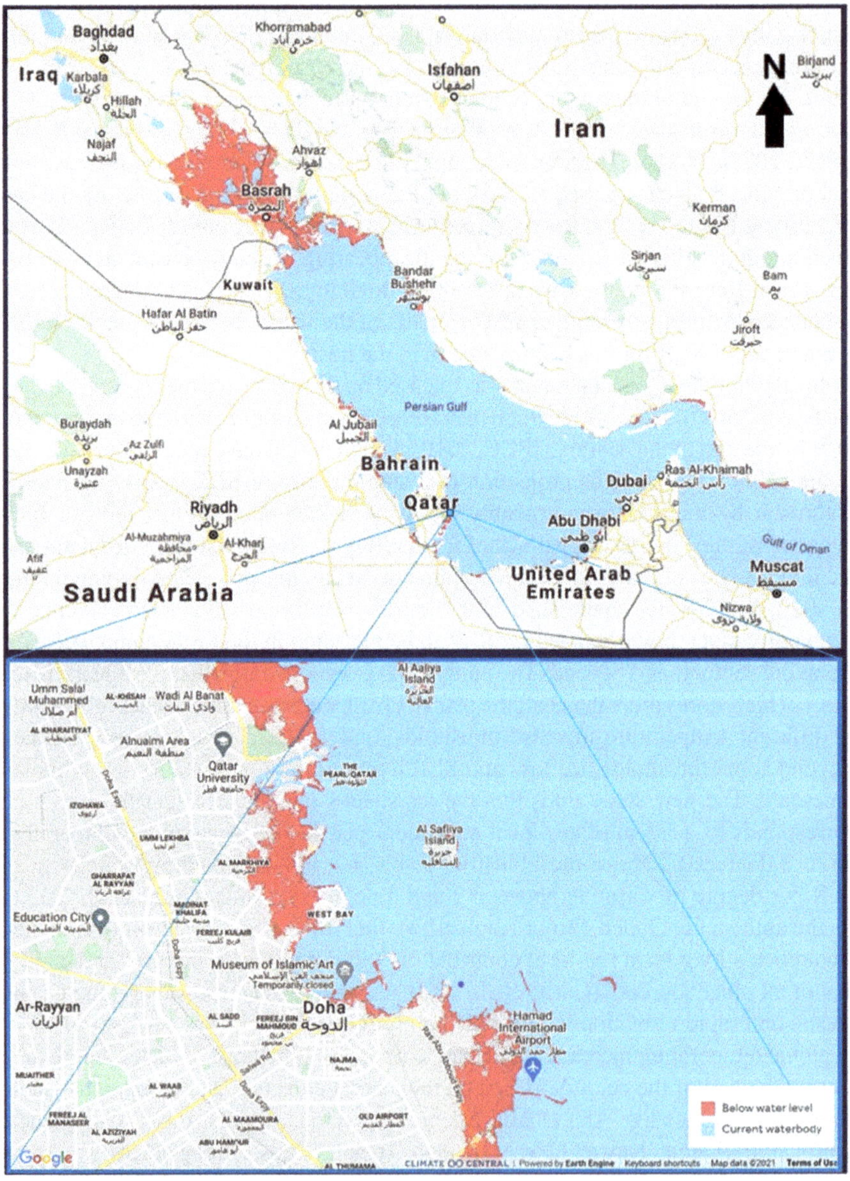

Fig. 9.1 Areas at risk in the Gulf Region and Doha's metropolitan area under 2 m of SLR (Laurent A. Lambert. Background sources: Climate Central, Google Maps, and Earth Engine)

since each sea surge (a particularly high wave generally created by a storm) could overwhelm these defenses and have a devastating impact on these cities. This is what happened when Hurricane Katrina's sea surge inflicted onto the New Orleans in 2005 and when Hurricane Harvey affected the coast of Texas in 2017, both killing scores of persons and generating damages approximately worth $125 billion and $161 billion US dollars, respectively (NOAA, 2022). Against this background, the city of New York is building elaborate seawalls to protect several vulnerable areas, for an initial cost of 1.41 USD billion, but potentially as part of a much larger program that could cost as much as 119 USD billion in total to protect the whole city for decades or more, provided the budget is accepted by all competent authorities (Barnard, 2020). If this figure seems far too high for the public budget of most Arab governments, it remains an appropriate investment for richer cities, with cities such as Shanghai, Saint Petersburg, and London, inter alia, having their expensive sea protections in place. And this is likely to become a feature in several other populated coastal urban centers of the USA due to the increasing costs and lethality of coastal extreme weather events. According to the US government's National Oceanographic and Atmospheric Administration (NOAA), tropical cyclones data from the past five years seems to indicate a sharp rise in multi-billion-dollar costs compared to previous periods.

> In 2021, there were 20 separate billion-dollar weather and climate disasters. The total cost for these events was $145 billion, making this the third most costly year on record, behind 2017 and 2005. (…)
>
> The total cost for the last five years ($742.1 billion) is more than one-third of the disaster cost total of the last 42 years ($2.155 trillion). (…)
>
> Of the 310 [separate] billion-dollar weather disasters between 1980 and 2021, tropical cyclones (or hurricanes) have caused the most damage: over $1.1 trillion total, with an average cost of $20.5 billion per event. They are also responsible for the highest number of deaths: 6,697 between 1980 and 2021. (NOAA, 2022)

Unless drastic and costly measures are taken and appropriate policies are co-designed with all relevant stakeholders, the joint dynamic of SLR, and increased extreme weather events, such as storms and tropical cyclones, mean that many coastal cities around the world could probably become very vulnerable to coastal hazards, with only a few decades left prior to becoming unsafe for urban life during the second half of the century.

9.3 Increasing Global Hazard Frequency, Exposure, and Vulnerability of Arab Coastal Cities

Across the world, 680 million people are directly vulnerable to SLR as they live in low-lying coastal zones, while an estimated 800 million people are living in more than 570 coastal cities that are vulnerable to only a 50 cm rise by 2050 (IPCC, 2019; WEF, 2019). From Nouakchott and Rabat in the West to Kuwait and Dubai in the East, including all North Africa and most of the Levant, the majority of the Arab region's capital cities and largest metropolises are located on at-risk coastal areas, deltas, or nearby, and their populations are becoming increasingly at risk to SLR and the extreme environmental events that make it more dangerous, like storm surges, extreme swale waves of several meters, and coastal flooding.

If the case of SLR has been increasingly documented in Egypt's coastal cities over the past decade, SLR is a particularly threatening yet relatively understudied problem in the Gulf Arab states (Agrawala et al., 2004; El-Raey & Doluschitz, 2010; Neelamani et al., 2021). More than 90% of that region's population lives in cities and, with the exception of Saudi Arabia, all GCC capital cities are sprawling coastal urban bodies with strong demographic growth patterns (Population Division of the Department of Economic & Social Affairs of the United Nations Secretariat, 2018). This trend includes some of the world's fastest-growing cities—such as Abu Dhabi, Doha, and Dubai—which happen to be located on low-lying coastal zones and islands, and are thus particularly vulnerable to even lower ranges of SLR as we will see.

Because warmer waters are more conducive to tropical storms and cyclones, the compounded effects of climate change and SLR will lead to more regular coastal flooding following tropical storms. In 2018, no less than three severe and very severe tropical storms (better known as 'tropical cyclones') have devastated parts of Yemen and, on the opposite side of the Gulf of Aden, created the worst cyclonic disaster ever recorded in Somalia (Lambert et al., 2021). There is now a strong and growing body of scientific publications showing that SLR risks of submersion are compounded by environmental processes such as high tides, coastal erosion, and extreme environmental events, like storms, swell waves, and sea surges, inter alia (Barnard et al., 2015; Serafin et al., 2017; Vitousek et al., 2017a). Coastal disasters can have profound social and economic implications. In 2007, for instance, severe tropical storm Gonu (generally called 'cyclone Gonu') directly killed more than 50 persons and affected 20,000 people and generated an estimated US$ 4 billion of direct damages in Oman alone and caused 28 deaths and US$ 216 million in damage in Iran (Al-Maskari, 2010). On May 18, 2018, the 'Sagar' tropical cyclone generated numerous casualties and heavy destruction in parts of the Arabian Peninsula and Horn of Africa, making history as the strongest tropical storm ever recorded in Somalia and also as the first of three tropical cyclones that brought devastation to war-torn Yemen in a single year. These tragic events are consistent with the literature on climatic changes in the North Indian Ocean that has been recording an increase in magnitude and frequency of tropical cyclones and other extreme weather events like coastal flooding (Vitousek et al., 2017a). In 2021, tropical cyclone Shahin was even more powerful and destructive than its predecessors, devastating areas in Oman and Iran, but also in the Northeast of the United Arab Emirates, a country generally off the reach of such extreme weather events. As the Northwestern part of the Indian Ocean where these cyclones form has featured several unusually stronger tropical cyclones in recent time, including powerful ones going far across the Gulf of Aden, this trend leaves the question open as to whether we are entering a period (rather than witnessing a series of anomalies) of increasing frequency and strength of cyclones battling the coasts of the Arab region (Lambert et al., 2021).

These events can impact essential services such as transport, logistics, schools, hospitals, as well as water and energy supplies. And if the toll and costs of such hazards keep on increasing, this may ultimately affect the levels of vulnerability of the populations, as repeated hazards can disrupt the resource base necessary for post-disaster resilience. Meanwhile, the vulnerability of several Arab cities is currently more acute than usual due to the COVID-19 pandemic toll on the national health systems and the economies of the region. This is especially true in coastal cities that have traditionally been depending on tourism.

The World Bank identified no less than 24 ports in the Middle East and 19 ports in North Africa at particular risk of SLR (WEF, 2019, p. 58). But the numbers could be higher. Until very recently, the global scientific literature on increased coastal flooding due to SLR generally did not integrate the risks posed by the additional factors of elevated water levels such as storms and sea surges, and thus underestimated the potential impact of SLR. In a watershed study, Vitousek et al. (2017a) quantified the risks of coastal flooding as doubling due to the joint risks of SLR and wave, tides, and sea surges. Their study calculated that even the limited 10 to 20 cm of SLR will more than double the frequency of extreme water-level events in the Tropics, a zone that includes many Gulf states and Red Sea countries. They concluded that even gradual sea level rise can rapidly increase the frequency and severity of coastal flooding by the mid-century and that it is the regions located in the Tropics that will suffer from the largest increases in flooding frequency (Vitousek, 2017b). The coastal cities, coastal groundwaters (increasingly salinized by seawater intrusion), as well as agricultural lands on and near the coasts and deltas in the MENA region, are expected to suffer from seawater flooding issues and the gradual salinization of soils and coastal water resources. In North Africa, the coastal areas (i.e., within 100 km from the sea) are where most agricultural production happens, as the south of the country is generally too arid for intensive production. These and other issues related to climate change are expected to significantly disturb—or in some cases, strongly disrupt—the growth of national economies of the MENA region and increase the vulnerability of most Arab countries.

Joint research by the international re-insurance company Swiss Re and the international NGO Oxfam has projected the economic impact of the world's regions by the mid-century under various scenarios of climate change. As it can be seen in Table 9.1, the Middle East and Africa region are expected to particularly suffer from the impacts, with an anticipated decrease of GDP by mid-century of −4.7% if the 2 °C temperature control target of the Paris Agreement is reached relative to a world without climate change (Swiss Re, 2021). However, the report also anticipates major decreases in GDP by mid-century, between −21.5 and −27.6% of GDP, relative to a world without climate change, if the global temperature increases between 2.0 °C to 2.6 °C by mid-century, a level which can be considered likely to happen given the world's current trajectory (Swiss Re, 2021, p. 2). Such a disruptive loss of economic growth would have important social and political consequences in a region already challenged by a young demography and high unemployment rates. Climate

Table 9.1 Expected impact on GDP loss by mid-century under various climate scenarios (Swiss Re, 2021)

	Temperature rise scenario, by mid-century			
	Well-below 2 °C increase	2.0 °C increase	2.6 °C increase	3.2 °C increase
	Paris target	*The likely range of global temperature gains*		*Severe case*
Simulating for economic loss impact from rising temperatures in % GDP, relative to a world without climate change (0 °C)				
World	−4.2%	−11.0%	−13.9%	−18.1%
OCED	−3.1%	−7.6%	−8.1%	−10.6%
North America	−3.1%	−6.9%	−7.4%	−9.5%
South America	−4.1%	−10.8%	−13.0%	−17.0%
Europe	−2.8%	−7.7%	−8.0%	−10.5%
Middle East & Africa	−4.7%	−14.0%	−21.5%	−27.6%
Asia	−5.5%	−14.9%	−20.4%	−26.5%
Advanced Asia	−3.3%	−9.5%	−11.7%	−15.4%
ASEAN	−4.2%	−17.0%	−29.0%	−37.4%
Oceania	−4.3%	−11.2%	−12.3%	−16.3%

change adaptation measures are thus urgently needed to protect societies, national economies, as well as local political stability.

9.4 Urban Planning, Proactivity, and Innovation Are Critically Important: Case of Doha, a Fast-Growing Middle East City

Everywhere in the world, classic adaptation strategies include restricting construction in the most at-risk urban areas, avoiding unplanned settlements, improving the reliability and durability of relevant existing real estate property (especially for collective housing in poorer areas), improving or constructing flood defenses, and preparing for contingency plans and potential relocation (similarly to what happens for cities at risks of earthquakes). Although these measures may affect some economic and financial interests in the short term, they may represent an undeniable necessity for the safety of the populations. All these policy measures assume that urban planning is key and operates with the premise that development control, by land use zoning generally, is enforceable. However, amid the impressive diversity of Arab cities, in terms of historical development and institutional culture, this is far from being the rule. The demographic growth of most Arab cities is often stronger in coastal cities and surrounding areas, which have become sprawling economic magnets in this era of globalization. The domestic demographic pressure and working migrations from Africa and Asia, with their consequent youth bulge, explains why many Arab coastal cities are expanding so rapidly (McKee et al., 2017). Qatar is a useful example when it comes to documenting the efforts and challenges related to SLR risks and adaptation in the contexts of the Gulf sub-region.

9.4.1 The Case of Doha

Qatar, a small peninsular state, features several important coastal urban areas, including its capital city of Doha, which is located on the Arabian Gulf coast. Its hinterland does not extend beyond 80 km and its most economically and politically important areas (e.g., West Bay, Al Corniche, and Al Dafna) are all adjacent to the waterfront. Furthermore, recent developments and mega-projects (e.g., the Pearl and Lusail neighborhoods) are all located along the coast. Against the background of the negotiations on the Paris Agreement, the Ministry of the Environment published in 2015 a startling statement in Qatar's official communication (its Intended Nationally Determined Contribution) to the United Nations Framework Convention on Climate Change:

> Qatar is extremely vulnerable to sea level rise as it is liable to inland flooding of 18.2% of its land area, at less than 5m rise in sea level, along with the associated adverse impacts on the population as 96% are living on the coastal areas. (State of Qatar, 2015)

This specific issue was also prominently featured in Qatar's follow-up Nationally Determined Contribution to the Paris Agreement:

> Sea level rise receives most of the government attention because of the large coastal population at risk of inundation (particularly during extreme sea level events). (…) Qatar's ecology, environmental resources, infrastructure and human systems are vulnerable to the adverse impact of climate change, the latter includes coastal and offshore installations such as power and water cogeneration facilities, and the oil & gas infrastructure. (State of Qatar, 2021)

Neighboring countries Bahrain, the United Arab Emirates, and the Kingdom of Saudi Arabia are also highly exposed to climate change and SLR. In 2010, The Abu Dhabi Environmental Agency had published an announcement which shows that 85% of UAE's population would also be affected by rising sea levels as well as 90 percent of its infrastructure (Luomi, 2014).

As a result of those identified climate risks, important urban planning efforts have been undertaken in the cities of Qatar over the past years, such as the Qatar National Master Plan and the Climate Change Strategy for the Urban Planning and Urban Development Sector.[2] But as these new plans and strategies will only provide substantial benefits via future designs and constructions, Qatar reiterated in clear terms this coastal vulnerability issue in its 2021 Intended Nationally Determined Contribution (INDC) to the UN:

> Given its geographic location, Qatar is likely to suffer from severe consequences of global warming. Under the major impact of climate change, Qatar is extremely vulnerable to rising

> sea level and air temperature increase, leading to inland flooding and heat exhaustion of its population, in addition to the potential risks to the marine biodiversity, food security, loss and damage due to climate change. Sea level rise receives most of the government attention because of the large coastal population at risk of inundation (particularly during extreme sea level events). Subsequently, climate change also poses a threat to the tourism industry of Qatar, largely due to the increase in temperature and frequency of dust storms. (State of Qatar, 2021, p. 10)

This statement reflects a deep understanding of the multi-dimensional climate risks and what is at stake for the country. It also reflects a more systemic perspective toward coastal risk management rather than the classic approach to hazards punctually disrupting the population's life, as in the INDC previously submitted to the UN. Worth mentioning, in the 2021 communication, the government of Qatar announced its ongoing studies of the coastal ecosystems and its plan to reforest with mangrove trees some coastal areas, thereby linking mitigation and the action of a known climate sink (mangroves' deep root system sequestrate large quantities of atmospheric carbon), to the coastal protection from erosion and storms that mangrove forests are known for (State of Qatar, 2021). In other words, this represents an ecosystem-based adaptation policy to climate change coastal risks with mitigation benefits (Fig. 9.1).

9.4.2 *Contributions from the City, State, and Other Stakeholders Are Essential*

In Doha, as in many cities of the MENA region, there are important economic challenges, such as the need to rapidly generate large numbers of jobs for a growing, young, and increasingly educated population, while continuously shifting the economy from a model of commodity-dependent rentierism and toward a more diversified national economy. Despite repeated grand projects and diversification policies over the past decades (Gray, 2011), reforms to give greater room to the private sector (Lambert, 2014), and cuts in the welfare state to decrease the citizenry's reliance on the generous state expenditure (Gengler & Lambert, 2016), the economies of hydrocarbons exporting countries of the Middle East and North Africa have largely failed to transform the structure of their national economy, which is still largely dependent on the export of a few commodities to provide the majority of government revenues and finance both the redistributive welfare state and state-led economic growth. Against that background of mixed results in economic reforms and diversification, short-term opportunities to develop tourism, including in at-risk areas, often tend to take precedence over sustainable coastal development. In Doha, Abu Dhabi, Dubai, or Manamah, the artificial islands for luxury hotels, villas, and resorts of the previous decades have been built on reclaimed areas which have depended on large quantities of sand dredging of fragile marine areas and thus have had an overall negative impact on the local marine environments, despite increasing remediation measures. Finding an appropriate balance between risks and economic activities will now have to be made with enough transparency for all stakeholders, for them to be able to assess, and eventually invest into, the appropriate environmentally friendly practices to start with, as well as the appropriate risk mitigation measures. Additionally, this information will be increasingly needed to contract insurance contracts to cover the climate risks. Unfortunately, most insurance in the Gulf States currently do not cover climate-related disasters. This is something national authorities will have to deal with, as the risks are rising, and the costs may become very substantial in the years and decades to come. New insurance products could be introduced, with the support of government entities initially to develop the market, for instance with government buildings being insured. The old model may not be appropriate anymore, and modern insurance services will be needed to increase Doha's climate resilience as well as that of other Arab cities, especially where powerful storms and/or cyclones might disrupt the economic activities and lead to corporate and individual bankruptcy.

Another economic challenge, which is more acute in low- and medium-income Arab states, is the planning and implementing (and staying committed to) the necessary but costly risk mitigation infrastructure developments given that national budgets vary significantly due to their heavy dependence on a few commodities, from phosphate and agricultural products in Morocco, to crude oil in Kuwait, and natural gas in Qatar. The government commitment to long-term infrastructure development, such as Doha has overall successfully managed over the past two decades, will be critically important for low-lying coastal cities as they seek to adapt to the formidable challenge of SLR.

Lastly, throughout the MENA region, the limited capacities of government agencies in charge of urban planning tend to generate problems of poor planning quality, underestimated risks, selection of inappropriate technologies, and sub-optimal returns on investment. In Qatar, for instance, despite strong efforts by the government to make its urban development more sustainable, some mega-projects have been assessed as not sufficiently taking into consideration SLR and other related risks to urban environments (Rizzo, 2014). The general lack of pro-activity on this matter across the region is likely to cost more over time in adaptation, climate disaster disruptions of the economy, and remediation measures than it would have initially with climate-smart designs, planning, and commitment.

9.5 South-South Transfers of Expertise and Innovation

The global phenomenon of SLR is undeniably a global threat for coastal populations around the world but is an even greater threat in the Arab region, wherein most of the population lives within 100 km from the sea, including in deltas and estuaries and in cities where rapid demographic growth already exacerbates urban planning challenges. In line with the United Nations Sustainable Development Goal 11 (Sustainable Cities and Communities), vulnerable cities, such as Amsterdam, Copenhagen, Queensland, Shanghai, and New York City, among others, have already accumulated a precious experience in assessing and deploying adaptation solutions at a large scale. Local and international ideas, projects, and cooperation will be needed to simultaneously face the formidable challenges of SLR, economic diversification, job creation, and, ultimately, poverty eradication.

The United Nations system provides numerous vehicles for facilitating and partly or wholly financing this process of transformation of MENA coastal cities. Yet, a review by the authors of the projects supported by the various climate finance facilities reveals that they have been under-utilized by Arab countries. For instance, the UN Climate Technology Centre and Network, in charge of green technology transfer to developing countries, has been under-utilized to facilitate technology transfers with UN funding and expertise, such as climate adaptation strategy designs, or for capacity building in climate-smart technologies, including solar desalination or projects of waste transformation into energy. As of June 2021, and out of the African continent's 154 funded projects (called "requests" in UN terminology), only 9 were called for by governments of North Africa.[3] This 5.84% share of total funded projects for Africa is particularly low as the population of North African states represents nearly 20% of the continent's total population, and with particularly daunting climate challenges in terms of SLR, water stress, droughts, food insecurity, and desertification, inter alia. Similarly, at the time of publishing this chapter, Arab countries have far from fully availed themselves of the UN's Clean Development Mechanism (CDM), the Adaptation Fund, the Green Carbon Fund (GCF), or the World Bank's Green Environment Fund (GEF). By contrast, some emerging countries, such as China, India, and Kenya, have been more proactive in requesting and obtaining multilateral resources for their sustainable development. This is something that most Arab countries will have to work on.

Sharing experiences, knowledge, lessons learned, and technological solutions and innovations among coastal cities in different parts of the world should be supported, as a relevant and cost-efficient way to adapt to climate change. This is especially relevant between and among metropolises of the Global South, as in Southeast Asia and Africa. This adoption of technology from outside the OECD economies holds the potential to greatly benefit Arab coastal cities, some of which must look for cost-efficient solutions, while others are widely known for developing and branding themselves as smart global cities. Against the backdrop of limited proactive action on climate change and SLR in many of the Global South's coastal cities, consistently investing in the management of sea-related urban risks could transform a major challenge for Arab cities into a structural competitive advantage and economic opportunity.

9.6 Conclusion

Based on the latest body of research, and using the case study of Doha, this chapter highlighted how the poorly documented hazard of SLR makes the populations, economies, and ecosystems of Arab cities, depending on their specific exposure and vulnerabilities, increasingly vulnerable to climate change and coastal hazards. Coastal risks are expected to directly affect the safety and standards of living of millions of inhabitants of the region's coastal cities in the years and decades to come, if appropriate policies are not devised and implemented in a rapid, consistent, and efficient manner. SLR is not the often-portrayed slow-onset and long-term risk that can wait; it is also a short- to medium-term major challenge for coastal cities, as it strongly increases coastal flooding risks and the frequency of several other important hazards, including the salinization of agricultural lands and water reserves, inter alia. The policy challenges and opportunities that could influence the nature of the risks (coastal flooding, erosion, and destructive storms) and the exposure of Arab coastal cities are first and foremost related to the urban planning strategy, risk transparency (to prepare and involve all stakeholders), the rule of law (implementing for all the same rules and regulations), greater cooperation with multilateral organizations, and locally relevant adaptation policies. By contrast with global trade regulations or global health issues, the global warming challenges constitute a policy field in which the governments of the region can be proactive and effective, if they decide so. Climate finance, though not yet at the scale decided in Copenhagen in 2009, is already available to most Arab states, but they need to process and submit the requests, something that most countries of the region have neglected so far.

Adequate technologies must be selected and appropriately used, and innovation will be necessary on a case-by-case scenario. State and city leadership involvement in SLR risks and measures are necessary, as well as stakeholders' participation, including the private sector, to ensure that the measures selected are appropriate to the cities and their communities. Regional and international collaboration is also critical to tackle the issue not only at the national level, but also collaboratively among the cities of the same region to ensure alignment and coordination in policies and actions. It must also be emphasized that SLR may also be an opportunity for new public–private partnerships for stakeholders to integrate global value chains in the blue economy perhaps, if opportunities are appropriately used and developed. Finally, the authors call for more anticipation in the region's urban planning of the general risks posed by climate change, which should be officially identified as, and treated like, any other national security challenges in Qatar's city of Doha, Arab countries and cities, and the world at large.

Notes

1. The topography of a city can be observed on https://en-us.topographic-map.com/maps/pdau/Abu-Dhabi/.
2. The Qatar National Master Plan can be accessed online at: www.mme.gov.qa/QatarMasterPlan/default.aspx, but the Climate Change Strategy for the Urban Planning and Urban Development Sector in the State of Qatar, completed in 2018, is not available to the public. The lead author of this chapter was consulted during its preparation and peer-reviewed parts of it, such as its executive summary.
3. See 'Requests by region' on the website of the United Nations' Climate Technology Centre & Network, https://www.ctc-n.org/technical-assistance/request-visualizations.

References

Agrawala, S., Moehner, A., El Raey, M., Conway, D., Van Aalst, M., Hagenstad, M., & Smith, J. (2004). Development and climate change in Egypt: Focus on coastal resources and the Nile. *Organisation for Economic Co-Operation and Development, 1*, 1–68.

Al-Maskari, J. (2010). How the National Forecasting Centre in Oman Dealt with Tropical Cyclone Gonu. In *Indian ocean tropical cyclones and climate change* (pp. 265–275). Springer Netherlands. https://doi.org/10.1007/978-90-481-3109-9_31

ARC Centre of Excellence. (2016). *Heat sickens corals in global bleaching event*. ARC Centre of Excellence for Coral Reef Studies. James Cook University. http://www.coralcoe.org.au/author/asojll

Barnard, A. (2020). The $119 billion sea wall that could defend New York... or not. *The New York Times*. https://www.nytimes.com/2020/01/17/nyregion/the-119-billion-sea-wall-that-could-defend-new-york-or-not.html

Barnard, P. L., Short, A. D., Harley, M. D., Splinter, K. D., Vitousek, S., Turner, I. L., Allan, J., Banno, M., Bryan, K. R., Doria, A., Hansen, J. E., Kato, S., Kuriyama, Y., Randall-Goodwin, E., Ruggiero, P., Walker, I. J., & Heathfield, D. K. (2015). Coastal vulnerability across the Pacific dominated by El Nino/Southern Oscillation. *Nature Geoscience, 8*(10), 801–807. https://doi.org/10.1038/NGEO2539

Clark, P. U., Shakun, J. D., Marcott, S. A., Mix, A. C., Eby, M., Kulp, S., Levermann, A., Milne, G. A., Pfister, P. L., Santer, B. D., Schrag, D. P., Solomon, S., Stocker, T. F., Strauss, B. H., Weaver, A. J., Winkelmann, R., Archer, D., Bard, E., Goldner, A., … & Plattner, G.-K. (2016). Consequences of twenty-first-century policy for multi-millennial climate and sea-level change. *Nature Climate Change, 6*(4), 360–369. https://doi.org/10.1038/nclimate2923

Climate Action Tracker. (2021). *Warming projections global update*. https://climateactiontracker.org/documents/853/CAT_2021-05-04_Briefing_Global-Update_Climate-Summit-Momentum.pdf

DeConto, R. M., & Pollard, D. (2016). Contribution of Antarctica to past and future sea-level rise. *Nature, 531*(7596), 591–597. https://doi.org/10.1038/nature17145

El-Nahry, A. H., & Doluschitz, R. (2010). Climate change and its impacts on the coastal zone of the Nile Delta, Egypt. *Environmental Earth Sciences, 59*(7), 1497–1506. https://doi.org/10.1007/s12665-009-0135-0

Garbe, J., Albrecht, T., Levermann, A., Donges, J. F., & Winkelmann, R. (2020). The hysteresis of the Antarctic ice sheet. *Nature, 585*(7826), 538–544. https://doi.org/10.1038/s41586-020-2727-5

Gengler, J., & Lambert, L. A. (2016). Renegotiating the ruling bargain: Selling fiscal reform in the GCC. *The Middle East Journal, 70*(2), 321–329.

Gray, M. (2011). *A theory of 'late rentierism' in the Arab states of the Gulf* (CIRS Occasional Paper [7]).

Hughes, T. P., Kerry, J. T., Baird, A. H., Connolly, S. R., Dietzel, A., Eakin, C. M., Heron, S. F., Hoey, A. S., Hoogenboom, M. O., Liu, G., McWilliam, M. J., Pears, R. J., Pratchett, M. S., Skirving, W. J., Stella, J. S., & Torda, G. (2018). Global warming transforms coral reef assemblages. *Nature, 556*(7700), 492–496. https://doi.org/10.1038/s41586-018-0041-2

Intergovernmental Panel on Climate Change. (2007). *Climate change 2007: The physical science basis. Contribution of working group I to the fourth assessment report of the intergovernmental panel on climate change* (S. Solomon, D. Qin, M. Manning, M. Marquis, K. Averyt, M. M. B. Tignor, H. L. Miller Jr., & Z. Chen, Eds.). Cambridge University Press.

Intergovernmental Panel on Climate Change. (2013). *Climate change 2013: The physical science basis. Contribution of working group I to the fifth assessment report of the intergovernmental panel on climate change* (T. F. Stocker, D. Qin, G.-K. Plattner, M. Tignor, S. K. Allen, J. Boschung, A. Nauels, Y. Xia, V. Bex, & P. M. Midgley, Eds.). Cambridge University Press.

Intergovernmental Panel on Climate Change. (2014a). *Climate change 2014: Impacts, adaptation, and vulnerability. Part A: global and sectoral aspects. Contribution of working group II to the fifth assessment report of the intergovernmental panel on climate change* (C. B. Field, V. R. Barros, D. J. Dokken, K. J. Mach, M. D. Mastrandrea, T. E. Bilir, M. Chatterjee, K. L. Ebi, Y. O. Estrada, R. C. Genova, B. Girma, E. S. Kissel, A. N. Levy, S. MacCracken, P. R. Mastrandrea, & L. L. White, Eds.). Cambridge University Press.

Intergovernmental Panel on Climate Change. (2014b). *Climate change 2014: Impacts, adaptation, and vulnerability. Part B: Regional aspects. Contribution of working group II to the fifth assessment report of the intergovernmental panel on climate change* (V. R. Barros, C. B. Field, D. J. Dokken, M. D. Mastrandrea, K. J. Mach, T. E. Bilir, M. Chatterjee, K. L. Ebi, Y. O. Estrada, R. C. Genova, B. Girma, E. S. Kissel, A. N. Levy, S. MacCracken, P. R. Mastrandrea, & L. L. White Eds.). Cambridge University Press.

Intergovernmental Panel on Climate Change. (2019). The ocean and cryosphere in a changing climate. *IPCC special report on the ocean and cryosphere in a changing climate* (H. O. Pörtner, D. C. Roberts, V. Masson-Delmotte, P. Zhai, M. Tignor, E. Poloczanska, & N. M. Weyer, Eds.). Cambridge University Press.

Intergovernmental Panel on Climate Change. (2021a). *Climate change 2021a: The physical science basis. Contribution of working group I to the sixth assessment report of the intergovernmental panel on climate change* (V. Masson-Delmotte, P. Zhai, A. Pirani, S.L. Connors, C. Péan, S. Berger, N. Caud, Y. Chen, L. Goldfarb, M.I. Gomis, M. Huang, K. Leitzell, E. Lonnoy, J.B.R. Matthews, T.K. Maycock, T. Waterfield, O. Yelekçi, R. Yu, & B. Zhou, Eds.). Cambridge University Press.

Intergovernmental Panel on Climate Change. (2021b). *Summary for policymakers*. In *Climate change 2021b: The physical science basis. Contribution of working group I to the sixth assessment report of the intergovernmental panel on climate change* (V. MassonDelmotte, P. Zhai, A. Pirani, S. L. Connors, C. Péan, S. Berger, N. Caud, Y. Chen, L. Goldfarb, M. I. Gomis, M. Huang, K. Leitzell, E. Lonnoy, J. B. R. Matthews, T. K. Maycock, T. Waterfield, O. Yelekçi, R. Yu, & B. Zhou, Eds.). Cambridge University Press.

Intergovernmental Panel on Climate Change. (2022). Climate change 2022: Impacts, adaptation and vulnerability. contribution of working group II to the sixth assessment report of the intergovernmental panel on climate change (H.-O. Pörtner, D. C. Roberts, M. Tignor, E. S. Poloczanska, K. Mintenbeck, A. Alegría, M. Craig, S. Langsdorf, S. Löschke, V. Möller, A. Okem, & B. Rama, Eds.). Cambridge University Press.

Lambert, L. A., Mahmood, A., & Mustafa, H. (2021). Climate change, humanitarian risks, and social-political (in)stability along the Gulf of Aden: Expert elicitation for the Case of Somalia and Yemen. In *EGU General Assembly Conference*, EGU21-7575. https://meetingorganizer.copernicus.org/EGU21/EGU21-7575.html

Lambert, L. A. (2014). *Water, state power, and tribal politics in the GCC: The case of Kuwait and Abu Dhabi* (CIRS Occasional Paper [15]).

Luomi, M. (2014). *The Gulf monarchies and climate change: Abu Dhabi and Qatar in an era of natural unsustainability*. Oxford University Press.

McKee, M., Keulertz, M., Habibi, N., Mulligan, M., & Woertz, E. (2017). *Demographic and economic material factors in the MENA region* (Middle East and North Africa Regional Architecture Working Papers 3). http://www.iai.it/sites/default/files/menara_wp_3.pdf

Neelamani, S., Al-Houti, D., Al-Ragum, A., Al-Salem, K., & Al-Saleh, A. H. (2021). Assessment of coastal inundation cost due to future sea level rise: A case study for Kuwait. *Marine Georesources & Geotechnology*, 1–18.https://doi.org/10.1080/1064119X.2021.1909195

National Oceanic and Atmospheric Administration. (2022). *Hurricane costs*. Office for Coastal Management. https://coast.noaa.gov/states/fast-facts/hurricane-costs.html

Poitou, J., & Braconnot, P. (2021). *Les Mots du Climat*. EDP Sciences.

Population Division of the Department of Economic and Social Affairs of the United Nations Secretariat. (2018). *World urbanization prospects*. https://population.un.org/wup/

Rignot, E., Mouginot, J., Scheuchl, B., van den Broeke, M., van Wessem, M. J., & Morlighem, M., Sub Dynamics Meteorology, & Marine and Atmospheric Research. (2019). Four decades of Antarctic Ice Sheet mass balance from 1979–2017. *Proceedings of the National Academy of Sciences—PNAS*, *116*(4), 1095–1103. https://doi.org/10.1073/pnas.1812883116

Rizzo, A. (2014). Rapid urban development and national master planning in Arab Gulf countries. Qatar as a Case Study. *Cities, 39*, 50–57. https://doi.org/10.1016/j.cities.2014.02.005

Serafin, K. A., Ruggiero, P., & Stockdon, H. F. (2017). The relative contribution of waves, tides, and nontidal residuals to extreme total water levels on U.S. West Coast sandy beaches. *Geophysical Research Letters*, *44*(4), 1839–1847. https://doi.org/10.1002/2016GL071020

Shepherd, A., Ivins, E., Rignot, E., Smith, B., van den Broeke, M., Velicogna, I., Whitehouse, P., Briggs, K., Joughin, I., Krinner, G., Nowicki, S., Payne, T., Scambos, T., Schlegel, N., Geruo, A., Agosta, C., Ahlstrom, A., Babonis, G., Barletta, V., … & Cullather, R. (2018). Mass balance of the Antarctic ice sheet from 1992 to 2017. *Nature, 558*(7709), 219–222.https://doi.org/10.1038/s41586-018-0179-y

Sieghart, L. C., & Betre, M. (2018). Climate change in the MENA: Challenges and opportunities for the world's most water stressed region. *MENA Knowledge and Learning* (p. 164). http://documents.worldbank.org/curated/en/894251519999525186/pdf/123806-REVISEDBLOG-CC-REGION-QN-002.pdf

State of Qatar. (2015). *Intended Nationally Determined Contributions (INDCs) Report*. www.unfccc.int/sites/submissions/INDC/Published%20Documents/Qatar/1/Qatar%20INDCs%20Report%20-English.pdf

State of Qatar. (2021). *Nationally Determined Contribution (NDC)*. www.unfccc.int/sites/ndcstaging/PublishedDocuments/Qatar%20First/Qatar%20NDC.pdf

Stone, R. (2007). A world without corals? *Science, 316*(5825), 678–681.

Swiss Re. (2021). *The economics of climate change: No action not an option*. https://www.swissre.com/dam/jcr:e73ee7c3-7f83-4c17-a2b8-8ef23a8d3312/swiss-re-institute-expertise-publication-economics-of-climate-change.pdf

The White House. (2021, April 23). *FACT SHEET: President Biden's Leaders summit on climate*. https://www.whitehouse.gov/briefing-room/statements-releases/2021/04/23/fact-sheet-president-bidens-leaders-summit-on-climate/

United Nations. (2015). *The Paris agreement*. https://unfccc.int/process-and-meetings/the-paris-agreement/the-paris-agreement

Vitousek, S., Barnard, P. L., Limber, P., Erikson, L. H., & Cole, B. (2017a). A model integrating longshore and cross-shore processes for predicting long-term shoreline response to climate change. *Journal of Geophysical Research. Earth Surface, 122*(4), 782–806. https://doi.org/10.1002/2016JF004065

Vitousek, Barnard, P. L., Fletcher, C. H., Frazer, N., Erikson, L., & Storlazzi, C. D. (2017b). Doubling of coastal flooding frequency within decades due to sea-level rise. *Scientific Reports*, *7*(1), 1399–1399. https://doi.org/10.1038/s41598-017-01362-7

World Economic Forum. (2019). *The global risks report 2019*. The World Economic Forum. http://www3.weforum.org/docs/WEF_Global_Risks_Report_2019.pdf

World Economic Forum. (2022). *The global risks report 2022*. The World Economic Forum. https://www3.weforum.org/docs/WEF_The_Global_Risks_Report_2022.pdf

World Resources Institute (WRI). (2011). *Reefs and risks in the middle east*. https://www.wri.org/resources/maps/reefs-risk-middle-east

Laurent A. Lambert is a faculty at the Doha Institute for Graduate Studies working on the Water-Energy-Climate Nexus, policies and politics in the Middle East and Africa. He is a former Board Member & Representative of the Research Community at the United Nations Climate Technology Center & Network (CTCN). He has been advising governments in Africa and the Middle East regarding climate change mitigation and adaptation for a decade and has published in English, French, and Arabic, including in leading publications of geography, such as Environmental Science & Policy, Energy Strategy Reviews, and Water. He served as expert reviewer for the International Panel on Climate Change Assessment Report 6. He holds bachelor and master's degrees in Geography from the University of Paris1 Sorbonne and a PhD from the Oxford University School of Geography and the Environment.

Cristina D'Alessandro is a Senior Fellow at the Centre on Governance at the University of Ottawa (Canada) and a Research Fellow at the PRODIG Research Centre, at the University of Paris 1 Sorbonne. She is widely published in English, French, and Italian. Her research focus is urban planning, management, and transformation; natural resources and environmental governance; political, economic, and territorial governance; institutional capacity building and leadership. With over 20 years of academic and policy research experience, she serves as an international consultant with international organizations, including the African Development Bank, UN-WOMEN, the United Nations Economic Commission for Africa, but also the BMW Foundation, and the Qatar National Research Found, inter alia. A former Scientific Manager at the French National Agency for Research and a former Professor at the Paris School of International Affairs (Sciences-Po, Paris, France), she holds a number of board membership positions, including the International Advisory Board of the Canadian Journal of Development Studies; the editorial board of the journals L'Espace Politique and of African Geographical Review.

Chapter 10
Assessing and Reporting Potential Environmental Risks Associated with Reefing Oil Platform During Decommissioning in Qatar

Radhouane Ben-Hamadou, Ahmad M. D. Mohamed, Sarra N. Dimassi, Mariam M. Razavi, Sara M. Alshuiael, and Muhammad O. Sulaiman

Abstract Decommissioning obsolete oil platforms in the upcoming decades in Qatar, through a complete removal, will imply a major loss of the fouling biodiversity and associated ecosystem functions and services. The rig-to-reef approach, successfully implemented in other seas of the world, provides an attractive alternative to the complete removal by reefing the underwater oilrig structures and theoretically provides positive environmental, economic, and social benefits to a wide range of stakeholders. This manuscript critically reports and analyzes the risks associated with the implementation of the rig-to-reef approach in Qatar oilfields by using a DPSIR conceptual framework focusing on the pressures, states, impacts, and responses. By portraying reefing approaches as an issue, several pressures were identified including potential navigational safety issue, pollution, and physical disturbances. These pressures are expected to modify the physical, biological, and chemical state of the ecosystems, potentially leading to several impacts, such as the loss of habitats, biodiversity, and associated ecosystem services. The identified hazards and their potential risks were analyzed, and these seemed to be outweighed by the potential benefits of reefing obsolete oil platforms. Nevertheless, the critical impairment to the rig-to-reef implementation, mainly associated to existing regulatory frameworks, should be addressed.

Keywords Marine environment · Marine biodiversity · Rig-to-Reef · Oil platform · Sustainability · Qatar

---

R. Ben-Hamadou (✉) · A. M. D. Mohamed · S. N. Dimassi · M. M. Razavi · S. M. Alshuiael · M. O. Sulaiman
Environmental Science Program, Department of Biological and Environmental Sciences, College of Arts and Sciences, Qatar University, Doha, Qatar
e-mail: benhamadou@qu.edu.qa

L. Cochrane and R. Al-Hababi (eds.), *Sustainable Qatar*, Gulf Studies 9,
https://doi.org/10.1007/978-981-19-7398-7_10

10.1 A Synthesis of Key Issues of the Marine Environment in Qatar

Boomed by a high economic development since the oil boom of the 1970s, all eight nations bordering the Arabian Gulf, including Qatar, have witnessed a major and sustained increase in their demographic indicators (van Lavieren et al., 2011). This economic and social growth did not come with no impacts to their marine environment (Burt et al., 2017). Coastal reclamation, pollution, and overexploitation of natural resources are the major pressures and have often outpaced environmental conservation, policy, and regulations (Sale et al., 2011). The resulting anthropogenic changes, mainly occurring in the coastal and nearshore zones, exacerbated the already naturally stressed environment. Indeed, the Arabian Gulf has harsh environmental conditions (i.e., shallow depths, high salinity and temperature regimes, slow hydrodynamics) that make its marine ecosystems unusually fragile and susceptible to impacts from human activities (Fanning et al., 2021).

While marine fauna and flora around Qatar are adapted to these conditions, the extreme nature of the environment puts this biota at the margins of their physiological tolerance, and any further stress—such as from human activities—can push these species over the edge, and result in mass mortality of organisms across whole ecosystems. Given the importance of coastal systems such as coral reefs and seagrass in supporting the biodiversity and economy of Qatar, it is critical that Ecosystem Based Management practices be instituted to limit the addition of man-made stressors to these fragile ecosystems.

Among major man-made environmental pressures in the region is the development of the oil and gas industry, relying mainly on extractions made offshore. The Arabian Gulf has the highest density of oil platforms among all seas of the world, with more than 800 oil platform structures (Sheppard et al., 2010), and the number of well-head platforms in the Arabian Gulf is estimated to be between 2000 and 3000 (Stachowitsch et al., 2002). These structures cause acute and chronic impacts on natural ecosystems during their drilling and installations (Albano et al., 2016); during operations due to chronic pollution caused by residual oil spills and discharges of produced waters (Bakke et al., 2013; Tornero & Hanke, 2016; Zhang et al., 2019); interaction with bird's migrations (Ronconi et al., 2015); or major loss of biodiversity and habitat destruction during decommissioning (Claisse et al., 2015). Nonetheless, several recent studies highlighted the positive effects of offshore oil platforms on the local and regional biodiversity and productivity (Meyer-Gutbrod et al., 2020; Van Der Stap et al., 2016) and recently in the Arabian Gulf around the Qatar EEZ (Torquato et al., 2017, 2021). Indeed, the growing body of evidence showing the useful ecosystem functions and services, habitat diversity, as well as the connectivity provided by these structures motivated the decision taken by some countries to allow these platforms to be converted into artificial reefs under the so-called Rig-to-Reef (R2R) program (Bull & Love, 2019).

In this chapter, the authors investigate the potential issues that may rise by implementing the R2R approach during decommissioning of Qatar's offshore oil platforms. We applied a systematic analysis framework, the DPSIR (Driver-Pressure-State-Impact-Response) (Gari et al., 2015; Lewison et al., 2016) to revise empirical scientific information and support understanding by stakeholders of appropriate actions to be taken, in order to maximize positive impacts of reefing obsolete oil platforms.

10.2 Justification, Benefits with a Historical Context

Facilities used to develop offshore oil and gas fields are found on continental shelves throughout the world's oceans. When the oil or gas reservoir reaches the end of its economic life the facilities are decommissioned, which usually requires the removal of all the installed facilities (platforms and wells). The world's offshore oil and gas infrastructure is aging, and the industry is rapidly approaching a decommissioning crisis (Fowler et al., 2014). In fact, a large proportion of the more than 12,000 active offshore installations is operating beyond or approaching the end of their designed life (Ars & Rios, 2017).

Nevertheless, oil and gas platforms are known to provide habitat for marine benthic species, as well as increase the diversity and productivity of marine ecosystems. They can also act as stepping-stones for these species, by connecting patches of natural habitats that were previously unconnected (Fowler et al., 2020; Torquato et al., 2019).

In recognition of the growing body of evidence showing the useful ecosystem functions, services, habitat diversity, and connectivity provided by these structures, some countries allow these platforms to be converted into artificial reefs under R2R programs. Subsequent international agreements introduced some exceptions to complete removal (the so-called partial removal options) if obligations associated with navigational safety and environmental protection were met. This allowed the R2R concept to be applied in the United States, Brunei, and Malaysia (Bull & Love, 2019).

10.3 Political, Economic, and Regulatory Framework Context

The decommissioning of redundant oil and gas facilities is a process which is regulated internationally, regionally, and nationally. Decommissioning regulations and guidelines in various countries and regions have been reported on and assessed extensively in the literature (Van Elden et al., 2019). Complete removal was first mandated in the 1958 Geneva Convention on the Continental Shelf to ensure that the oil and gas industry was liable for their infrastructure following cessation of production. The

OSPAR Commission's Decision 98/3 requires complete removal of offshore installations, with some exceptions that fulfill purely technical criteria (Fowler et al., 2018). In Qatar, as well as the whole Arabian Gulf region, the current regulatory framework provides no other alternative, apart from the complete removal of obsolete offshore platforms. The intention of these regulations is clear, namely, to protect the marine environment and ensure proper management of redundant resources.

10.4 Reefing Obsolete Oil Platforms, a DPSIR Analysis

10.4.1 Rationale

Over 60% of global oil reserves are found within the jurisdiction of the Arab Gulf Nations (OPEC, 2018). In addition to oil, Arab countries just a decade ago were home to approximately 55 trillion m^3 of reserves (Fattouh & Darbouche, 2011). These huge reserves have allowed unparalleled economic growth and development in the region (Din, 1990; Calder, 2015). This oil and gas rich economy has led to the construction of 800+ offshore platforms as well as 25 large terminals containing some of the largest oil and gas infrastructures to date (Sheppard et al., 2010). While some of this oil is transported through underwater pipelines to Harbor's in the Red Sea and Mediterranean regions, most of it leaves through the Strait of Hormuz, transported via tankers which distribute this oil to different parts of the globe, such as Western Europe and Japan (Din, 1990; Calder, 2015).

Qatar is no different from the other Gulf countries; according to the US Energy Information Administration (2015), Qatar's mainly hydrocarbon economy was responsible for almost half of its revenue in 2014. Its large oil and gas reserves are the main driver behind the construction of multiple oil platforms in the North and East of the Peninsula. Since some of them are reaching the end of their lifetime, they need to be decommissioned.

Recent studies however revealed, using remotely operated vehicles (ROV) footages, some promising findings, where the artificial reefs seem to support a very well-established ecosystem with a variety of fouling macroinvertebrates (Torquato et al., 2021) and rich fish assemblages (Torquato et al., 2017). These studies open the doors to explore the possibility of reefing the rigs. While it is widely known that coral reefs are among the most diverse and productive ecosystems in the world, they have been degraded, and their coverage is only a fraction of what they used to be. However, amidst the Gulf's extreme environmental conditions of ocean surface temperatures above 36° during the summer months and salinity levels above 45 psu (Range et al., 2018), the fact that these corals are able to survive at different depths on these platforms is very promising. According to Torquato et al. (2021), the videos were taken et al.-Shaheen oilfield during the monitoring surveys conducted from 2007 to 2014. Using underwater Remotely Operated Vehicles, they analyzed 4510 videos and found 17 functional groups categorized by morphology, among which they found hard corals. They conducted this study from the surface to a depth of 60 m, where they found that the coral communities increased in abundance at depths exceeding 30 m as well as on older platforms (Torquato et al., 2021).

The most interesting finding was the identification of a new species of coral belonging to the family *Caryophylliidae* which was seen for the first time in Qatar. As a result, this study is of paramount importance as it shows that not only there is a well-established ecosystem attached to the oil platforms consisting of reef building corals, but also that they can potentially compensate for the loss of reefs and associated flora and fauna elsewhere (Range et al., 2018; Torquato et al., 2021).

These ecosystems provide various functions and services. Approximately 912% of global fisheries are associated with coral reefs (Smith, 1978), and up to 25% in certain regions (Cesar, 1996).

Scientists have also discovered anticancer and anti-inflammatory substances in corals and their associated flora and fauna for the pharmaceutical industries (Carté, 1996). In addition, coral associated flora, such as seaweed, is used to extract agar (Birkeland, 1997), as well as to produce manure (Craik et al., 1990). Corals themselves seem to be useful in bone surgery (Spurgeon, 1992). According to Tsounis et al. (2010), red corals were sold for up to 50.000 US$ kg^{-1}.

The presence of corals prevents coastal erosion caused by natural events such as strong currents and storms. In Indonesia, Cesar (1996) estimates that up to a million dollars are lost per kilometer of eroded coast, while in the Maldives, it cost twelve million dollars to replace the lost reefs with an artificial barrier (Weber, 1993).

Reefs are also beneficial to fisheries by exporting nutrients into the pelagic food web, which in turn provides nutrients to plankton, leading to increased productivity and helping fisheries (Sorokin, 1990). Moreover, they clean the water by sequestering human waste and detoxifying it, such as breaking down hydrocarbon pollution to CO_2 and H_2O (Peterson & Lubchenco, 1997).

The decommissioning of these rigs would result in loss of the biodiversity, ecosystems, and their provided services (Torquato et al., 2021) justifying the prospect of their conservation through the R2R approach instead of a complete removal during decommissioning.

10.4.2 Qatar Oilfields

The Arabian Gulf is recognized as a youthful sedimentary basin which is found in the subtropical area. The Gulf area ground was nearly entirely exposed in the last glacial maximum (LGM) since the level of the sea was about 120 m less than the current level (Sheppard et al., 2010; Torquato et al., 2021). Presently, the Arabian Gulf seawater floods show a maximal width of 350 km with 35-m mean depth, hardly surpassing 100 m of depth (Seibold, 1973; Vaughan & Burt, 2016). The exchange of water with the open ocean in that region is restricted to the Hormuz strait. The climate in the Arabian Gulf is allied with the desert condition since the area is encircled with arid land, which generates extreme environmental conditions resulting in hypersaline conditions where water which usually exceeds 42 psu (Swift & Bower, 2003) and extremely seasonal variations in sea surface temperatures (Nesterov et al., 2021). During the summer months, the Gulf is characterized as being the hottest sea on earth with sea surface temperatures exceeding 36 °C (Vaughan & Burt, 2016). However, during the winter season, the temperature can drop as low as 13 °C (Coles, 2003; Torquato et al., 2021).

This manuscript is critically analyzing risks associated with the implementation of the R2R to the oil platforms pertaining to the different oilfields into the Exclusive Economic Zone of Qatar.

Qatar's offshore oilfields are located mainly toward the northeast and East of the peninsula. They comprise, from South to North, El Bunduq, Al-Karkara, Idd El Shargi, Bul Hanine, Maydan Mahzam, Najwat Najem, Al-Khalij, Al-Shaheen, and Al-Rayyan oilfields. The latter two fields are inserted in the Qatar's North Field, the world largest single natural gas field (Fig. 10.1). All oilfields under operation are located at 60 km or more off Qatar's mainland and are therefore away from direct exposure of anthropogenic pressures generated at the coastal zone. Each oilfield will comprise 5– 33 oil rigs or platforms, each weighing several thousands of tons and the submerged rig is usually made of austenitic stainless steel.

Most of the agreements signed by the different operators of these offshore structures with the State of Qatar last for more than 20 years, and thus, oilfield platforms play a crucial role as a long-lasting artificial habitat for several marine organisms in the offshore environment in the Arabian Gulf (Torquato et al., 2021).

10.4.3 The Rig-to-Reef Approach

Rig-to-reef (R2R) is considered as the reusing and conversion process of offshore submerged platform structures into artificial reefs that consist of similar characteristics and functions of natural reefs. The main objective of this process is to protect and regenerate the production of marine organisms to enhance the conservation of aquatic environment and fisheries (Verbeek, 2013). The conversion of oil platforms into the environmental and economic valuable structure after decommissioning can be achieved through the re-utilization of the rig structure and its associated fouling biological communities which in turn enhance the biological production, preserve the marine ecological resources, and form a well-established habitat for marine organisms (Bull & Love, 2019; Kaiser, 2019; Nugraha et al., 2019).

The increase in offshore oil exploration, the construction of oil platforms, and the decommissioning after reaching the end production period have led to the exploitation of the artificial reefs resulting from the structure of the rig, left after the platforms closure (Nugraha et al., 2019). This would provide environmental and economic benefits:

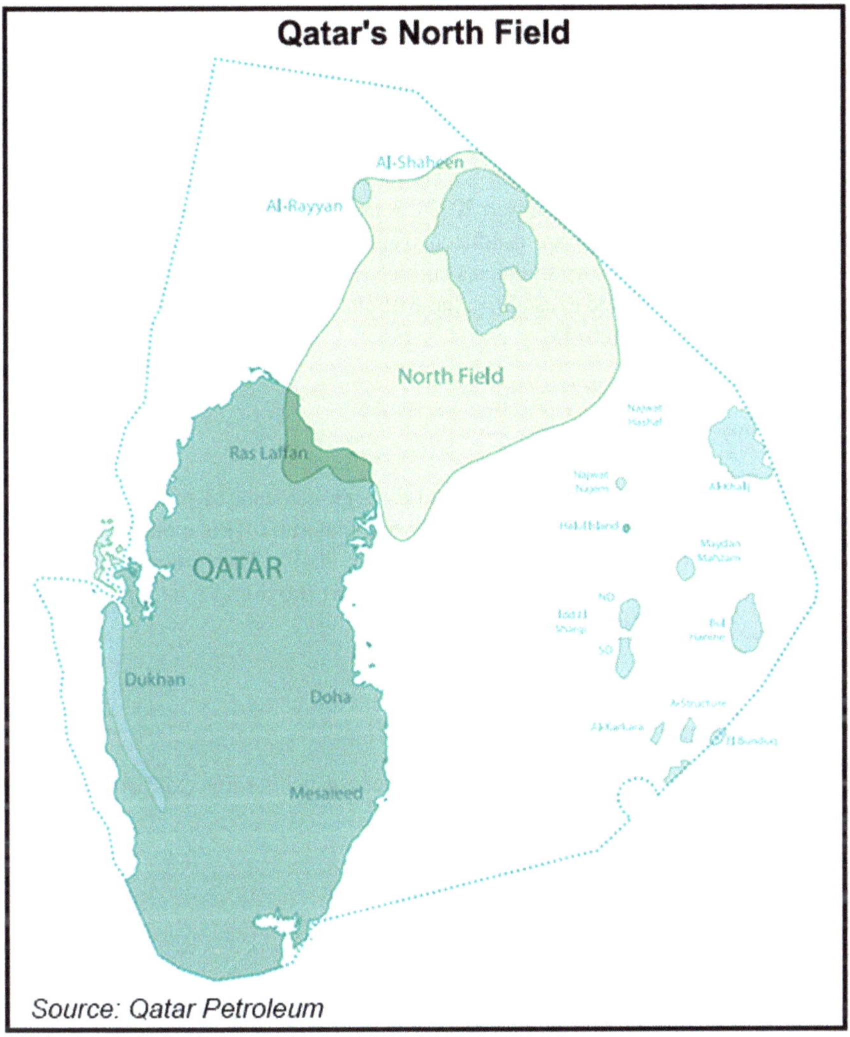

Fig. 10.1 Map of the offshore oil and gas fields in the Exclusive Economic Zone of Qatar (delimited by the dotted border). *Source* Qatar Petroleum

- Habitat restoration, loss compensation, and resources conservation: rig-to-reef supports the production and development of different marine organisms that enhance the ecological connectivity.
- Fisheries: production of valuable marine organisms such as algae, crustaceans, molluscs, and pelagic and benthic fishes.
- Recreational activities: attract divers and recreational fishermen and increase the economy of the tourism sector.

10.4.3.1 Methods of Rig-to-Reef

The conversion of oil platforms into artificial reefs is considered as a sustainable solution, and it occurs following three different methods: (1) partial removal, (2) topple-in-place, and (3) tow and place (Fig. 10.2). Each method requires different levels of sea work interventions on the rig jacket, including use of explosives and heavy mechanical processes (Kaiser, 2019). A jacket represents the support structure of the oil platforms which are produced from steel and placed offshore where it extends from the seafloor up to the sea surface (Bull & Love, 2019). The following methods were mainly used to establish the artificial reefs (Bull & Love, 2019; Kaiser, 2019; Macreadie et al., 2011).

Partial Removal Method

The upper part of the rig jacket is severed through mechanical process and placed next to the bottom part without using explosive materials. This method can reduce the disturbance of marine habitat (Bull & Love, 2019). The bottom part remains in the same place of the rig without any movement providing a habitat for marine organisms (Kaiser, 2019).

Topple-in-Place Method

The oil rig jacket is laid down in a horizontal position using some explosive materials on the seafloor.

Tow-and-Place Method

The oilrig jacket is pulled up from the sea floor and towed using heavy lifting vessels after which it is placed vertically or horizontally on the sea floor (Kaiser, 2019).

10.4.3.2 Outcomes of the Rig-to-Reef Approach

Benefits of Rig-to-Reef

Rig-to-reef is considered as a beneficial project since it provides a habitat for marine organisms to spawn, grow, and aggregate which improves the carrying capacity of the ecosystem which in turn increases biomass production (Macreadie et al., 2011; Verbeek, 2013). Also, it develops and augments the marine ecosystem functions that provide energy and nutrients for the aquatic environment (Verbeek, 2013). In addition, it offers several ecosystem services such as provisioning services, cultural services, and supporting services, where rig-to-reef improves the marine biodiversity

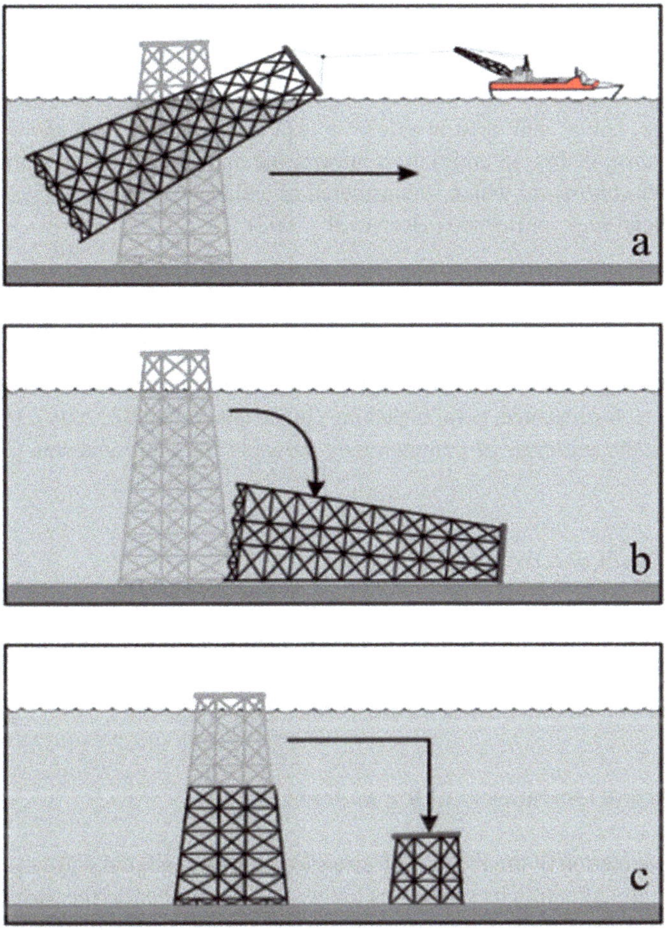

Fig. 10.2 Platform reefing methods. **a**. Tow-and-place platform reefing. **b**. Topple-in-place platform reefing. **c**. Partial removal platform reefing. Methods **a** and **b** often use explosives to severe steel jacket legs below the seafloor. Methods **b** and **c** often use mechanical tools to severe steel jacket legs either below or above the seafloor. Method **c** may or may not include placement of shallow water severed jacket on the seafloor as additional reef material (Bull & Love, 2019)

and biomass production through the change of food webs, and provides recreational activities (Jagerroos & Krause, 2016). Furthermore, artificial reefs play a significant role in the development of new habitats in the aquatic environment, restoration, protection, and conservation of the habitats (Becker et al., 2018; Jagerroos & Krause, 2016). The main goals of rig-to-reef approach are:

Creation of New Habitat

Rig-to-reef can produce new habitat that develops and matures with time which in turn increases the abundance of biofouling communities and marine organisms including macro-algae, fishes, and most invertebrate taxa such as corals and sponges (Jagerroos & Krause, 2016). In addition, it attracts the endangered species, for instance *Eretmochelys imbricata* which is considered as endangered turtle that was obtained in the artificial reefs in Borneo (Tisen et al., 2002).

Restoration of Habitat

Rig-to-reef can restore the damaged sites through the maximization of larval recruitment and its distribution. Moreover, it provides an alternative habitat for the impacted marine areas, for instance, coral bleaching (Jagerroos & Krause, 2016). In addition, it enhances the presence of commercially valuable marine organisms (Jorgensen, 2009).

Protection and Conservation of Habitat

Rig-to-reef represents a barrier to prevent active fishing, for instance, trawl fisheries that will enhance the conservation of marine organisms, and to achieve this, regular monitoring should be conducted in artificial reefs area to avoid any illegal fishing methods and reduce overfishing (Jagerroos & Krause, 2016).

Implementation Determinants of Rig-to-Reef

The implementation of the rig-to-reef scenario could be hindered with several obstacles potentially affecting its success including, legal compliance, the overall costs of cleaning, severance of platforms, maintenance, and regular surveillance (Jagerroos & Krause, 2016; Ounanian et al., 2020). According to the Gulf of Mexico case, 10% of the oil platforms have been reefed; however, some decommissioned platform's structure has been fully removed and brought onshore because of the prohibitive costs of maintenance (Kaiser, 2019). Moreover, the liability problems of platforms leakage, the pollution level of the artificial reef installation site, and the abundance of invasive species can motivate their complete removal (Ounanian et al., 2020).

Additional criteria should be also considered for the implementation of the R2R approach in the Qatar oilfields such as the location of the different oil platforms, determining their connectivity with natural, and other artificial reefs and its exposure to differential hydrodynamic forcings (Jagerroos & Krause, 2016).

Perhaps the most significant obstacle toward the implementation of the R2R approach in Qatar is the lack of a regulatory framework for such decommissioning scenario. The United Nations Convention on the Law of the Sea (UNCLS), to which Qatar is a signatory, recognizes the need for artificial structures developed on the sea, either for commercial purposes or protection of the territorial integrity of a state.

According to Byrd et al. (2018), the state of Qatar established regulations that sought to ensure the safety of all those accessing the country's seabed which may impact the future implementation of the R2R approach. Accordingly, reefing oil platforms in the EEZ of Qatar will lead to several legal challenges (da Fonseca et al., 2020). This is due to the possible disruption of marine transport due to the many accidents likely to be reported by having reefs beneath the sea. In Qatari law, oil and gas companies that fail to comply with the set of regulations on marine operations risk being decommissioned. Any move by the operators of oil and gas companies to establish rig-to-reef is seen as a violation of a country's regulations on marine transport (Van Elden et al., 2019). This is in accordance with Law No. 29 of 1966 together with Law No. 8 of 2017 on the conduct of marine works in Qatar which collectively require oil and gas companies to provide seamless passages to marine transportation companies using both the inland and offshore waters.

Lim (2021) noticed that one key challenge at offshore Qatar oilfield is the lack of a natural reef that can provide a safe breeding ground for fish. In this case, the fundamental idea is that once the entire oil exploration is completed, there will be a need to remove the top part so that the reef underneath will not be affected. However, according to da Fonseca et al. (2020), this represents a set of challenges to maritime navigation. In most cases, such ambitious projects are prone to accidents due to the reefs already existing under the water hindering the free movement of cargo across the sea (Van Elden et al., 2019).

On the technological front, there are logistical challenges of time, manpower, and resources, both financial and non-financial, to fully realize the objective of a rig-to-reef system (Lim, 2021). Moreover, not all the platforms can be used for reefing. Transforming a platform to a permanent artificial reef requires engineering and environmental criteria. The size, complexity, structural integrity, and location of the platform are all important factors to consider when assessing its reefing potential. Platforms that are complex, stable, durable, and clean are good prospects for reefing. Reefing is not an option for platforms that have tipped over owing to structural breakdown (BSEE, 2021).

Rig-to-reef is meant to offer favorable environmental conditions for fish and other macroinvertebrate communities to thrive but with the increased oil and gas exploration, the project would pose significant legal, logistical, and environmental challenges to stakeholders (Lim, 2021). Byrd et al. (2018) noted that the development of artificial reefs on any oilfield requires the removal of the top structures while leaving behind those buried deep in the sea.

In the following sections, we are considering the DPSIR framework to critically analyze the risks associated with reefing oil rigs in Qatar from different perspectives, such as leaving an artificial structure in place, implementing marine protected areas, pollution that is likely to occur, the effect on fisheries, and expert knowledge regarding the case.

10.4.4 *The Driver-Pressure-State-Impact-Response (DPSIR) Framework*

The complexity of the marine environment is mainly generated from the interaction of its morphological and physical structures which creates a continuous variation between the physico-chemical processes and the ecological structure and function (Elliott et al., 2017).

Recently, several comprehensive assessments based on conceptual models have been utilized for problem structuring and facilitating empirical research in coastal regions (Lewison et al., 2016). The Drivers-Pressures-State-Impact-Response (DPSIR) framework remains one of the most widely adopted comprehensive approaches in some of the coastal regions around the globe. Although there are various frameworks which are available and accessible for exploring the components interactions and the integrated assessments of the different systems (Lewison et al., 2016; Potschin, 2009). DPSIR is a powerful framework which was initiated by the economic co-operation and development organization (OECD, 1993) in order to offer an integrated reporting strategy related to the environmental assessment (Kelble et al., 2013). This framework is effective in organizing and compiling various sets of information to manage the system; therefore, it unambiguously illustrated the causal relationship to the stakeholders (Maxim et al., 2009). In addition, this conceptual model can distinguish addressed issues by associating and recognizing the causal relationship, which will facilitate the definition and the study of the system, and to further analyze the issues in order to generate potential solutions for the same (Daniels, 2010). DPSIR framework can be successfully implemented for the evaluation of environmental variations in the coastal regions and marine ecosystems, to make predictions for the future challenges, and to enhance the managemental practices (Goble et al., 2017; Kaur et al., 2020; Lin et al., 2007; Miranda et al., 2020; Newton & Weichselgartner, 2014). Moreover, this straightforward approach has the ability to form a link between pressures and impacts for evaluating objectives by mainly directing the risk assessment and the major pressures which are altering the state and thus impacting the ecosystem services and benefits which will eventually impact "us" as humans (Atkins et al., 2011; Smith et al., 2016; Smyth & Elliott, 2014). Within the marine context, managing marine ecosystems by implementing the DPSIR framework is coherent with the ecosystem approach (Cooper et al., 2013; de Jonge et al., 2012; Elliott et al., 2017). Therefore, due to the effectiveness of this conceptual model, the risks associated with the rig-to-reef approach in Qatar's platforms at the different oilfields will be evaluated and critically analyzed via A DPSIR framework analysis.

10.4.4.1 Evidence Search and Review

A systematic literature review (Booth et al., 2021) was performed in order to critically analyze the risks associated to reefing approach of the Qatari platforms by using A

DPSIR framework. Science Direct and Scopus databases were used, by searching specific syntaxes:

- "DPSIR framework, AND marine, (artificial structures OR sustainability)". Total papers found were 98.
- "MPA Marine DPSIR" (yielded 73 results out of which the relevant papers were chosen).
- "Rig-to-Reef" (yielded 7 results).

No refinement was done based on any specific titles, abstracts, or date limitations. Additional references were also manually added due to their relevance to the topic.

10.4.4.2 DPSIR Framework Set-Up and Analysis

A DPSIR framework was generated to identify, evaluate, and critically analyze the risks associated with the rig-to-reef approach implementation in Qatar oilfields (Fig. 10.3). This approach is potentially applicable to all similar oilfields in the Arabian/Persian Gulf. Potential risks of impacts are reported here as a result of a cascade of effects starting from a set of pressures driven by primarily societal changes resulting in changes of environmental features triggering impacts for which policy- and decision-makers are required to remedy with technical and non-technical set of responses.

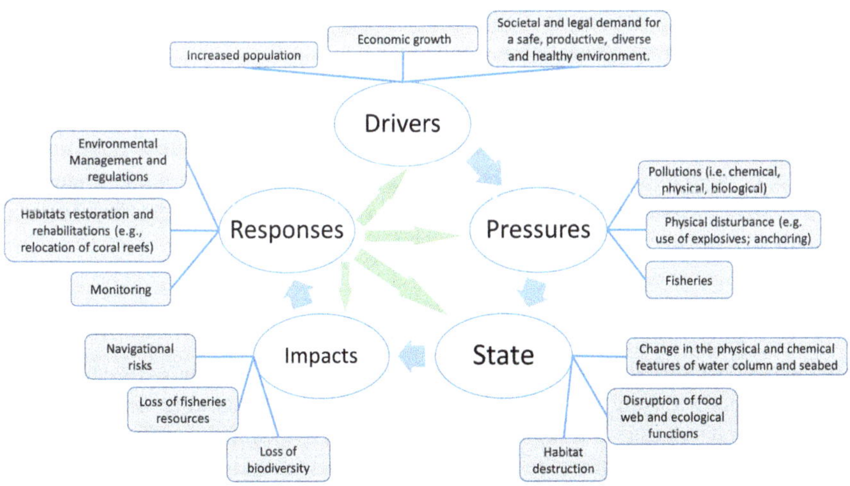

Fig. 10.3 Schematic representation of the generated DPSIR framework

Pressures

Chemical Pollution

The driver for this activity is that pipelines attached to the main structure may be classified as remaining materials left on site and can be considered as litter since they are alien objects to the natural environment and may eventually breakdown and disperse into the local environment. There are two options regarding the issue. First is pipeline abandonment, which, as the name suggests, is the activity of leaving the pipeline within the environment after disconnection and purging. Second is the physical removal of a pipeline installation from within the environment intended to be turned into a marine-protected area. The first issue that arises is in terms of chemical and radioactive pollution in the form of naturally occurring radioactive materials (NORMs), since a buildup of barium and strontium compounds within pipelines is commonplace. This may be due to transportation of crude and incompatible water injection which results in a barium/strontium, sulfate, and calcium carbonate co-precipitate (Hamlat et al., 2001). Trace amounts of precipitate may leak into the environment via degradation of pipeline if it is left within the environment. Trace amounts may also be released into the environment during purging and decommissioning process which can lead to adverse effects.

NORM as well as other elements can indeed be made directly accessible to life-forms on site or via discharge into the adjacent seabed ecology by hydrocarbon granulation (Ossai et al., 2020). Although pipeline precipitate is not a soil pollutant, it will settle on and combine with the surface layer of soil after the pipeline erodes. Through nutritional absorption of particulates and soluble particulates or physical adsorption through carapace, benthic and pelagic ecosystems may well be subjected to related pollutants. The relationship of marine animals with NORM precipitate-based pollutants is influenced by suspended particulate quantities, as well as the dietary habits and biochemistry of the biota. Organisms living in close proximity to pipelines or other hydrocarbon sediments linked toxins can bioaccumulate chemicals and experience eventual ecotoxicological impacts due to the precipitates biochemical and radioactive characteristics (Kennish, 1997).

Physical and Biological Pollutions

Coral reefs provide habitat for aquatic creatures but determining the ecosystem functions or impacts on the local plant and animal life is challenging. As artificially deployed coral reefs age and expand, this introduced environment leads to an increase in the population of creatures by luring surrounding species to the suitable environment, as well as the creation of biomass. The first problem created may be introduced in the form of alien and potentially invasive species as individuals that do not belong to the original habitat potentially leading to a shift in the trophic structure. Furthermore, the substrate provided by the rig is responsible for attraction of invasive species as it forms a perfect anchoring point. Secondly, the richness of local marine fauna that can also be accessed by fishermen is undeniably increased by increased coral habitats. As a result, increased fishing pressure may result in lower fish populations in the long term. Artificial reefs attract fish and other species because they provide protection from waves and predation, reproductive and incubation locations, and a substrate that allows for more accessibility to phytoplankton and sunlight.

The habitat now existing on and around the rig will undoubtedly be impacted by the total removal of subsea structures following dismantling, but the long-term environmental effects of artificial reef installation are less obvious. Most sessile marine creatures, in contrast to migratory species, are unable to be attracted to artificial reefs owing to their confined or sessile lives (Reed et al., 2004). As a result, colonization and expansion of such individuals will almost certainly result in only minor gains in biomass at first (Reed et al., 2004). Acquisition of invertebrates via larval colonization contributes to improvements in secondary production of these organisms as the reef grows and matures.

However, there are multiple forms of pollution that can invariably affect creation and operation of a rig to a coral reef that may have varying effects on the local community that has formed over the course of the rig's lifetime with one of these stressors being noise pollution. The activity of converting an oilrig to a reef is bound to introduce large amounts of sound into the immediate environment. Since most aquatic species utilize sound as a vital function of their physiology, the excessive input can likely result in deformations, juvenile, and clutch mortality, slowed growth and furthermore it is determined that zooplankton are directly affected with a higher mortality rate when exposed to excessive auditory input (Weilgart, 2007). This will result in key ecosystem services being provided by inhabitants of the existing ecosystem to be disrupted. In a similar manner, a higher exposure to sunlight into an otherwise benthic community that is used to a lowered light concentration can also result in a modification of the physiological performance of the community. Artificial illumination has been shown to introduce changes in species abundance since it disrupts the natural predator prey interactions that an ecosystem develops which results in a disruption of the trophic equilibrium (Davies et al., 2012).

Fisheries

From a fishing perspective, the main causes of pollution are fishing gears such as nets and anchorage. Mangi et al. (2007) applied the DPSIR analysis on the reef fisheries management in Kenya and found out that the destructive fishing gear was one of the pressures affecting the reefs. Similarly, Ojeda-Martínez et al. (2009) in their study of the conceptual framework for the integral management of marine protected areas found that the gears effect, gear lost, and wastes were the main pressures for the management of the protected areas. This is also expected in the rig-to-reef case, as many nets may be lost while fishing and the use of fishing gears (e.g., gargoor or fishing traps) that falls on the seafloor, affect the living communities, or may be lost where the fishermen cannot locate it again.

State of the Environment

All the previously mentioned pressures resulting in the implementation of the R2R approach will cause several changes in the structure and functions of the natural ecosystems and thus modifying their state. These pressures and human activities have adverse effects in altering the marine ecosystems since it mainly creates a potential biological loss and damage to the ecology, hydrodynamics, and ecosystem services (e.g., suffocation of the benthos, sediments resuspension, and thus the release of the contaminants) (Li & Hu, 2021). In addition, there is a potential change in the physical and chemical nature of the water column and seabed (Tarr, 2014). All of these changes will eventually lead to a disruption in the food web due to habitat destruction and loss of the biological diversity.

Regarding fisheries activities, studies have shown that there can be changes in the species abundance, species richness, and habitats (Mangi et al., 2007; Ojeda-Martínez et al., 2009). In the case of the rig-to-reef of Qatari oilfield platforms, we are expecting these changes to be in the contrary affecting the status of environment in a net positive manner, due to the relatively low productivity recorded nearshore and offshore of the EEZ of Qatar in most of the seasons (Rakib et al., 2021).

Impacts

Since several pressures are altering the state of natural ecosystems by causing changes in their structure and functions, these changes may lead to various impacts. Maintaining the rigs in place will change the originally existing seabed's nature affect the benthic populations and their predators. The physical and chemical changes in the water column can affect sea mammals and fishes. Furthermore, marine mammals in general are potentially endangered via offshore operations (i.e., boat propeller accidents) leading to a loss of biodiversity (Tarr, 2014). In addition, food web disruption due to pollution will potentially lead to a loss of food sources and medicine (e.g., corals mortality). Moreover, changing the structure and functions of the natural ecosystems will eventually lead to the loss of the ecosystem services provided by these ecosystems (Smyth et al., 2015).

Previous studies found that some impacts on the fisheries industry are the changes in the habitat's spatial structure and declining fish catch, which will show reduced livelihood benefits that will be a change in socioeconomic relationship (Gebremedhin et al., 2018; Mangi et al., 2007; Ojeda-Martínez et al., 2009). Since the rig-to-reef is expected to boost fisheries, its implementation in the Qatari oilfields would inversely benefit the fisheries and tourism.

Climate change plays a negative role in the success of rig-to-reef, namely, due to the impacts resulted, increase of temperature, and acidification rate that accelerate physical breakdown of the artificial reefs (Jagerroos & Krause, 2016).

Responses

In order to tackle the potential navigational safety issue of the remaining structure, many regulations were developed for the achievement of an equilibrium between international agreements which tackle navigational safety, protection, and preservation of the created marine ecosystem (Smyth et al., 2015). Navigational aid (NAVAID) may be installed to assist navigators in determining safe course and/or warn of the existence of underwater structures. In addition, a set of regulations and environmental management solutions needs to be set in order to protect the biodiversity and connectivity with adjacent natural reefs, such as restoring and rehabilitating habitats (e.g., relocation of coral reefs) and thus preventing the loss of habitats, food source, and ecosystem services. Implementing all of these responses will tackle the main risks associated with the rig-to-reef approach in Qatar oilfields, taking into consideration the monitoring process in order to modify these measures based on their efficiency.

As a response to the previously mentioned impacts, past experiences have suggested that certain steps can be implemented to mitigate potential impacts. These mitigation measures include proper policy implementation, changes in the legislation, environmental impact assessments, education, and awareness especially for fishermen, restriction on the number of recreational and professional fishermen and fishing gears, and monitoring and control for the newly established artificial reefs (Gebremedhin et al., 2018; Mangi et al., 2007; Ojeda-Martínez et al., 2009). Furthermore, a regulatory management needs to be developed including policies and monitoring plans that should be implemented by the governmental regulation system to ensure the sustainability and successful work of artificial reefs over long term (Jagerroos & Krause, 2016).

In the case of Qatar, there is already a good environmental law that is constantly updated and that should include the new established rig-to-reef and would have a dedicated policy and legislation. Also, all the fishing activities are licensed, and monitoring using Automatic Identification System (AIS) using radar technology would allow a prompt control of the access to reefed rigs and activities around.

An ecosystem-based approach for management of the newly artificial reefs is necessary to efficiently manage their uses. Declaring these areas as Marine Protected Areas (MPAs) is to be considered. MPAs provide protected zones for fish reserves, underwater parks, as well as wildlife sanctuaries (National Research Council, 2001).

As a result, management responses are required to tackle the changes that can take place due to applied pressures, changes in the status of the environment, and their associated impacts. Regarding decommissioning of oil platforms in these potential marine protected areas, the impacts of decommissioning activities on the conservation objectives need to be included, by incorporating mitigation, adaptation, and compensation measures to minimize negative consequences, as well as to utilize management measures to further enhance gains in goods and services (Burdon et al., 2018).

The selection of the type of MPA as fully protected; highly protected; lightly protected; or minimally protected (Grorud-Colvert et al., 2021) for the reefed rigs should be purely based on the desired outcome from the MPA. Since the aim here is to protect the ecosystem but at the same time promote human development, the precautionary approach would consist of keeping the area fully protected until studies can be conducted to understand the carrying capacity of the established system to human utilization. Once the capacity of the system to cope with disturbances is assessed then an increasing usage can be implemented at a later stage.

10.4.5 Conclusion

The prosperous economy of the State of Qatar, as all other Arabian Gulf neighboring countries, is based on a foremost oil and gas industry established in the second half of the twentieth century. Most of today's production is conducted offshore using oil platforms, that during the last decades have been colonized by a thriving biological fouling assemblages and visited by rich fish and megafauna communities. Decommissioning these man-made infrastructures in the upcoming decades, through a complete removal, will imply a major loss of the established biodiversity and associated ecosystem functions and services. The rig-to-reef approach, successfully implemented in other seas of the world, provides an attractive alternative to the complete removal by reefing the underwater oilrig structures and theoretically provide positive environmental, economic, and social benefits to a wide range of stakeholders. We thoroughly assessed and reported environmental and technical risks associated to the implementation of the rig-to-reef in Qatar, following the Driver-Pressure-State-Impact-Response (DPSIR) framework. Considering risks associated with navigation, pollution, fisheries, and enforcement of the protection status of the reefed structures, potential impacts and needed mitigation response measures have been identified and linkages with pressures revealed. Obstacles to the effective implementation of the rig-to-reef approach in Qatar have been reviewed and appeared to be mainly connected with the existing regulatory framework and the limited knowledge of conservation value of the oil platforms to be reefed. Hazardous outcomes of the implementation of the rig-to-reef approach in Qatar, and potentially in the region, are of low risk and should be outweighed by the potential benefits achieved by analogous projects and logically expected by the conservation of functional ecosystems, contributing to the maintenance of local biodiversity and regional productivity and connectivity.

10.5 Future Options

The rig-to-reef alternative to complete removal during decommissioning has a potential successful outcome if implemented to Qatar and the Arabian Gulf oil platforms. Its successful implementation will be contingent of the following considerations:

- Ensure stakeholder engagement with clear definition of liability and compensation schemes under a consistent policy and regulatory framework.
- Consider asset integrity and risks from corroding structure to the safety of maritime navigation, surrounding habitat destruction and release of contaminants.
- Thoroughly assess how important are these platforms to ecological productivity and diversity, fisheries sustainability, or the development, use, and enjoyment of marine fisheries in Qatari waters. The State of Qatar should consider this in the proposal of new MPAs toward the 30-by-30 (protection of 30 percent of national marine environment by 2030) as a commitment for the Sustainable Development Goals (SDGs).
- Anticipate and mitigate risks associated with spread of invasive species and changes in the marine food webs.
- Implement monitoring programs and enforcement actions to ensure sustainability of the artificial reef and its safe use.

All these actions should be integrated through space (single or multiple oil platforms and/or oilfields) and time (a timeline for decommissioning and reefing) for the effective implementation of a national rig-to-reef strategy roadmap.

Acknowledgements This work was conducted within the framework of the NPRP project "Assessing the contribution of oil platforms and the impact of decommissioning scenarios to marine biodiversity, productivity and connectivity (PlatDecom)—NPRP13S-0208-200302" funded by the Qatar National Research Fund (a member of The Qatar Foundation) and TotalEnergies, Qatar. The statements made herein are solely the responsibility of the authors.

References

Albano, P. G., Filippova, N., Steger, J., Kaufman, D. S., Tomašových, A., Stachowitsch, M., & Zuschin, M. (2016). Oil platforms in the Persian (Arabian) Gulf: Living and death assemblages reveal no effects. *Continental Shelf Research, 121*, 21–34.

Ars, F., & Rios, R. (2017, May). *Decommissioning: A call for a new approach*. Offshore Technology Conference. OnePetro.

Atkins, J. P., Burdon, D., Elliott, M., & Gregory, A. J. (2011). Management of the marine environment: Integrating ecosystem services and societal benefits with the DPSIR framework in a systems approach. *Marine Pollution Bulletin, 62*, 215–226.

Bakke, T., Klungsøyr, J., & Sanni, S. (2013). Environmental impacts of produced water and drilling waste discharges from the Norwegian offshore petroleum industry. *Marine Environmental Research, 92*, 154–169.

Becker, A., Taylor, M. D., Folpp, H., & Lowry, M. B. (2018). Managing the development of artificial reef systems: The need for quantitative goals. *Fish and Fisheries, 19*(4), 740–752.

Birkeland, C. (1997). *Life and death of coral reefs*. Springer Science & Business Media.

Booth, A., Sutton, A., & Papaioannou, D. (2021). Systematic approaches to a successful literature review.

BSEE. (2021). *Rigs-to-reefs*. Rigs-to-Reefs | Bureau of Safety and Environmental Enforcement. Retrieved December 9, 2021, from https://www.bsee.gov/what-we-do/environmental-compliance/environmentalprograms/rigs-to-reefs

Bull, A. S., & Love, M. S. (2019). Worldwide oil and gas platform decommissioning: A review of practices and reefing options. *Ocean & Coastal Management, 168*, 274–306.

Burdon, D., Barnard, S., Boyes, S. J., & Elliott, M. (2018). Oil and gas infrastructure decommissioning in marine protected areas: System complexity, analysis and challenges. *Marine Pollution Bulletin, 135*, 739–758.

Bureau of Safety and Environmental Enforcement. (2018). *Rigs to Reefs*. https://www.bsee.gov/what-we-do/environmental-compliance/environmental-programs/rigs-to-reefs

Burt, J. A., Ben-Hamadou, R., Abdel-Moati, M. A., Fanning, L., Kaitibie, S., Al-Jamali, F., Range, P., Saeed, S., & Warren, C. S. (2017). Improving management of future coastal development in Qatar through ecosystem-based management approaches. *Ocean & Coastal Management, 148*, 171–181.

Byrd, R. C., Smith, J. B., & Spease, S. J. (2018). The challenges facing the industry in offshore facility decommissioning on the California Coast. Offshore Technology Conference, 2018. Onepetro.

Calder, K. E. (2015). *The United States, Japan, and The Gulf Region*. Washington, DC: Reischauer Center for East Asian Studies.

Carté, B. K. (1996). Biomedical potential of marine natural products: Marine organisms are yielding novel molecules for use in basic research and medical applications. *Bioscience, 46*, 271–286.

Cesar, H. (1996). *Economic analysis of Indonesian Coral Reefs*. http://web.worldbank.org/archive/website00672/WEB/PDF/ECONOM-3.PDF

Claisse, J. T., Pondella, D. J., Love, M., Zahn, L. A., Williams, C. M., & Bull, A. S. (2015). Impacts from partial removal of decommissioned oil and gas platforms on fish biomass and production on the remaining platform structure and surrounding shell mounds. *PLoS ONE, 10*, E0135812.

Coles, S. L. (2003). Coral species diversity and environmental factors in the Arabian Gulf and the Gulf of Oman: A comparison to the Indo-Pacific Region. *Atoll Research Bulletin*. No. 507. Smithsonian Institution. Washington, DC, USA.

Cooper, K., Burdon, D., Atkins, J. P., Weiss, L., Somerfield, P., Elliott, M., Turner, K., Ware, S., & Vivian, C. (2013). Can the benefits of physical seabed restoration justify the costs? An assessment of a disused aggregate extraction site off the Thames Estuary, UK. *Marine Pollution Bulletin, 75*, 33–45.

Craik, W. A., Kenchington, R. A., & Kelleher, G. (1990). Coral-reef management. *Ecosystems of the World, 25*, 453–467.

da Fonseca, A. P., Banet, C., Hall, K. B., Pereira, E. G., & Trischmann, H. (Eds.). (2020). *The regulation of decommissioning, abandonment and reuse initiatives in the oil and gas industry: From obligation to opportunities*. Kluwer Law International BV.

Daniels, P. L. (2010). Climate change, economics and Buddhism—Part 2: New views and practices for sustainable world economies. *Ecological Economics, 69*, 962–972.

Davies, T. W., Bennie, J., & Gaston, K. J. (2012). Street lighting changes the composition of invertebrate communities. *Biology Letters, 8*, 764–767.

De Jonge, V. N., Pinto, R., & Turner, R. K. (2012). Integrating ecological, economic and social aspects to generate useful management information under the EU directives' 'ecosystem approach.' *Ocean & Coastal Management, 68*, 169–188.

Din, M. A. E. (1990). The transport importance of the Arabian (Persian) Gulf. *Transport Reviews, 10*(2), 127–148.

Elliott, M., Burdon, D., Atkins, J., Borja, A., Cormier, R., De Jonge, V., & Turner, R. (2017). "And DPSIR Begat DAPSI (W) R (M)!"—A unifying framework for marine environmental management. *Marine Pollution Bulletin, 118*, 27–40.

Fanning, L. M., Al-Naimi, M. N., Range, P., Ali, A. S. M., Bouwmeester, J., Al-Jamali, F., Burt, J., & Ben-Hamadou, R. (2021). Applying the ecosystem services-EBM framework to sustainably manage Qatar's coral reefs and seagrass beds. *Ocean & Coastal Management, 205*, 105566.

Fattouh, B., & Darbouche, H. (2011). *The implications of the Arab uprisings for oil and gas markets*. Oxford Institute for Energy Studies. Retrieved from https://ncusar.org/blog/2013/03/basic-facts-about-oil-and-gas-in-thearab-world/

Food and Agriculture Organization. (2020). *The state of world fisheries and aquaculture 2008*. FAO Rome.

Fowler, A. M., Jørgensen, A. M., Coolen, J. W., Jones, D. O., Svendsen, J. C., Brabant, R., Rumes, B., & Degraer, S. (2020). The ecology of infrastructure decommissioning in the North Sea: What we need to know and how to achieve it. *ICES Journal of Marine Science, 77*(3), 1109–1126.

Fowler, A. M., Jørgensen, A. M., Svendsen, J. C., Macreadie, P. I., Jones, D. O., Boon, A. R., Booth, D. J., Brabant, R., Callahan, E., & Claisse, J. T. (2018). Environmental benefits of leaving offshore infrastructure in the Ocean. *Frontiers in Ecology and the Environment, 16*, 571–578.

Fowler, A., Macreadie, P., Jones, D., & Booth, D. (2014). A multi-criteria decision approach to decommissioning of offshore oil and gas infrastructure. *Ocean & Coastal Management, 87*, 20–29.

Gari, S. R., Newton, A., & Icely, J. D. (2015). A review of the application and evolution of the DPSIR framework with an emphasis on coastal social-ecological systems. *Ocean & Coastal Management, 103*, 63–77.

Gebremedhin, S., Getahun, A., Anteneh, W., Bruneel, S., & Goethals, P. (2018). A drivers-pressure-state-impact-responses framework to support the sustainability of fish and fisheries in Lake Tana, Ethiopia. *Sustainability, 10*, 2957.

Goble, B. J., Hill, T. R., & Phillips, M. R. (2017). An assessment of integrated coastal management governance and implementation using the DPSIR framework: Kwazulu-Natal, South Africa. *Coastal Management, 45*, 107–124.

Grorud-Colvert, K., Sullivan-Stack, J., Roberts, C., Constant, V., Horta e Costa, B., Pike, E. P., Kingston, N., Laffoley, D., Sala, E., & Claudet, J. (2021). The MPA guide: A framework to achieve global goals for the Ocean. *Science,* 373, EABF0861.

Hamlat, M., Djeffal, S., & Kadi, H. (2001). Assessment of radiation exposures from naturally occurring radioactive materials in the oil and gas industry. *Applied Radiation and Isotopes, 55*, 141–146.

Jagerroos, S., & Krause, P. R. (2016). Rigs-to-Reef; Impact or enhancement on marine biodiversity. *Journal of Ecosystem & Ecography, 6*(2), 1000187.

Jorgensen, D. (2009). An oasis in a watery desert? Discourses on an industrial ecosystem in the Gulf of Mexico Rigs-to-Reefs program. *History and Technology, 25*(4), 343–364.

Kaiser, M. J. (2019). *Decommissioning forecasting and operating cost estimation: Gulf of Mexico well trends*. Gulf Professional Publishing.

Kaur, M., Hewage, K., & Sadiq, R. (2020). Investigating the impacts of urban densification on buried water infrastructure through DPSIR framework. *Journal of Cleaner Production, 259*, 120897.

Kelble, C. R., Loomis, D. K., Lovelace, S., Nuttle, W. K., Ortner, P. B., Fletcher, P., Cook, G. S., Lorenz, J. J., and Boyer, J. N. (2013). The EBM-DPSER conceptual model: Integrating ecosystem services into the DPSIR framework. *Plos One, 8*, E70766.

Kennish, M. J. (1997). *Pollution impacts on marine biotic communities*. CRC Press.

Lewison, R. L., Rudd, M. A., Al-Hayek, W., Baldwin, C., Beger, M., Lieske, S. N., Jones, C., Satumanatpan, S., Junchompoo, C., & Hines, E. (2016). How the DPSIR framework can be used for structuring problems and facilitating empirical research in coastal systems. *Environmental Science & Policy, 56*, 110–119.

Li, Y., & Hu, Z. (2021). A review of multi-attributes decision-making models for offshore oil and gas facilities decommissioning. *Journal of Ocean Engineering and Science, 7*(1), 58–74.

Lim, J. (2021). *Industry challenges as basis for repurposing oil rigs and barges*. Springer.

Lin, T., Xue, X.-Z., & Lu, C.-Y. (2007). Analysis of coastal wetland changes using the "DPSIR" model: A case study in Xiamen, China. *Coastal Management, 35*, 289–303.

Macreadie, P. I., Fowler, A. M., & Booth, D. J. (2011). Rigs-to-Reefs: Will the deep sea benefit from artificial habitat? *Frontiers in Ecology and the Environment, 9*(8), 455–461.

Mangi, S. C., Roberts, C. M., & Rodwell, L. D. (2007). Reef fisheries management in Kenya: Preliminary approach using the Driver–Pressure–State–Impacts–Response (DPSIR) scheme of indicators. *Ocean & Coastal Management, 50*, 463–480.

Maxim, L., Spangenberg, J. H., & O'connor, M. (2009). An analysis of risks for biodiversity under the DPSIR framework. *Ecological Economics, 69*, 12–23.

Meyer-Gutbrod, E. L., Love, M. S., Schroeder, D. M., Claisse, J. T., Kui, L., & Miller, R. J. (2020). Forecasting the legacy of offshore oil and gas platforms on fish community structure and productivity. *Ecological Applications, 30*, E02185.

Miranda, M. N., Silva, A. M., & Pereira, M. F. R. (2020). Microplastics in the environment: A DPSIR analysis with focus on the responses. *Science of the Total Environment, 718*, 134968.

National Research Council (NRC). (2001). *Marine protected areas: Tools for sustaining ocean ecosystems*. Committee on the evaluation, design and monitoring of marine reserves and protected areas in the United States. Ocean Studies Board, National Research Council, Washington, DC, USA.

Nesterov, O., Temimi, M., Fonseca, R., Nelli, N. R., Addad, Y., Bosc, E., & Abida, R. (2021). Validation and statistical analysis of the group for high resolution sea surface temperature data in the Arabian Gulf. *Oceanologia, 63*, 497–515.

Newton, A., & Weichselgartner, J. (2014). Hotspots of coastal vulnerability: A DPSIR analysis to find societal pathways and responses. *Estuarine, Coastal and Shelf Science, 140*, 123–133.

Nugraha, R. B. A., Basuki, R., Oh, J. S., Cho, I. H., Naibaho, N., Secasari, Y., & Mbay, L. O. N. (2019, March). Rigs-to-Reef (R2R): A new initiative on re-utilization of abandoned offshore oil and gas platforms in Indonesia for marine and fisheries sectors. In *IOP Conference Series: Earth and Environmental Science* (Vol. 241, No. 1, p. 012014). IOP Publishing.

Organisation for Economic Co-operation and Development, P. (1993). OECD core set of indicators for environmental performance reviews. *OEDC Environment Monographs*.

Ojeda-Martínez, C., Casalduero, F. G., Bayle-Sempere, J. T., Cebrian, C. B., Valle, C., Sanchez-Lizaso, J. L., Forcada, A., Sanchez-Jerez, P., Martín-Sosa, P., & Falcón, J. M. (2009). A conceptual framework for the integral management of marine protected areas. *Ocean & Coastal Management, 52*, 89–101.

Organization of the Petroleum Exporting Countries, O. (2018). Share of world crude oil reserves. OPEC Annual Statistical Bulletin. Vienna, Austria.

Ossai, I. C., Auwalu, A. A. H., & Shahul, H. F. (2020). Remediation of soil and water contaminated with petroleum hydrocarbon: A review. *Environmental Technology & Innovation, 17*, 100526.

Ounanian, K., Van Tatenhove, J. P., & Ramírez-Monsalve, P. (2020). Midnight at the oasis: Does restoration change the Rigs-to-Reefs debate in The North Sea? *Journal of Environmental Policy & Planning, 22*(2), 211–225.

Peterson, C., & Lubchenco, J. (1997). On the value of marine ecosystems to society. In *Nature's services. Societal dependence on natural ecosystems* (pp. 177–194). Island Press.

Potschin, M. (2009). Land use and the state of the natural environment. *Land Use Policy, 26*, S170–S177.

Range, P., Riera, R., Omerspahic, M., Bouwmeester, J., Bach, S. S., Møller, P. R., & Ben-Hamadou, R. (2018). Fouling assemblages on offshore oil platforms: Diverse vertical Reefs in the Arabian Gulf. Qatar Foundation Annual Research Conference Proceedings Volume 2018 Issue 1. Hamad Bin Khalifa University Press (HBKU Press), EEPP894.

Rakib, F., Al-Ansari, E. M. A. S., Husrevoglu, Y. S., Yigiterhan, O., Al-Maslamani, I., Aboobacker, V. M., & Vethamony, P. (2021). Observed variability in physical and biogeochemical parameters in the central Arabian Gulf. *Oceanologia, 63*(2), 227–237.

Reed, D. C., Schroeter, S. C., & Raimondi, P. T. (2004). Spore supply and habitat availability as sources of recruitment limitation in the giant kelp *Macrocystis pyrifera* (Phaeophyceae)1. *Journal of Phycology, 40*, 275–284.

Ronconi, R. A., Allard, K. A., & Taylor, P. D. (2015). Bird interactions with offshore oil and gas platforms: Review of impacts and monitoring techniques. *Journal of Environmental Management, 147*, 34–45.

Sale, P. F., Feary, D. A., Burt, J. A., Bauman, A. G., Cavalcante, G. H., Drouillard, K. G., Kjerfve, B., Marquis, E., Trick, C., Usseglio, P., & Van Lavieren, H. (2011). The growing need for sustainable ecological management of marine communities of the Persian Gulf. *Ambio, 40*(1), 4–17.

Seibold, E. (1973). *Biogenic sedimentation of the Persian Gulf*. Springer.

Sheppard, C., Al-Husiani, M., Al-Jamali, F., Al-Yamani, F., Baldwin, R., Bishop, J., Benzoni, F., Dutrieux, E., Dulvy, N. K., & Durvasula, S. R. V. (2010). The Gulf: A young sea in decline. *Marine Pollution Bulletin, 60*, 13–38.

Smith, C. J., Papadopoulou, K.-N., Barnard, S., Mazik, K., Elliott, M., Patrício, J., Solaun, O., Little, S., Bhatia, N., & Borja, A. (2016). Managing the marine environment, conceptual models and assessment considerations for the European marine strategy framework directive. *Frontiers in Marine Science, 3*, 144.

Smith, S. V. (1978). Coral-reef area and the contributions of reefs to processes and resources of the world's oceans. *Nature, 273*, 225–226.

Smyth, K., & Elliott, M. (2014). Develop risk assessments leading to best practice: Resource.

Smyth, K., Christie, N., Burdon, D., Atkins, J. P., Barnes, R., & Elliott, M. (2015). Renewables-to-Reefs? Decommissioning options for the offshore wind power industry. *Marine Pollution Bulletin, 90*, 247–258.

Sorokin, Y. I. (1990). Aspects of trophic relations, productivity and energy balance in coral-reef ecosystems. *Ecosystems of the World, 25*, 401–418.

Spurgeon, J. P. G. (1992). The economic valuation of coral reefs. *Marine Pollution Bulletin, 24*, 529–536.

Stachowitsch, M., Kikinger, R., Herler, J., Zolda, P., & Geutebrück, E. (2002). Offshore oil platforms and fouling communities in the Southern Arabian Gulf (Abu Dhabi). *Marine Pollution Bulletin, 44*, 853–860.

Swift, S. A., & Bower, A. S. (2003). Formation and circulation of dense water in the Persian/Arabian Gulf. *Journal of Geophysical Research: Oceans, 108*, 4-1–4-21.

Tarr, D. (2014). *From exploration drilling to decommissioning: Where does responsibility lie for regulating and monitoring offshore oil platforms in Liberian waters?* (World Maritime University, Dissertations 452).

Tisen, O.B., Bali, J., Mosier, A., Foley, A., & Brost, B. (2002). Current status of marine turtle conservation programmes in Sarawak, Malaysia. In *Proceeding of the Twentieth Annual Symposium on Sea Turtle Biology and Conservation. Orlando, Florida, USA* (pp. 12–14).

Tornero, V., & Hanke, G. (2016). Chemical contaminants entering the marine environment from sea-based sources: A review with a focus on European seas. *Marine Pollution Bulletin, 112*, 17–38.

Torquato, F., Jensen, H. M., Range, P., Bach, S. S., Ben-Hamadou, R., Sigsgaard, E. E., Thomsen, P. F., Møller, P. R., & Riera, R. (2017). Vertical zonation and functional diversity of fish assemblages revealed by ROV videos at oil platforms in The Gulf. *Journal of Fish Biology, 91*, 947–967. https://doi.org/10.1111/jfb.13394

Torquato, F., Range, P., Ben-Hamadou, R., Sigsgaard, E. E., Thomsen, P. F., Riera, R., Berumen, M. L., Burt, J. A., Feary, D. A., Marshell, A., D'agostino, D., Dibattista, J. D., & Møller, P. R. (2019). Consequences of marine barriers for genetic diversity of the coral-specialist yellowbar angelfish from the Northwestern Indian Ocean. *Ecology and Evolution, 9*, 11215–11226. https://doi.org/10.1002/ece3.5622

Torquato, F., Omerspahic, M. H., Range, P., Bach, S. S., Riera, R., & Ben-Hamadou, R. (2021). Epibenthic communities from offshore platforms in the Arabian Gulf are structured by platform age and depth. *Marine Pollution Bulletin, 173*, 112935.

Tsounis, G., Rossi, S., Grigg, R., Santangelo, G., Bramanti, L., & Gili, J. M. (2010). The exploitation and conservation of precious corals. *Oceanography and Marine Biology: An Annual Review, 48*, 161–212.

US Energy Information Administration. (2015). *Qatar: International energy data and analysis*. Retrieved from https://www.eia.gov/beta/international/analysis_includes/Countries_long/Qatar/qatar.pdf

Van Der Stap, T., Coolen, J. W., & Lindeboom, H. J. (2016). Marine fouling assemblages on offshore gas platforms in the Southern North Sea: Effects of depth and distance from shore on biodiversity. *Plos One, 11*, E0146324.

Van Elden, S., Meeuwig, J. J., Hobbs, R. J., & Hemmi, J. M. (2019). Offshore oil and gas platforms as novel ecosystems: A global perspective. *Frontiers in Marine Science, 6*, 548.

Van Lavieren, H., Burt, J., Feary, D., Cavalcante, G., Marquis, E., Benedetti, L., Trick, C., Kjerfve, B., Sale, P. (2011). *Managing the growing impacts of development on fragile coastal and marine ecosystems: Lessons from the Gulf*. A policy report, UNU-INWEH.

Vaughan, G. O., & Burt, J. A. (2016). The changing dynamics of coral reef science in Arabia. *Marine Pollution Bulletin, 105*, 441–458.

Verbeek, S. (2013). Worldwide Rigs-to-Reefs experiences: Considerations and motivation for a living North Sea. Would the North Sea Benefit from Reusing Offshore Materials for Artificial Reefs. IMSA, Amsterdam, Netherlands.

Weber, P. (1993). Reviving coral reefs. *State of the World*, 42–42.

Weilgart, L. S. W. S. (2007). The impacts of anthropogenic ocean noise on cetaceans and implications for management. *Canadian Journal of Zoology, 85*, 1091–1116.

Zhang, B., Matchinski, E. J., Chen, B., Ye, X., Jing, L., & Lee, K. (2019). Marine oil spills—Oil pollution, sources and effects. In *World seas: An environmental evaluation*. Elsevier.

Radhouan Ben-Hamadou (Ph.D. and Env. Eng.) Associate Professor of Marine Sciences and Lead of the Marine Conservation and Restoration Research Group at Qatar University, is an ecohydrologist and oceanographer with an Aquaculture Engineer and an MSc and a PhD degrees from the Sorbonne Universities (UPMC-Paris 6) in Biological Oceanography. Prior to joining Qatar University, from 2006 to 2013 he was Research Associate Professor at the Centre of Marine Sciences and Assistant Professor at the University of Algarve. He also acted as Deputy Director of the UNESCO International Centre for Coastal Ecohydrology (UNESCO-ICCE) from 2009 to 2013. He participated in the launching of the EU-ERASMUS-MUNDUS Master on Ecohydrology. He served also as consultant to several companies and governmental agencies in the GCC and Europe. At Qatar University, he was successful in securing since 2013 over 6.8 M$ funding for projects mainly dealing with marine resources sustainability, biodiversity conservation, coral reefs ecology and restoration and impacts of man-made structures on the marine environment. His major R&D project aimed to understand the ecological functioning of offshore oil and gas platforms, their connectivity with natural reefs, and their contribution to support local productivity, fisheries, and regional biodiversity. He was appointed Head of the Department of Biological and Environmental Sciences (2017–2019) and to the task force to develop Qatar University Research Strategy in 2014, 2016, and 2019. He acted up to 2016 as President of the Arabian Gulf Branch of the SETAC society and is co-founder of the International Society of Ecohydrology (IS-ECOHYD). He was awarded in 2016 the QU-CAS Excellence Research Award. He co-authored 105+ peer-reviewed publications in international journals and books and has supervised 17 M.Sc. and 6 Ph.D. students.

Ahmad Mujthaba Dheen Mohamed AMIEnvSc, AIEMA, is an Environmental Professional and an Associate Member at the Institute of Environmental Management and Assessment (IEMA), UK, with experience in various environmental sectors. Upon attaining his undergraduate degree in Environmental and Marine Sciences from Qatar University where he conducted extensive research on Qatar's coral Communities and got awarded 1st place for the departmental poster presentation titled "Bleaching dynamics of Coral Assemblages of Inshore sights in Qatari Waters," he started his career with the oil and gas industry where he worked for 2 years in QAPCO as a sustainability and environment inspector for the Sustainability and Environment Department. He then joined Qatar University and worked as a Research Assistant on corals and fisheries with the Marine Conservation and Restoration Group (MARESCO). He has also collaborated with Qatar Museums and the Smithsonian Institution's National Museum of Natural History in Washington DC on paleontological projects related to fossilized dugongs. Currently, he is pursuing his MSc in Environmental Sciences and works as a Research Assistant at the Centre for Sustainable Development where he conducts research on Wastewater Treatment using Algal Technologies.

Sarra N. Dimassi is currently part of the research team at the Sustainable Development Center. She is pursuing her M.Sc. in Environmental Science at Qatar University. She received her B.Sc. in Chemistry also from Qatar University, where her thesis focused on the usage of bio-waste as an efficient adsorbent for dyes from wastewater. She was an active undergraduate student, where she worked with the Anti-Doping Lab, Qatar, in the preparation and testing of samples during the Asian Games in 2018. Currently, her research interests focus on understanding plastic waste degradation-fragmentation under marine conditions. During her graduate career, she has extensively worked and possesses expertise in several research areas including the synthesis and characterization of lithium-ion batteries, as well as the isolation, differentiation, and biodiversity of ureolytic soil bacteria.

Mariam M. Razavi is a masters student majored in Environmental Sciences at Qatar University. She received a bachelor's degree in Environmental Sciences from Qatar University. Her current work field placement is with the educational system where she worked as a science teacher who teaches 7th grade students. Moreover, she is passionately interested in doing scientific research. Her research work focuses specifically on the remediation of wastewater and contaminated soil.

Sara M. Alshuiael graduated from Qatar University recently in 2022 with the degree of Master of Science in Environmental science with research distinguished honors. Her research work was concentrated on environmental chemistry, specifically focusing on adsorbents and analytical chemistry. She is also currently working as a research assistant in the environmental science center (esc) under the chemistry department, focusing on monitoring the marine environment contaminants.

Muhammad Omair Sulaiman graduated from Qatar University with a bachelor's in Environmental Sciences and is currently active as a research assistant and student for master's in Environmental Sciences within Qatar University. With multiple research and field knowledge of flora within Qatar, the current focal point is irrigation utilizing impure water sources and aims for the contribution of a sustainable methodology of water management.

Chapter 11
The Domestic Water Sector in Qatar

Jenny Lawler, Annamaria Mazzoni, and Sa'd Shannak

Abstract Qatar is considered one of the most highly water stressed countries in the world, with little in the way of natural water resources. It is almost entirely reliant on the desalination of seawater to provide for municipal and industrial needs, while the growing agricultural sector has historically been dependent upon the desalination of brackish groundwater. This chapter outlines the sustainability initiatives, national strategies, policy instruments, and drivers for efficiency improvements in the water sector in Qatar. The Qatar National Vision 2030 has a particular focus on reducing consumption, improving conservation, and on the circular water economy, highlighting the value of research, development, and innovation (RDI) in implementing the Sustainable Development Goals (SDGs). Significant investment has taken place toward developing a sustainable water industry, where a transition toward renewable energy resources can support desalination, wastewater treatment, and reuse policies for a sustainable Qatar.

Keywords Water resources · Water security · Groundwater · National strategies · Sustainable development · Water sector

11.1 Background Information

Qatar is surrounded by the Arabian Gulf, while it shares its only land border of about 60 km with the Kingdom of Saudi Arabia and it is classified as a hyper-arid environment (Aridity Index—A.I. <0.05) (UNEP, 1992). The World Bank classified Qatar as one of the wealthiest countries in terms of GDP per capita, ranking among the top ten wealthy countries in the world. This prosperity has been driven by revenues

J. Lawler (✉) · A. Mazzoni · S. Shannak · S. Shannak
Qatar Environment and Energy Institute, Hamad Bin Khalifa University, Doha, Qatar
e-mail: jlawler@hbku.edu.qa

A. Mazzoni
e-mail: annamaria.mazzoni@unito.it

S. Shannak
e-mail: sshannak@hbku.edu.qa

L. Cochrane and R. Al-Hababi (eds.), *Sustainable Qatar*, Gulf Studies 9,
https://doi.org/10.1007/978-981-19-7398-7_11

from fossil fuel exports. The country's economy is highly dependent on oil and gas production which accounts for more than 50% of GDP, 85% of export earnings, and 70% of government revenues (MOFA, 2021). The land surface is dominated by Eocene limestone and gypsum rocks with very little soil. The available soil is characterized by high salinity, mainly in coastal areas, and areas with heavy irrigation application using saline groundwater. Qatar largely consists of flat rocky surfaces, with some hills reaching an altitude of 100 m above sea level. Nearly, all of the country's land is sandy desert covered with scrub plants and loose gravel. Shifting sand dunes, with an average height of about 40 m, are found in the southern part of the country and on the northeastern coast near Ras Laffan. The climate in the peninsula is an arid climate, characterized by high temperature in the summer (as high as 50 °C), with very mild winters and the common wind is moist and blows from the north-west, more frequently in the summer months. The temperature averages around 17 °C in both spring and autumn months and precipitation in the winter is scarce with less than 100 mm annually (World Bank, 2022). Most water supply resources are affected by climatic conditions and are mainly determined by precipitation. This phenomenon is evident in arid regions where precipitation is less than evaporation, and the availability of freshwater from natural resources is limited. The Countries of the Gulf Cooperation Council (GCC) are no exception, as they are located in an arid zone, they are endowed with few water resources, and they have experienced tremendous economic and population growth over the past few years. This growth has led several countries to search for alternative water resources. For example, seawater is one alternative that is not affected by climatic conditions, and it has a substantial potential to supply freshwater in dry regions. Removing salts from seawater is a process known as "desalination." Despite the incredible development in desalination technologies, Qatar's natural freshwater resources are highly depleted, and the country's water security is currently one of the key pillars of Qatar's agenda for sustainable development. Reduction of water subsidies, diversification of the water resources portfolio, and R&D in the water sector are crucial steps for a more sustainable water governance. In this chapter, the challenges and opportunities revolving around Qatar's water management will be reviewed. In Sect. 11.2, the main challenges of the water sector are outlined; in Sect. 11.6.2 the economic development and the climate change objectives linked with the water resources are highlighted. In Sects. 11.3 and 11.4, a brief outline of water resources status and water demand is presented. Sections 11.5 and 11.6 highlight the current country's strategy and the planned efforts toward sustainability in the water sector, respectively, while Sect. 11.7 also includes the policy implications of the future plans and strategies toward sustainability.

11.2 Water Sector Challenges

The lack of surface renewable freshwater sources, with the only natural resources being the fossil aquifers, makes Qatar severely constrained by water scarcity. Alongside erratic precipitation, high temperature, and evapotranspiration the water resources' portfolio is limited to three main sources: brackish groundwater, seawater desalination, and treated sewage effluent. As such, the water sector faces several challenges, which interest different actors and distinct levels of the water governance system. Given the resources constraints, the increasing demand for water presents a matter of contention, as does limited groundwater resources, while desalination activities remain energy intensive and directly affect other sustainability aspects of the water sector. These three main issues are briefly outlined in this sub-section.

Water demand increase: Fast industrial and economic development as well as population growth creates competition among water users and sectors. The water demand has risen steadily, reaching 1.9 million m^3 in 2019 and this number is expected to grow up to 2.2 million m^3 in 2022 (Kahramaa, 2022). In addition, the country has one of the highest levels of domestic water consumption globally, of about 430 L of water per person per day (Ismail, 2015), a figure which is a mean estimate from the municipal desalinated water production (approx. 600 L/capita/day) and the wastewater treated (approx. 250 L/capita/day). The government highly subsidizes both water and electricity for its citizens and residents, and therefore there are not many economic incentives in place to decrease such demand while from a behavioral perspective, more initiatives should be promoted to reduce consumption.

Groundwater resources: This water source is heavily relied upon to meet the needs of the agricultural sector. The fossil aquifers are mainly recharged through limited rainfall and underground water flow from Saudi Arabia. The rate of exploitation exceeds by three to five times the rate of replenishment (Baalousha, 2016; Planning & Statistics Authority, 2018a), making the use of these resources unsustainable. Aquifer over-abstraction poses risks not only in terms of water availability, i.e., quantity, but it also allows for seawater and saline intrusion, further deteriorating quality. At the same time, extensive use of fertilizers and pesticides can further pollute the groundwater through percolation.

Seawater Desalination: As the main source for municipal water supply, desalination is critical to the water system. The first plant was established in 1953, with a capacity of 682 m^3/day (Kahramaa, 2022). Currently, there is a total eleven desalination plants plus two under construction, of which one will be finalized in 2022 and the other in 2027 (for more details see Table 11.1, Sect. 11.3). Still, real and apparent water losses in the network cause a volume decrease. In order to overcome this issue, Kahramaa (Qatar General Electricity and Water Company, responsible for distribution and billing for both water and electricity in the country) has invested, since the first National Development Strategy (NDS-1) 2011–2016, in decreasing the rate of water leakage and loss in the desalinated water network. Currently, the real loss is ~5.28%. Along with the dependence on desalination, water storage is also a major concern. As of 2022, Qatar has only two days of strategic water stock in the network in case of an accident. To overcome this challenge, the country is developing a mega reservoirs project that will increase the water storage network to seven days. The first phase, which started in 2015, consists of building 24 concrete reservoirs by 2026, with a total storage volume of about 10 million m^3. The second phase, targeted to 2036, will add further storage, reaching 40 concrete reservoirs for ~17 million m^3 of water storage (Kahramaa, 2020). Another major challenge associated with the desalination processes is the production and consequently the management and disposal of the brine. Brine is the main by-product of all the desalination processes and its amount depends on the water source that needs to be desalinated. Usually, in seawater desalination plants the brine constitutes approximately 50–60% of the quantity of seawater that is pumped from the sea, while in brackish desalination plants the brine constitutes between 15 and 30% of the brackish water pumped to the desalination plant. Regardless of the source, brine has a negative environmental impact and must be disposed of appropriately (Ogunbiyi et al., 2021). Qatar, together with Saudi Arabia, UAE, and Kuwait account for 55% of the total global share of brine production, with Qatar having a 5.8% share (Jones et al., 2019). Strategies for improving brine management and disposal to reduce the negative externalities as well as its economic costs need be considered and implemented to decrease the impacts of desalination processes. Lastly, desalination plants mainly rely on fossil fuels for co-generation of water using the heat from gas turbines. As such, the water sector still has a considerable carbon emission footprint and in order to contribute to the national target of emission reduction of 25% by 2030, shall look toward the use of renewable energies for powering the desalination plants.

11.3 Water Resources Status

Qatar's only natural freshwater resources are rainfall and groundwater. The country does not have surface water and the solely natural freshwater available is coming from the fossil transboundary aquifers. Qatar is currently relying heavily on desalination for meeting the needs of the municipal and industrial sectors, while the groundwater resources are employed in the agricultural sector for crop irrigation and cooling; lastly, the treated sewage effluent (TSE) is used mainly for landscaping purposes and fodder irrigation. The cheap energy prices (and lack of natural freshwater resources) make the desalinated water the preferred and main water resource for the country. At the same time, this comes at environmental costs in terms of marine pollution and carbon emissions. On the other hand, water portfolio diversification should become a pillar of the new water sector strategy shifting away from groundwater extraction and toward higher value uses of TSE and treated produced water (TPW) from the oil and gas industry, as well as the adoption of the latest most energy efficient desalination technologies, and coupling desalination with renewable energies.

Qatar's groundwater mainly feeds the agricultural sector (Baalousha & Ouda, 2017). Several formations overlap constituting the different aquifers of Qatar. The Dam and Dammam Formations constitute the top layers, with shallow aquifers.

Table 11.1 Municipal desalination plants in Qatar (*Gulf Times*, 2021)

No	Plant	Location	Commissioning year	Ownership	Technology used	Capacity (MIGD)
1	Ras Abu Fantas	Al Wakrah	1977–98	QEWC	MSF	55
2	Ras Abu Fantas B2	Al Wakrah	2008	QEWC	MSF	29.2
3	Dukhan	Dukhan	1997	QEWC	MED	2
4	Ras Abu Fantas A1	Al Wakrah	2010	QEWC	MSF	45
5	Ras Abu Fantas A2	Al Wakrah	2015	QEWC	MSF	36
6	Ras Abu Fantas A3	Al Wakrah	2016	QEWC	RO	36
7	Ras Laffan A	Ras Laffan	2003–04	Ras Laffan Power	MSF	40
8	Ras Laffan B	Ras Laffan	2006–08	Qatar Power	MSF	60
9	Ras Girtas	Ras Laffan	2010–11	Ras Girtas Power	MED	63
10	Umm Al Houl	Umm Al Houl	2016–17	Umm Al Houl Power	MSF RO	76 60
11	Pearl of Qatar	Doha	2008	United Development Company	RO	8.7
In Construction and Pipeline Projects						
1	Umm Al Houl	Umm Al Houl	2021–22	Umm Al Houl Power	RO	64
2	Ras Abu Fantas A4*	Al Wakrah	2024–27	QEWC	Unknown	100

Below, the Rus Formation developed during the Lower Eocene, while the Umm er Radhuma Formation, belonging to the Paleocene represent the deeper aquifer with the most saline content. The main basins are the northern, central, and southern basins, which present salinity levels from low to high while progressing toward the South. The Northern Basin has a salinity of 500 to 3000 ppm and up to 10,000 near the coastal areas (drinking and crop irrigation water should be <200–400 ppm). It covers ~19% of Qatar at 10–40 m of water depth. The Southern Basin covers about half of the country and is also moderately saline.

The groundwater safe yield (amounts of groundwater withdrawn from a basin over a period of time without exceeding the long-term recharge of the basin) amounts to approximately 56 million m^3 per year. However, the current groundwater abstraction reached just over 250 million m^3 per year, leading to depletion of aquifers, low groundwater levels, and increased salinity. The total increase in stocks (total recharge from rainfall, inflow from Saudi Arabia, artificial recharge, and irrigation returns) increased from 108 million m^3 in 2008 to 258 million m^3 in 2019. However, the water balance decrease remained relatively unchanged between 267 and 269 million m^3 per year for the same period. Consequently, there has been an annual water deficit ranging from 166 million m^3 and 11 million m^3 during the period 2008–2019 (Planning & Statistics Authority, 2021a). The artificial recharge and irrigation returns represent the largest source of additions to water reserve. The decrease in water reserve is attributed to the water withdrawal for agricultural purposes.

Until the 1970s, most of the desalination plants installed were based on thermal desalination processes. The main technologies available in the desalination sector are multi-stage flash (MSF) distillation and multi-effect distillation (MED), and membrane-based processes, such as reverse osmosis (RO) (Table 11.1). While at the beginning of the desalination development MSF was preferred due to the availability of cheap fuel and the high salinity of the feed water in the Gulf, recently RO is becoming increasingly employed due to its lower costs of production, lower energy consumption, lower temperature brine, and higher recovery rates (i.e., less rejected brine output and therefore better environmental standards). Of key benefit is the ability to utilize PV electricity directly to operate the plant, no longer relying on co-generation electricity and water plants. In 2020, the desalinated water produced in Qatar among the Independent Power and Water Producers, namely QEWC, Ras Laffan, Umm Al Houl and The Pearl amounted to 691.5 million m^3, which translates into ~50 to 63 million m^3 of water per month (Kahramaa, 2021). The current landscape of municipal desalination in Qatar is outlined in Table 11.1.

While RO is indeed lower energy demand than traditional thermal desalination plants, research efforts are ongoing to improve efficiencies in the thermal desalination sector. This is particularly important for Qatar; as the seawater characteristics of high salinity (up to 57,000 in the west of Qatar) and high boron content pose a particular challenge to seawater RO. Qatar Environment and Energy Research Institute (QEERI) has implemented a multi-effect distillation pilot plant at Dukhan on the west coast which is currently producing 25,000 L/day of freshwater, and exhibiting up to a 22% reduction in unit water cost (with up to 40% improvement in energy efficiency) in comparison with that of a conventional Thermal Vapor Compressor along with Multi-Effect Distillation (MED-TVC) desalination plant. This is based on improvements in feed and pumping energy efficiencies as well as implementation of an absorption vapor compressor, and also a reduction in the plant footprint and capex by virtue of the innovative design features. The innovation roadmap includes the transition to power the plant using solar PV (Aly et al., 2021, 2022).

TSE is the only water source that is in surplus in the country (and only in the winter, currently not meeting the demand in summertime). This water is primarily used for fodder irrigation but can play a more significant role, especially for industrial use. Ashghal, the Public Works Authority, is responsible for the design, construction, operation, and management of all major infrastructural projects in Qatar, including the wastewater and sewage drainage and treatment. The use of TSE for fodder started in 2004 at 5% and increased to 16% in 2013 (Planning & Statistics Authority, 2018a).

In 2015, the TSE volume was ~194 million m^3, which corresponds to 98.2% of total wastewater. Of these, ~66 million m^3 were used in the agricultural sector (fodder), ~31 million m^3 for landscape irrigation, and ~57 million m^3 in deep injection of non-freshwater aquifers. The remaining 40 million m^3 were instead discharged into lagoons (General Secretariat for Development Planning, 2011; Planning & Statistics Authority, 2018b). Already in 2017, TSE reuse rose to 229 million m^3 and only about 15% was discharged into lagoons unused, while in 2019, it further declined to below 12%, while usage in green space and agriculture increased by about 1% in each (Planning & Statistics Authority, 2019). This indicates that there is a possibility to make a better and more sustainable use of these resources. For example, the change in regulation to manage TSE has attracted additional industries beside agriculture to utilize this resource. District cooling industry utilizes TSE to reduce capital and operational expenditures, while decreasing the demand on fresh water.

Similarly, produced water (PW) is the major wastewater stream generated in the oil and gas industries and while one of the challenges of this industry is to reduce the PW water volumes injected in the disposal wells to achieve long-term reservoir sustainability, due to the needs of diversification of water sources within Qatar, alternative uses for this resource could potentially be helpful for the future of the water sector, with the implementation of appropriate treatment strategies for reuse.

11.4 Water Demand

Rising population and economic growth are commonly associated with increasing water demands. The chief driving factors of water demand and its availability are population growth, climate conditions, water tariff, economic status (i.e., GDP), creeping urbanization, and water regulations such as metering. The Intergovernmental Panel on Climate Change (IPCC) has projected that freshwater resources are susceptible and could be impacted significantly by climate change, with broad-ranging implications on the different sectors. The newly published report (IPCC, 2022) further suggests, with medium to high confidence, that the projected water cycle changes will affect agriculture, energy production, and urban water uses. In particular, worldwide depletion of non-renewable groundwater storage will increase due to higher water demand.

One approach that has been applied to rationalize water demand and secure supply is demand-side management. This approach has evolved as an important component of a total water cycle management approach and complements the supply management side. In other words, any actions that result in reducing the amount of water used or allowing water to be used more efficiently are part of the water demand management approach. It is worth noting that in the past, the focus was mainly on the supply side, where the concern was to build infrastructure to collect, purify, transport, and recycle water. The recognition that water demand should be tackled from both sides of supply and demand is essential to better understand the scope of the challenge and better inform the policymaking process. The significance of water demand management is indicated by the IPCC as a "no-regrets option" to manage growing demand in the face of climate change effects (Bates et al., 2008). Demand-side management is differentiated from supply-side management in that it concentrates on the amounts and forms of water used by consumers (Bates et al., 2008; Brooks, 2006). This way, demand management includes as much attention on water use behavior as it does on technology or infrastructure (Baumann et al., 1984; Brooks, 2006).

In Qatar, as it is the case for the countries of the Gulf Cooperation Council, sustaining water resources has been a concern due to the lack of freshwater resources and high dependency on energy-intensive alternatives such as desalination. The country developed a 30-year master plan in 2009, which accounts for mega-investment in desalination plants, wastewater treatment plants, and other water infrastructure projects, equivalent to more than 5 billion USD between 2010 and 2015. On the other hand, and from a demand side, the agricultural industry and the government sectors are regarded as the largest water users (Kamal et al., 2021). In 2015, the water used in agriculture amounted to 230 million m^3 abstracted from groundwater, while 66.29 million m^3 from treated wastewater. The industrial sector, which accounts for mining and quarrying, manufacturing, electricity and water and construction, relies on three main sources of freshwater: the water supplied by Kahramaa (desalinated seawater), desalinated water from groundwater wells (usually for the agricultural sector), and seawater desalinated by the industrial establishment itself (these latter type of data are not available). In 2019, the total water use for this sector, excluding the desalinated industrial water, amounted to 34.18 million m^3. In the same year, the water use in the commercial/municipal sector, which is solely supplied by Kahramaa, amounted to 85 million m^3. The water use in the government sector that includes both waters supplied by Kahramaa and TSE used for the irrigation of green spaces, was ~156.65 million m^3 in 2019. Lastly, the water use in the domestic sector in 2019 amounted to 291.10 million m^3 (Planning & Statistics Authority, 2021a).

Considerable effort has been devoted to meeting water demand, and demand management is considered a crucial element in the government's sustainable development vision, which focuses on managing demand for water by maximizing its efficiency, controlling leaks, and optimizing its use. The current water strategy along with the country's vision has placed clear structures to focus on developing, exploring, and employing environmentally sustainable techno-economic viable, and socially acceptable options to meet water demand.

11.5 Qatar's Water Strategies

Both NDS-1 and NDS-2 along with the QNV 2030 plans have strong focus on water security and water sector sustainability. Through these overarching frameworks, over the past decade, Qatar has highly invested in developing sustainable strategies for the water sector, improving water network and infrastructure, enhancing sustainable water use and launching water awareness and conservation campaigns. The QNV 2030, launched in 2008, defines the long-term development outcomes for Qatar along four interrelated pillars: human, social, economic and environmental development. The document also outlines a framework within which different plans and strategies can be prepared and implemented. The main goal is to achieve balance between development and natural resources conservation for future generations, i.e., amid the environmental pillar and the other three dimensions. The need for sound water management to sustain the country's growth despite water resources scarcity is recognized in the vision, and the nation's Permanent Water Resources Committee (PWRC) mandated the development of the Qatar Water Security Policy and Qatar Water Strategy and Implementation Plan to provide the needed strategic vision and action plan toward a water-secure future.

Looking at water as a human right, and therefore the rights to water and sanitation, the water sector policy is in compliance with QNV 2030 and is able to ensure continuity of water supply and pressure, highest water standards, water storage, and water security for future generations. NDS-1 covers the period 2011–2016 and for what concerns the water sector it first highlights the needed reforms for the sector, which span from leakages and losses in the desalination network, aquifers over-abstraction, and large amounts of wastewater that are uncollected, untreated, and unused. As such, it identifies a range of initiatives to tackle these challenges from technological fixes to reduce network water losses to introducing water savings devices and user charges (initially targeted for non-citizens) that reflect more accurately the full economic cost, or at least the operation and management of the water resources. Changing water consumption patterns in the agricultural sector is also part of the vision, coupled with the adoption of the irrigation technology and crop mixing. Lastly, new regulatory approaches shall be implemented to push for the needed reforms in the water sector. Along this line, a comprehensive National Water Act is elaborated to merge the fragmented system of laws and regulations. NDS-2, outlined for the period 2018–2022 builds on the achievements and the outcomes of NDS-1 to support the implementation of the QNV 2030. In particular, it focuses on developing treated wastewater networks and stimulating its use. There are three specific targets for TSE, which include:

1. Provide infrastructure to use 70% of the TSE produced in different projects by 2022.
2. Establish an integrated management of water and of accompanying contaminants in industrial zones by 2022.
3. Reduce loss of drinking water rate (actual + administrative) to 8% by 2022.

The Third National Development Strategy (NDS-3) 2023–2027 is currently under preparation and will aim at ensuring that such targets are achieved and building also on the efforts of the Voluntary National Review 2021 and the Qatar 2020 Census.

11.5.1 Specific Initiatives, Best Practices

Kahramaa launched the National Campaign for the Conservation and Efficient Use of Water and Electricity *Tarsheed* in 2012 with the aim of educating Qatar's community about energy and water conservation, sustainability, and environmental awareness to induce behavioral changes and practices oriented toward environmental-friendly habits and lifestyle. The Tarsheed campaign supports the efforts of preserving and sustaining natural resources for future generations through setting a target for decreasing the per capita consumption of electricity (−10%) and water (−6%) by 2022. The overall goal of Tarsheed is to place Qatar as a regional leader in the reduction of electricity and water by involving citizens, residents, and businesses in reduction practices. The program has several initiatives and mechanisms in place to reach its goals, ranging from awareness and education campaigns, financial incentives, as well as regulations and legal penalties. One example is the Conservation Law No. 20/2015, which halts the use of potable water for car wash or yard cleaning, leaving outdoor lights turned on between 7.00am and 4.30 pm, and neglecting to fix broken or damaged taps and pipes. Tarsheed also aims at providing technical support to the country's renewable energy efforts. In 2019, the Tarsheed Program contributed to water savings of 33.76 MCM (Kahramaa, 2021). In 2017 Kahramaa also established the *Kahramaa Awareness Park* (KAP), which is an awareness museum to educate the public about water and electricity consumption, use, and production complemented with the showcase of the latest technologies in this domain. KAP museum is a member of the Water Museums Global Network, which is a global initiative to encourage people to use water responsibly and with respect to the Sustainable Development Goals (SDGs) framework.

The *District Cooling (DC)* system is an urban utility service that centralizes the production and distribution of chilled water for air conditioning in different neighborhoods in Doha. This system is developed and often privileged over the traditional air conditioning system as it is more efficient, environment-friendly and helps reducing both water and electricity consumption thus supporting the efforts of Kahramaa in the water resources conservation and sustainability. By the target date, DC is expected to reach its full exploitation by providing the country with an array of benefits spanning from economic, to social and environmental. DC is aiming to reach 100% usage of non-potable water in DC plants, saving 520 MCM of potable water, similarly this will allow 10% savings in potable water capital demand. This effort dates back to 2013, when a ban on the use of potable water in district cooling, which is highly water-intensive, was approved. Ashghal, who is responsible for TSE operation and management has been expanding its network to increase TSE recycling and use for DC. In 2019, DC reduced potable water usage for cooling by 65% with a target reduction of 85% in 2020. In parallel, TSE utilization increased with a percentage of 67% in 2019 compared to 60% in 2018. Further help from DC to Kahramaa entails the implementation of new absorption chilling systems to increase efficiency as well as the utilization of residual heat sources from industry and bio-heat from underground sources.

11.6 Pathways Toward Sustainability

QNV 2030 and the NDSs both recognize the need for sound water management to sustain Qatar's fast growth despite scarce natural freshwater. Those documents remain a pillar for sustainable development, but at the same time, new endeavors and goals shall be outlined for long-term strategies in the water sector. In particular, in the way ahead, water sector quantity and quality monitoring and assessment should be integrated into the new strategies. Similarly, indicator-based frameworks should be developed from existing ones, and the involvement of the different stakeholders operating within the water sector is the base line for successful strategies and goals definition. In addition, new strategies for achieving a higher level of food security and meeting the increasing food demand shall be coupled with sound efforts toward water resources conservation in the agricultural sector.

11.6.1 Qatar Voluntary National Review (2021)

The Qatar Voluntary National Review (Planning & Statistics Authority, 2021b) shows the progress that has been made in achieving the goals of the Sustainable Development Agenda 2030 and also includes the future work needed to meet the SDGs targets by 2030, which will be aligned with the forthcoming NDS-3. This document is extremely relevant as it collects data and information at the national level on the actual progress and achievement of Qatar for the SDGs, involving several stakeholders from ministries, government agencies, the private sector, NGOs, civil society, academia, and research centers. Looking closer at the progress in the SDGs directly linked to the water sector, for what concerns SDG 2 "Zero Hunger" and in particular the target 2.4, related to ensuring sustainable food production system and implementing resilient agricultural practices has been achieved, and there are several examples of best practices that have been introduced to meet this target, such as the development of smart agricultural technologies (hydroponics and aquaponics) that ensure savings in irrigation and minimize the water usage, also for the cooling system. At the same time, there remain several structural and natural obstacles that hinder Qatar's agricultural system. First of all, the productivity and the agricultural revenues against the GDP remain low, while water and arable land scarcity remain an issue for the projected food demand. Population growth and urban expansion have led to a higher demand for food that can only be met at the current status through imports. Likewise, natural resource challenges imposed by the scarcity of irrigation water and land can be addressed by the adoption of unconventional practices in food production, such as hydroponics, greenhouse construction, and the use of advanced technologies in agriculture, livestock and fish production. The SDGs 6 "Clean Water and Sanitation" shows steady progress and achievements of the sanitation-related targets, while the use of treated wastewater for agricultural irrigation has seen a decline from 86.1 MCM/yr. in 2016 to 61.7 MCM/yr. in 2019 (Planning & Statistics

Authority, 2021b). SDG 12 "Responsible Consumption and Production" refers to the achievement of better consumption and production patterns while minimizing the use of natural resources, waste, and pollution. As such, progresses toward water resources rationalization are also captured in this SDG, in particular, toward the minimization of groundwater depletion, reduction in total losses of the drinking water network as well as per capita consumption, and promotion of integrated water and electricity management.

Similarly, the sustainability reporting tools employed by the different public and private companies could be adopted across the entire water sector (not just in Kahramaa) and improved to capture all the aspects of the water realm. This approach could ensure consistent monitoring and data collection and the effective achievement of water resources management targets. Furthermore, it will allow to compile together information and data coming from different stakeholders operating in the water sector, granting that all the relevant aspects are considered and measured against the goals. Lastly, in addition to the alignment to the SDGs, *ad-hoc* indicators could be developed to respond to policy needs and enhancement in the water realm.

11.6.2 Ministry of Environment and Climate Change Policies

Under a government reshuffle in October 2021, a new Ministry for the Environment and Climate Change (MoECC) was created. This decision came along with the participation of Qatar in the COP26 in Glasgow and further stresses the country's interests in environmental issues, to achieve comprehensive and sustainable development and that climate change-related policies have new prominence in the State of Qatar. The efforts toward environmental sustainability have been translated into the goals of the Qatar National Environment and Climate Strategy (QNE) (Qatar Government Communications Office, 2022). Such goals spread in different priority areas, but for what concerns the water sector, the aims include:

- Regular and effective monitoring of all water sources (groundwater, seawater, and potable water).
- More than 55% desalination from RO or more sustainable technology.
- 60% reduction in groundwater extraction.
- Reduce and maintain total loss to <8%, real loss to <4%.
- Conduct a household water survey and develop roadmap to reduce per capita household water consumption by 33% from 2019 level to ~310 L/capita/day.
- 100% of recycled water reuse.
- Complementary to the very ambitious targets above, one other target from the Land Use priority areas is related to the water realm is: 40% improvement in water consumption/tons of crop produced from 2019 levels.

The actions and goals of the newly created ministry are extremely ambitious and indeed address the majority of the challenges that the water sector is experiencing.

Those plans will be realized in collaboration with the efforts of the Ministry of Municipalities and the Environment and Kahramaa, as well with multiple actors within the water sector.

11.6.3 Food Security Strategies Sustainability

While there is a separate chapter in this book covering sustainability and food security (see Chap. 12), since food security is inextricably linked with water, it is worth mentioning here. The geographic and climatic conditions of Qatar hinder the development of standard agricultural practices. Therefore, the expansion of the food supply has relied on imports from foreign markets, which made Qatar's supply and demand more susceptible to exogenous disruption. Local food production takes a toll on domestic groundwater sources, while importing foods puts pressure on the so-called virtual water sources (i.e., the total volume of water used to produce a specific good in the country of origin of the same). At the same time, the global food crisis in 2008 and the war in Syria in 2011 caused food price spikes and deprived Qatar of key suppliers of fresh fruit and vegetables. As a response to those events, in 2008, the government decided to launch the first National Food Security Program (QNFSP). This program aimed to reduce Qatar's reliance on food imports through the development of several initiatives oriented toward self-sufficiency, to achieve 40% of food supplied domestically by 2030. The plan included the expansion of the agricultural sector and its endowment with the latest technologies. Despite a few examples of successful implementation of the QNFSP strategies, Qatar continued to depend on imports for a very significant portion of its food needs. The blockade starting in June 2017, however, exposed the country to an unprecedented level of food insecurity. Two years after the beginning of the blockade, a new Strategic Food Security Project for 2018–2023 was introduced by the Food Security Department of the Ministry of Municipality and Environment (MME) (Ministry of Municipality & Environment, 2020). Unlike the first plan of 2008, the current plan is better organized from a logistics perspective and the food security targets are well outlined. The strategy is based on four main pillars: (1) boosting local production; (2) increasing the strategic storage that aims to provide non-produced goods in the country for covering its needs for up to six months; (3) keeping international trade as a cornerstone; and (4) starting local market studies. The new plan has a much shorter time horizon, with objectives to be achieved as early as 2023, and very ambitious autarkic aims. These new strategies were disclosed only in March 2019, although the preparation of these plans started immediately after the blockade. The action plan was rapidly implemented, and local production started to increase, providing again the local market with livestock and vegetables, fruit and dairy products—this time, produced domestically. The increase in domestic food production also increased the pressure on water resources.

Despite efforts in the agricultural sector to increase its efficiency by developing "smart" agricultural technologies such as hydroponic, aquaponic, and aeroponic, at least in the major farms, the pressure on groundwater resources remains high and even if sustainable water uses and reducing net depletion of groundwater are a key aspiration of the food security strategies, this sector still relies heavily on these resources. In order to meet the future food demand requirements, the current food security strategies, set to expire in 2023, need to be updated. The new strategies should include more detailed and targeted research and investment programs for sustainable agricultural management in which food security is developed along with water security and sustainability. Crop selection and diversification of the water sources for food production are the key areas of intervention.

11.7 Conclusions and Policy Implications

Water security and water management are at the heart of sustainable development in Qatar. Their achievement is highlighted throughout the numerous strategies in place since 2008 from the QNV 2030 to the NDS-1 and NDS-2, the future NDS-3 (2023–2027) and the multiple local, regional, and international projects in which the country is involved. Water in hyper-arid environments as Qatar remains high on the political agenda and calls for further actions are needed to move Qatar's economy and development forward. The impressive achievements in the water sector span from scientific and technological development to policy strategies, regulations, and laws are complemented by a growing public awareness and involvement in water conservation projects by the civil society and the local communities. As Qatar continues to flourish and to aim for greater self-reliance of food production, industry as well as expanding the workforce, further demands on water resources are projected for the future. As such, old and new challenges await, and sustainable solutions should be planned and implemented, targeting a holistic approach that involves all the actors and segments of the water sector. Sustainable water management can be approached from an allocation efficiency perspective, as well as from a societal point of view in terms of quantity and access. At the same time, monitoring and regulations enforcement is crucial, as it ensures water conservation and safeguards that targets and goals are met and accounted, and adjusted along the way should parameters, thresholds, and dynamics change. From a water demand perspective, initiatives to reduce the per capita water demand in the municipal sector should be further encouraged and complement *Tarsheed* initiatives to push the reduction efforts forward. This would become crucial, especially for the service sector in light of the FIFA 2022 World Cup events and should be realized through a coordinated and collaborative campaign among the different stakeholders. In addition, revision and updates of the current water tariff system, in a way that better reflects the scarcity of the natural water resources and can at least recover the costs of operations and management of the water sector should be encouraged. On the supply side, the actions toward the reduction of leakages and losses along the water network shall be pushed forward together with the technological advancements in wastewater management, produced water management and the infrastructures needed to step up and enhance and diversify the water portfolio. Similarly, a huge potential lies in the desalination sector, which is environmentally unsustainable because of its use of fossil fuels for power generation as well as for the elevated discharge of highly concentrated brine. Lastly, water sector digitalization, which focuses on the integration of the spectrum of digital solutions and technology into Qatar's water management system can become a key component for water sustainability and the creation of smart cities. To conclude, revision of current KPIs and indicators for the evaluation of water utility operations can pave the way for designing regulatory and managerial incentives to improve performance in the water sector. As of now, Kahramaa's and Agenda 2030 SDGs appear as the main KPIs, but acquisition and recollection of additional indicators from different actors within the water sector can prove useful for providing more accurate and reliable data and metrics as well as linking together different agents in the water realm to maximize the collaboration, improve service quality, expand water access, and diminish water losses and demand.

References

Aly, S., Jawad, J., Manzoor, H., Simson, S., Lawler, J., & Mabrouk, A. N. (2022). Pilot testing of a novel integrated multi effect distillation—Absorber compressor (MED-AB) technology for high performance seawater desalination. *Desalination, 521*, 115388. https://doi.org/10.1016/j.desal.2021.115388

Aly, S., Manzoor, H., Simson, S., Abotaleb, A., Lawler, J., & Mabrouk, A. N. (2021). Pilot testing of a novel multi Effect Distillation (MED) technology for seawater desalination. *Desalination, 519*, 115221. https://doi.org/10.1016/j.desal.2021.115221

Baalousha, H. M. (2016). Development of a groundwater flow model for the highly parameterized qatar aquifers. *Modeling Earth Systems Environment, 2*, 67. https://doi.org/10.1007/s40808-016-0124-8

Baalousha, H. M., & Ouda, O. K. M. (2017). Domestic water demand challenges in Qatar. *Arabian Journal Geoscience, 10*, 537. https://doi.org/10.1007/s12517-017-3330-4

Bates, B., Kundzewicz, Z. W, Wu, S., & Palutikof, J. (2008). *Climate change and water* (Intergovernmental Panel on Climate Change Technical Paper; 6).

Baumann, D. D., Boland, J. J., & Sims, J. H. (1984). Water conservation: The struggle over definition. *Water Resources Research, 20*, 428–434. https://doi.org/10.1029/WR020i004p00428

Brooks, D. B. (2006). An operational definition of water demand management. *International Journal of Water Resources Development, 22*, 521–528. https://doi.org/10.1080/07900620600779699

General Secretariat for Development Planning. (2011). National Development Strategy 1.

Gulf Times. (2021). QEWC plans new plant in 2027 with 2,600 MW, 100 MIGD capacity [WWW Document]. *Gulf-Times*. https://gulf-times.com/story/686250. Accessed 30 September 2021.

Intergovernmental Panel on Climate Change. (2022). Climate change 2022: Impacts, adaptation, and vulnerability. contribution of working group II to the sixth assessment report of the intergovernmental panel on climate change.

Ismail, H. (2015). Food and water security in Qatar: Part 2—Water resources 1.

Jones, E., Qadir, M., van Vliet, M. T. H., Smakhtin, V., & Kang, S. (2019). The state of desalination and brine production: A global outlook. *Science of the Total Environment, 657*, 1343–1356. https://doi.org/10.1016/j.scitotenv.2018.12.076

Kahramaa. (2020). *KAHRAMAA—Water security mega reservoirs project*. http://www.watermegareservoirs.qa/. Accessed 16 April 2020.

Kahramaa. (2021). *Annual statistics report 2020.*

Kahramaa. (2022). *Qatar general electricity & water corporation*. https://www.km.qa/AboutUs/Pages/History.aspx. Accessed 21 April 2022.
Kamal, A., Al-Ghamdi, S. G., & Koç, M. (2021). Assessing the impact of water efficiency policies on Qatar's electricity and water sectors. *Energies, 14*, 4348. https://doi.org/10.3390/en14144348
Ministry of Municipality and Environment. (2020). *Qatar national food security strategy 2018–2023*.
Ministry of foreign Affairs. (2021). *Oil and gas sector*. https://mofa.gov.qa/en/qatar/economy-today/oil-and-gas-sector. Accessed 21 April 2022.
Ogunbiyi, O., Saththasivam, J., Al-Masri, D., Manawi, Y., Lawler, J., Zhang, X., & Liu, Z. (2021). Sustainable brine management from the perspectives of water, energy and mineral recovery: A comprehensive review. *Desalination, 513*, 115055. https://doi.org/10.1016/j.desal.2021.115055
Planning and Statistics Authority. (2021a). *Water statistics in the State of Qatar 2019*.
Planning and Statistics Authority. (2021b). Qatar *National voluntary review 2021b—Report on the implementation of the 2030 agenda for sustainable development*.
Planning and Statistics Authority. (2019). *Environment statistics bulletin 2019*. https://www.psa.gov.qa/en/statistics/Statistical%20Releases/Environmental/EnvironmentalStatistics/Environment_Statistics_bulletin_2019_AE.pdf
Planning and Statistics Authority. (2018a). *Water statistics in the State of Qatar 2017*.
Planning and Statistics Authority. (2018b). *National development strategy 2*.
Qatar Government Communications Office. (2022). *Environment and sustainability*. https://www.gco.gov.qa/en/focus/environment-and-sustainability/. Accessed 21 April 2022.
United Nations Environment Programme. (1992). *World atlas of desertification*.
World Bank. (2022). *World bank climate change knowledge portal*. https://climateknowledgeportal.worldbank.org/. Accessed 21 April 2022.

Jenny Lawler is the Senior Research Director of the Water Center at Qatar Environment and Energy Research Institute (QEERI), part of Hamad Bin Khalifa University (HBKU). She is a chemical engineer with a PhD in Membrane Separations. Her research focuses on the development of novel technologies and membrane-based strategies to solve environmental problems including those encountered in drinking water and wastewater treatment systems. She is particularly interested in the targeting of emerging pollutants such as pharmaceuticals and nanomaterials, and their impact on public health and the environment. QEERI Water Center research, innovation, and development strategy focus on the core water-related issues for Qatar, desalination, and wastewater treatment, with key insights into water management and the future of the water sector for arid regions.

Annamaria Mazzoni is a Researcher at the Qatar Environment and Energy Research Institute, where she works as an environmental economist, sustainability specialist, and policy analyst in the Environment and Sustainability Center. She has over 7 years of experience in international research institutions. She obtained her PhD in Economics from the University of Turin and Collegio Carlo Alberto, Italy. She has extensive practice in water resources management in hyper-arid environments and her main study area is the Middle East and North Africa Region. Before joining QEERI, she acquired direct experience in project management and administration, collaborating in international project involving US institutes, Middle East Universities, and with international agencies involved in social development and environmental sustainability (UNDP, USAID).

Sa'd Shannak is a scientist at the Qatar Environment and Energy Research Institute, part of Hamad Bin Khalifa University. He is an experienced environment and energy professional with quantitative and qualitative research skills such as econometric modeling, short- and long-term forecasting, model optimization, financial analysis, and risk analysis. He has a strong track record in energy and water policy and technology, gained over 15 years of work in research institutions, consultancy, local authorities, and NGOs in the USA and the Middle East.

Chapter 12
Contribution of Non-profit Organizations to Food Security and Sustainability in the State of Qatar

Sana Abusin, Ebaidalla M. Ebaidalla, and Maryam F. Al-Thani

Abstract Despite the relatively long experience of charitable organizations in social and humanitarian aspects in Qatar, little is known about their role in food security and sustainability. This chapter sheds light on the issue of food waste and assesses the role of non-profit organizations in food security and sustainability in the State of Qatar. The analysis is based on data sourced from two active non-profit organizations working on food surplus and redistribution in Qatar, namely, Hifz Al Naema and Wahab. The results indicate that the organizations under study are exerting a considerable effort to maintain food security and to reduce food waste, and their intervention increase over time. However, despite the sizable efforts exerted by charitable organizations to sustain food security, the issue of food surplus in Qatar still is a puzzling problem. The chapter recommends that collective cooperation between academics, government entities, and civil society would help in designing an effective strategy that aims to sustain food security. Moreover, policymakers need to support and encourage charitable organizations to participate in managing, sustaining, and achieving food security. Therefore, allocating grants and facilities to organizations that operate in food waste management would stimulate the charitable societies to contribute to food sustainability.

Keywords Non-profit organizations · Food security · Sustainability · Food waste management · Qatar

S. Abusin (✉) · M. F. Al-Thani
Social and Economic Survey Research Institute, Qatar University, Doha, Qatar
e-mail: sabusin@qu.edu.qa

M. F. Al-Thani
e-mail: maralthani@qu.edu.qa

E. M. Ebaidalla
Ibn Khaldon Center for Humanities and Social Sciences, Qatar University, Doha, Qatar
e-mail: ebaidalla@qu.edu.qa

L. Cochrane and R. Al-Hababi (eds.), *Sustainable Qatar*, Gulf Studies 9,
https://doi.org/10.1007/978-981-19-7398-7_12

12.1 Introduction and Motivation

In a world where consumption is increasing exponentially due to modern lifestyle and sophisticated marketing campaigns, the waste of products from different types is increasing too. On the one hand, almost one-third of the food in the world goes to waste, which is equal to 1.3 billion tons of food and costs an estimated $1 trillion. On the other hand, many people around the world are suffering from insufficient food and hunger. According to FAO's (2021) report, the number of undernourished people in the world was around 800 million in 2020.

Food waste (FW) is defined as food that is appropriate for human consumption being discarded or left to spoil at the retail or consumption phases. Food waste occurs when consumers dispose food waste intentionally without using it or storing it until the expiration date. Often, food waste happens at the distribution and consumption stage due to technical difficulties concerning storage, processing, packaging, and marketing systems.

The organic waste leads to several environmental consequences. For instance, decomposed waste food in landfills releases methane gas and carbon dioxide. The first accumulation for a long time causes climate change. Climate change, in turn, causes the sea levels to rise putting lives and livelihoods at risk. At the same time release of these gases, causes air pollution and contributes to serious health problems, such as mortality, chronic health conditions, health deterioration, behavior problems, and poor mental health (FAO, 2011). The chronic conditions include asthma, bronchitis, and other lung diseases. In addition, food loss results in excessive greenhouse gas (GHG) emissions and unnecessary utilization of scarce resources (Luo et al., 2021).

Food waste has been a crucial topic that gained continuous attention from researchers and policymakers. Numerous regional and international initiatives have been launched in recent decades, aiming at reducing food loss and waste, and attaining food security. For instance, the United Nations Sustainable Development Goal (SDG) 12.3 aims to "halve per capita global food waste at the retail and consumer levels and reduce food losses along production and supply chains, including post-harvest losses" by 2030.

Following the Gulf Arab rift in 2017, Qatar paid considerable attention to food security and made it a top priority. Consequently, the country has expanded and supported the local agricultural producers, aiming at achieving self-sufficiency and satisfying the increasing demand for food that resulted from rapid population growth and economic development (Karanisa et al., 2021). The Ministry of Municipality (MM) statistics show that the local agricultural production increased from 523 million Qatari Riyals (QR) in 2015 to 1.22 billion QR in 2019 (MM, 2020). Specifically, the volume of vegetable production increased from 53,000 tons in 2016 to 91,000 tons in 2019, with a growth percentage of 72% (MM, 2020). However, the expansion in agricultural production is associated with remarkable food loss and waste due to the small size of the domestic market and the sizable exports from trading partners (MM, 2020). Indeed, food waste is regarded as a prominent phenomenon in the Gulf Cooperation Council (GCC) countries (El Bilali & Ben Hassen, 2020). In Qatar, food waste is considered a major problem, especially during the holy month of Ramadan, estimated at around 1.4 million metric tons in 2012 (mentioned by EcoMENA 2020, a volunteer-driven initiative to create mass environmental awareness).

The issue of food waste has received considerable attention from charitable organizations and associations in recent years. In this regard, many organizations engage in collecting food surplus and distributing it to needy people. It is worth to mention that Hifz Al Naema and Wahab are the most active charity organizations, working exclusively in food redistribution. These charities contribute to achieving food justice and social cohesion among members of the community by distributing valid food from producers to those who are in need. In addition, Wahab aims to reduce food waste by finding sustainable management solutions to compost wastage as well as raising awareness. However, despite the relatively long experience of these organizations in reducing food waste by redistributing food surplus, the information available on their size and contribution to food security sustainability is unknown. Therefore, this chapter aims to shed light on the issue of food waste in Qatar and the role of Hifz Al Naema and Wahab in food security.

The contribution of this chapter is fourfold. First, this chapter provides a new and significant contribution to the existing literature on the role of non-profit organizations in food security in Qatar; as to the best of the authors' knowledge, this is the first study that addresses this issue in Qatar. Second, examining the contribution of non-profit organizations to food security would provide policymakers and key stakeholders (e.g. government, NGOs, and regional organizations) with useful insights to adopt measures that help in sustaining the food sector. These measures may contribute effectively to improving food security, attaining self-sufficiency, and mitigating food waste. Third, the chapter provides an in-depth review of the potential contribution of non-profit organizations to the achievement and sustainability of food security, hence, informing policymakers on the importance of supporting and funding charities. Four, this study is timely, relevant, and quite consistent with Qatar's National Vision 2030 and the United Nations SDGs, especially, SDG 12.3. Therefore, the study's outcomes will promote sustainable food practices and improve economic efficiency to achieve national and global development strategies.

The chapter is organized into four sections, the first presents the chapter's motivation and justification by providing some stylized facts about food waste. The second section provides a background and historical context of the food sector in Qatar. How society and government contribute to food management is presented in section three with more focus on Hifz Al Naema as a charity and Wahab as a business company doing also significant charitable work. Finally, the chapter concludes by providing policy input on innovative solutions toward sustainable food security in Qatar by widening the role of non-profitable organizations and engaging them in the food sector system.

12.2 Social, Economic, and Environmental Impacts of Food Loss and Waste

There are social, economic, and environmental impacts of food loss and food waste in the society. From a social perspective, and according to FAO in 2020, there are between 720 and 811 million hungry people (FAO, 2021). With the increasing number of people in need of food, there is an expanding amount of food loss and food waste (Sarkozi & Rukikaire, 2020). Thus, managing food loss may reduce the amount of food waste and consequently achieve food security.

The economic impact of food waste can be highlighted through waste management costs and money wasted, given the amount of edible food disposed yearly. Cost includes the maintenance of landfills, transport costs, operation costs in treatment plants, and separation costs in some cases (El Bilali & Ben Hassen, 2020). There is a gap in the literature in analyzing the economic impact of food waste in the region, and even a bigger gap when discussing Qatar. Optimized composting facilities can bring about significant economic benefits. A study conducted in Saudi Arabia found that diverting food waste from landfills to composting facilities will feed to the Saudi economy about $70.72 million per year (El Bilali & Ben Hassen, 2020). In Qatar, more research needs to be done on the economic impact of food waste/loss and surplus.

Food waste and food loss can have serious negative effects on the environment. For instance, a study conducted in Europe to illustrate the environmental impact of food waste found that food waste is attributed in Europe with '186 million tonnes of CO2 equivalent (Mt CO2-eq), 1.7 Mt SO2-eq. and 0.7 Mt PO4-eq' and that around three-quarters of the impacts of all food waste for global warming are created from GHG emissions throughout the production stage (Scherhaufer et al., 2018). For example, 6% of the emissions come from food disposal and 7% from retail and distribution (Scherhaufer et al., 2018). Similarly, Alruqaie and Alharbi (2012) investigated the environmental impact of food waste recycling in Riyadh, Saudi Arabia and found that methane gas was higher than the other types of gaseous pollutants with 99% of the total gaseous emission in the experiment.

12.2.1 Historical Context of Food Security in Qatar

In the past five years, there have been huge shifts in Qatar's food security policies. In 2017, Saudi Arabia, Bahrain, the United Arab Emirates, Egypt, Yemen, and Libya banned Qatar's diplomatic ties and imposed an economic blockade. Prior to the blockade, Qatar relied heavily on imported food, medicine, and by-products from the neighboring countries. Before 2017, Qatar imported 90% of its food, 40% of which enters through its border with Saudi Arabia, therefore, shutting air and land borders had implications on the food supply. In addition, multiple studies from different countries have found a significant link between food insecurity and psychological disorders such as stress, anxiety, depression, and sleep problems (Arenas et al., 2019; Huddleston-Casas et al., 2009; Kirang & Frongillo, 2007; Leung et al., 2015; Martin et al., 2016; Myers et al., 2020; Yang & Matthews, 2010). The Qatari government took immediate action to address the food shortage.

The blockade compelled Qatar to restructure its food security. Qatar shifted the imports from Turkey and Iran to compensate for the blockage of the sea and air routes. Furthermore, Qatar imported thousands of cows from the United States and Europe, which has allowed it to supply 40% of its milk requirements within months. The country allocated a significant amount of financial resources to quickly achieve nearly 100% self-sufficiency. As a result, the production of vegetables, dates, red meat, poultry, eggs, fish, and green fodder has increased by 400% in one year (Abusin et al., 2020).

Local farms are putting tremendous effort to improve their agricultural production by using modern production methods such as greenhouse structure, efficient water-conserving irrigation techniques, and advanced cooling technologies to increase the shelf life of products. As a result, they have achieved high yields and met food security goals (Karina Enikeeva & Carol Khadra, 2020; Qatar National Food Security Strategy 2018–2023, n.d.). Qatar is also witnessing increasing demand for by-products, particularly vegetables and fruits.

Several Qatari farms now produce vegetables and fruits at affordable prices (Global Organic Trade Guide for Qatar, 2020). Qatar's market for packaged food and beverage consumption was estimated at 14.4 million tons in 2020. The number of farms and farmers is increasing in the country (Qatar Guide, 2020). The government encourages Qataris to invest in food growing and provides many facilities to ease farming. Qatar National Food Security Strategy 2018–2023 (QNFSS) gives special attention to the farmer support program through the farmer representative entities such as centralized procurement of inputs, crop advisory services, market and trading, warehousing, farm credit, and product traceability. It also focuses on increasing the productivity of local farms, providing an efficient go-to-market channel for local farmers, and reforming market price regulation. This creates overproduction of food, and the latter results in food surplus loss/waste.

Food surplus is an issue that affects the sustainable achievement of food security in Qatar. The per capita income has increased the quality and quantity of food that the population demand (Adema, 2016). Based on the existing literature, excessiveness in food consumption is a common trend in Qatar. When it comes to traditions and customs that revolve around the food industry, Qatari culture revolves around hospitality, and Qataris are known for their generosity. Edelstein (2011) notes that the culture of generosity is largely felt across Qatari supports the sentiments. Food plays a central role when hosting parties such as weddings or any other form of communal dining, demonstrating generosity and hospitality. This act of hospitality and generosity is extended beyond private settings and into restaurants and cafes. For instance, according to Sillitoe and Al-Misnad (2014), it is highly welcoming to dine with the Qataris; a Qatari will always insist that the visitor eat or take the meal or drink respectively to the last piece or drop. Similar to other nations and cultures, family life has an influence on food consumption patterns in Qatar. All these mentioned behavior and practices are responsible for the food surplus created because of the misunderstanding of generosity. Therefore, this surplus could be one of the key reasons for food waste in Qatar (Abusin et al., 2020; Adema, 2016; Edelstein, 2011; Sillitoe & Al-Misnad, 2014).

Consumer protection is one of the requirements to sustain food security. Consumer protective policies in Qatar relating to food quality and expiry dates are strict, with special teams designated to outtake the monitoring system. Therefore, the chapter will not go into detail on this.

12.3 Food Self-Sufficiency and Domestic Markets (Government's Efforts to Achieve Food Security)

The international economic crisis of 2007–2008 shed light on the importance of food self-sufficiency (FSS) due to the food crisis and extreme food price volatility. Many countries expressed interest in improving their levels of food self-sufficiency. The FAO (2015) defines it as "a country closing its borders to all food trade—both imports and exports—and concentrating its resources on its agriculture sector so as to be able to produce all of its food requirements domestically". Others have defined FSS as a country's ability to produce the total calorie required to meet its demand (Pradhan et al., 2014).

At the turn of the millennia, Qatar started to focus on different approaches to improve food security. The government had decided to lease or purchase arable lands in countries like Kenya and Ghana to produce food for the people in Qatar (Amery & Series, 2019). Domestically, the country invested in hydroponics and other innovative farming technologies to produce food that would not be vulnerable to supply disruptions or price volatility. In 2008, the Qatar National Food Security Program (QNFSP) was founded in order to consider strategic risk scenarios and develop a system that would improve the country's resilience to water and food supply shocks. It aimed to expand food storage capacity, diversify the food supply chains, and increase domestic food production. For example, Qatar's production of red meat and poultry skyrocketed from 10,792 tons in 2012 to 183,988 in 2016 (Amery & Series, 2019).

In January 2020, an updated QNFSS was established with the aim of building a robust food security strategy. The main target is to increase local production of perishables to secure 30–70% self-sufficiency in strategic commodities. The government has outlined four different approaches in order to achieve this goal. First, increasing vegetable production by establishing a hydroponics greenhouse cluster to reach 70% self-sufficiency in greenhouse vegetables, including tomatoes, peppers, cucumber, and squash. This is to be achieved by finalizing a greenhouse cluster infrastructure plan and developing bid guidelines, including subsidy programs for private sector operators. Secondly, expansion and improvement of production capacity for fish and red meat by fattening units and breeding farms for sheep and goats. Thirdly, to cap production of fresh milk and poultry to 100% self-sufficiency by shifting poultry surplus to egg production by suspending any new project tenders and shifting additional capacity toward milk derivatives or frozen poultry to egg production. Finally, to reduce groundwater-based fodder production by switching to treated sewage effluent (TSE) by developing a transition plan for existing fodder producers.

As mentioned previously, the blockade against Qatar in 2017 has increased domestic markets' production. There was a 20% increase in commercial farming in Qatar, from 364 farms in 2016 to 437 in 2019. Vegetable production has increased from 58,627 in 2016 to 1,055,488 in 2019 (Awadh, 2020). There was also a direct positive correlation in the sales of local vegetables in the Central Market. Qatar has also focused on organic farming and the production of domestic markets. The number of farms and farmers' markets in Qatar is increasing. Some local farms hold international organic certifications like USDA Certification by Agrico Agricultural Development and IFOAM certification by Al Safwa farms (Karanisa et al., 2021). Karanisa et al. (2021) suggest that there are several opportunities for Qatar to improve its domestic markets by focusing on exports. The winter season in Qatar allows for the production of high-quality agricultural products that could not only satisfy local demands but also to export fresh high-quality agricultural products to countries in the Northern Hemisphere. Although, in 2019, Qatar exported products like dates, cereals, eggs, milk, and fish, the number of exports is lower in comparison to exports (Planning and Statistics Authority, 2018).

12.3.1 Food Types and Management

For the sake of this chapter, food is divided into three categories "surplus, waste or loss". The strategic commodities or perishable food include red meats, fruits and vegetables, and fresh dairy. Understanding the types of foods available simplifies the food management process. According to the literature, foods classified as perishable are those that spoil, decay, or become unsafe for consumption if not kept refrigerated at 4.4 °C or below, or frozen at −17.8 °C or below. Examples of foods that must be kept refrigerated for safety include meat, poultry, fish, dairy products, and all cooked leftovers. Foods that are perishable can only be stored for one or two days at room temperature, i.e. they have a shelf life of one or two days (Kumar & Samadder, 2017). On the other hand, non-perishable foods have a long shelf life and do not require refrigeration such as canned food, dry foods, and dehydrated foods (Kumar & Samadder, 2017).

The Qatar Second National Development Strategy 2018–2022 (NDS-2) has set two main targets in relation to waste reduction and prevention. The first target focuses on restricting the generation of domestic waste to less than 1.6 kg per capita/day between 2018 and 2022. The second target is recycling 15% of solid waste by the end of 2022. The Qatari government approved QNFSS with the aim of reducing food waste by a target of 5%, to be measured against a baseline that will be established as part of the program.

Fig. 12.1 Food management by food type

A question may be raised about the role of non-profitable organizations and private sector in food waste management while the Domestic Solid Waste Management Centre (DSWMC) in Mesaieed is the largest composting facility in Qatar with a capacity of 550 tons of waste per day. The process in which the food waste is collected by this center does not involve the segregation of food. Therefore, charitable efforts by non-profitable organizations and private sector can assist MM, but not as a replacement.

Figure 12.1 "elaborated by authors", links the types of food and the proposed management method in order to sustain the food sector.

Surplus food could be perishable and non-perishable to be allocated to donation to those who need it through registered and professional charities that take the responsibility of all health, legal, and quality food conditions that made it suitable for human consumption. If the non-perishable food, meets all conditions except of marketing, then the food can be allocated to processing into by-products. Food processing is very important, in this stage after agricultural expansion, there are a large amount of fruits and vegetables that are valid for consumption but not accepted by marketing. The large amount is due to overproduction. Unfortunately, there are no transformative industries to produce jams, tomato paste, etc.

Food waste that is segregated can be composted into organic fertilizers by a professional recycling company or MM Centre. After that, the composters take the decision of donating or selling it.

12.3.2 Non-profit Organization Contribution to Food Security

On the international scale, food rescue and allocation optimization have received little attention from waste management researchers until recently, regardless of its emergence as a legitimate form of food waste diversion (Reynolds et al., 2015). The logistics of Meals-on-Wheels programs mark the earliest studies on food rescue costs and optimization focused. Moreover, the work of Youn et al. (1999) discusses the food rescued by the United States' operations; similarly, Cotugna and Beebe (2002) give a historic background to the rise of food rescue in the United States. More recently, deliveries in foodbank networks have received increasing attention, due to the complexity of their operations (Davis et al., 2014; Gunes et al., 2010; Solak et al.

2012). According to Reynolds et al. (2015), food rescue is seen as more economically costly than landfill or composting, however, is a cheaper method of obtaining food for the food insecure rather than directly purchasing with every dollar spent on food rescue enabling $5.71 of edible food to be rescued. A few studies have focused on optimizing delivery and collection schedules (Davis et al., 2014; Gunes et al., 2010; Nair et al., 2017; Solak et al., 2014).

Charities collect foods that are still edible but are either close to becoming unfit for consumption or are classified as excess (i.e. leftovers from events) (Nair et al., 2017). Generalizations are possible although each charity operates in its own unique operations. The common factor is that these organizations receive donated foods mainly from the service sector (hospitality and events) followed by the manufacturing sector. Donated foods are transformed by the charities into meals or food parcels that are given directly to recipients in need or through secondary charities or religious organizations (Reynolds et al., 2015).

12.4 Role of Non-profit Organization in Achieving Food Security in Qatar

There is a well-established relationship between societal responsibilities, the services provided by charities, and the SDGs, especially SDG 12.3 on food waste reduction. Charitable societies have become concerned with their responsibilities toward societal issues, including environmental issues, in an effort to reach a better and environmentally sustainable society. Charitable organizations seek to contribute to achieving SDGs through community activities and events that they practice.

Environmental sustainability has become at the forefront of priorities, as the environment is a common denominator for living among members of society as a whole. This also includes the actors within society, such as the government, the private sector, and civil society organizations represented by charities, which have become concerned about communities and adopt local environmental issues such as pollution, waste, clean air, and conservation.

Non-profit organizations and associations are well-known for their ability to volunteer work and to identify the needs of society and the challenges they face in all aspects, the most important of which is the field of environmental development and methods of overcoming these problems and challenges. From a religious perspective, wasting food can create guilt from extravagance and is sometimes unintentional. Food waste during Ramadan is not accepted and is becoming a significant ethical dilemma in Qatar (Abusin et al., 2020). It has been identified that the larger the household, the greater the chance for food wastage (Koivupuro et al., 2012; Parizeau et al., 2015; Van Garde & Woodburn, 1987; Williams et al., 2012). A household that spends more on food purchases tends to have bigger food wasters (Parizeau et al., 2015). Consumers feel bad about food-wasting (Evans, 2012; Watson & Meah, 2012) and are concerned when they throw food away (Abeliotis et al., 2014; Graham-Rowe et al.,

2014). The Qatari culture is known for its customs that reflect people's generosity. A culture of hospitality is widely acknowledged and practiced; however, generosity may lead to overproduction. The first step to control overproduction is balancing the production with demand. Overproduction of food creates food surpluses that require redistribution by charities within a short amount of time.

In Qatar, there are two active organizations that work exclusively with food redistribution, Hifz Al Naema and Wahab. The aforementioned charities contribute to achieving food justice and social cohesion among members of the community by distributing valid food from producers to those who are in need. In addition, Wahab aims to reduce food waste by finding sustainable management solutions to compost wastage as well as raising awareness.

12.4.1 The Role of Hifz Al Naema in Achieving Food Security

Qatar has a total of fourteen charitable organizations in which some are working on food management, including Red Crescent, Qatar charity, Hifz Al Naema, and many other initiatives such as ones in Education City and Georgetown University, but because it is beyond the capacity of the chapter to review each one separately, we will concentrate on Hifz Al Naema for surplus management and Wahab for surplus and raise awareness (a non-profitable activity they carry out), recycling and composting service (profitable activity).

Hifz Al Naema is a charity established in Qatar in 2008 with the main goal of redistributing food surplus to those in need. It is considered the first food bank in Qatar that reprocess the access food to deliver it to people in need. Its main role is to collect food surplus from all different sources and distribute it inside Qatar. There are two main services in the center (a) food bank and (b) charity exhibition. The food bank collects the food surplus from the restaurants, banquets, and funerals and then redistributes it to the people in need. It also accepts individual donations such as Zakat al-Fitr and wraps them appropriately to be distributed later as meals. The number of registered families in 2018 was approximately 700.

The organization contacts the beneficiaries and brings the donations to classify them based on their type and situation. In addition, it sells the surplus at low prices and donates the financial return to low-income families. In 2018, Hifz Al Naema has participated in helping the people in need during Ramadan. It has distributed about 1000 meals on a daily basis as a fasting breakfast meal. Moreover, the center supports low-income families on certain occasions, such as Eid Al-Adha. For instance, in 2021, it distributed 700 sacrifices to 1400 families (*The Peninsula*, 2021) (Fig. 12.2).

The first figure explores the trend in number of meals distributed during the period 2010–2020. The amount of food or meal distributed increased dramatically from the year 2010 up to 2019. Interestingly by 2020, the number of meals distributed was halved which reflects the government efforts that followed 2017 Blockade on the achievement of self-sufficiency in food.

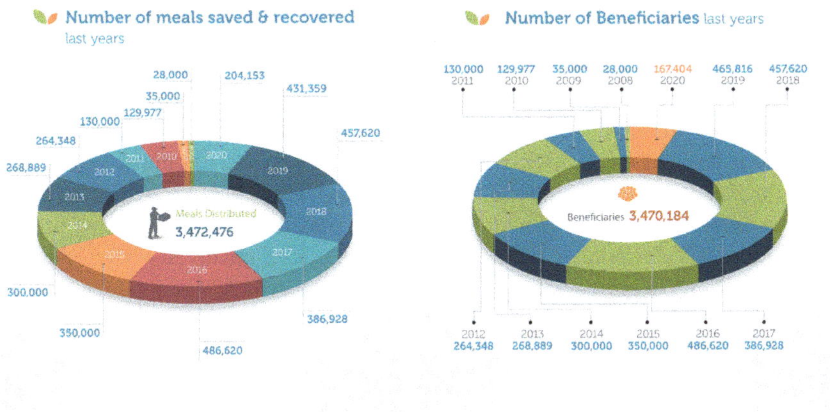

Fig. 12.2 (Hifz Al-Naema Charity, 2018) numbers of meals distributed and beneficiaries since its foundation

The second figure presents the number of beneficiaries during 2012–2020, the figure shows the positive increase of beneficiaries up to 2019. After that, the number decreased to less than half. It is clear that the self-sufficiency strategy and increase in local food production is the main reason.

Hifz Al Naema has helped to improve life in Qatar in many aspects. From one perspective, it is helping the people in need by providing food. It is also assisting in solving one of the biggest problems that affect life in Qatar, which is the environmental issue. Food waste along with the other types of waste are extremely harmful to the environment. Hifz Al Naema is doing a type of recycling the products "other than food" and reusing them.

The above statistics reflect the great role that Hifz Al Naema plays in controlling food surplus and reducing the amount transported to landfills. Additionally, reducing the Methane emission and therefore contribute to fighting climate change. The main challenge that Hifz Al Naema faces is that they work at full capacity, but the donated foods are a lot more. They are left with two options: to increase their resources in terms of the number of food cars and human resources and to start recycling the amount of food that gets spoiled intentionally due to temperature or handling. The compost then is distributed for free to grow food and hence contributes to achieving food sustainability.[1]

[1] The authors are very thankful to Ali Al-Gahtani, the Executive manager of Hifz Al Naema Center, for being very generous and cooperative to provide all information in this session.

12.4.2 The Role of Wahab in Achieving Food Security

Wahab is an award-winning company that fights food waste and climate change in Qatar by implementing the 3 R's of food waste—Reduce, Reuse, and Recycle. To address the issue of food waste, Wahab believes that first and foremost, awareness of achieving food security is important. In this regard, it has held more than 20 awareness sessions for a combined audience of 4000 plus individuals, targeting schools, universities, and hospitality industries and events, both onsite and online (Wahab, 2021). Over the last three years, it has helped to redistribute more than 500 tons of surplus food to the community, by tying up with food banks and local NGOs. Instead of hiring paid workers, Wahab fully relies on community volunteers. Through its 150 plus group of volunteers, they have demonstrated a well-knit organized group that has worked with various establishments including McDonalds, Siemens, KEO, Lulu, Mall of Qatar as well as schools. They were also nominated by Qatar tourist Authority (QTA) to be the official partners to redistribute the food surplus during Qatar International Food Festival 2018 and 2019 (Wahab, 2021).

In the absence of efficient and cost-effective food waste recycling solutions, Wahab has stepped in to fill the gap in food waste management solutions in Qatar. It has introduced food waste recycling solutions of various capacities and technologies, ranging from machines intended for large-scale commercial institutions to small compost bins for urban homes (see Table 12.1 for recycling machines capacity). By composting inedible food like vegetable scraps and chicken bones and returning valuable nutrients back into the soil, which is key to growing the next generation of crops. Wahab's actions come in line with QNV 2030, which aims to achieve self-sufficiency in food production. However, the community organization lacks a wide infrastructure as it depends fully on volunteers. Moreover, the food recycling industry in Qatar is still new, therefore, there is no clear guidance and conditions in which a company like Wahab can operate under.[2]

Despite the fact that government support and attention are increasing toward sustainability and food security in Qatar, and despite the sizable efforts exerted by charitable organizations to sustain food security, only little attention is given to them regarding their challenges and difficulties in continuing to provide and maintain this great services.

There are some measurements that need to be taken into consideration to sustain food security which may include, cooperation between governmental entities and food actors in both private and public sectors. Facilitating coordination between academics, ministries, and charities is needed to achieve sustainable ways of managing food redistribution and food waste. In addition, government support for economic diversification through opening opportunities for recycling companies and the control of food production, i.e. balancing food supply and demand is vital to improve food security. Charities, responsible ministries, and specialized academics must set a clear goal to reduce food waste and discourage sending food waste to

---

[2] The authors are very thankful to Wardah Mamukoya, the Executive Director of Wahab Company, for being very generous and cooperative to provide all information in this session.

Table 12.1 Wahab recycling machines capacities and specifications summary 2021 (Wahab, 2021)

	EP-25	EP-50	EP-100	EP-250	EP-500	EP-750
Nominal capacity per day (kg)	25	50	100	250	500	750
Machine weight when empty (kg)	600	650	1,000	1,700	2,200	3,400
Size (L x W x H) cm	115 × 80 × 108	171 × 102 × 145	224 × 117 × 132	297 × 134 × 170	350 × 155 × 195	470 × 176 × 205
Power rating (KW)	3	5	7	10	15	18

EP stands for EcoProbe, the name of the food recycling machine. EP-50 means the machine can compost 50 kg of organic waste daily

landfills. Charities help to reduce the number of those who need quality food and raise awareness about the donation of surplus food and recycling of wastes. The role of academics is to provide data and statistics about quantities of food and identify gaps in research on food security.

12.5 Policy Implications

The huge efforts and participation of charities and non-profit organizations though are appreciated inside and outside Qatar, no literature or documentation of these significant works in terms of achievement and contribution to food security. This chapter tries to provide some useful information on the issue of food waste and present some of the contribution of non-profit organizations to food security and sustainability in the State of Qatar. The chapter benefited from the experiences of two active non-profit organizations working on food surplus and redistribution in Qatar, namely Hifz Al Naema and Wahab. It has been noted that the organizations under consideration are paying a lot of efforts to maintain the food security and to reduce the food waste, and their intervention increase over time to try to cover the increasing number of beneficiaries taking all health, legal and humanitarian responsibilities on dealing with those in need.

The way forward to help improve the sustainability of the food sector in Qatar should consider the following:

- Encouraging other charities to engage in waste management and providing them opportunities to contribute to sustainable food security in the country will add a lot of value given their long experience.
- Segregation of organic food from total waste to be composted and reused for growing food is vital to sustaining food security.

- Establishment of two collection and distribution points is necessary to manage food waste on a national level. The collection point will take the responsibility of classifying the food into two categories: surplus valid for human consumption to be sent to people in need and food loss to be sent to transformative industries and transferred to another valuable products such as animal feed or canned products. The collection point could work within the ministry of public health section and/or department of Standardization. This will take the total health and legal responsibilities of food redistribution.
- The distribution point will save effort and time by providing food cars that are well-prepared and take into consideration, health risks that might happen during food distribution. Also from legal aspect, manage illegal markets that may appear from food distributors by selling food items at a low price.
- The responsibility of food waste should be taken by the Ministry of Municipality (MM), Centre for waste treatment and recycling by offering a prepared truck to take segregated food waste for composting.
- Engagement of charities in food management will help in providing daily statistics on the amount of food waste, loss, and surplus. Therefore, support the calculation of the footprint of Qatar. It will also help to reduce the risk associated with donating unsellable, edible food.
- Finally, raising awareness is a very important step toward food management. Civil society and organizations in Doha could enhance their efforts in engaging with volunteers for awareness-raising and recycling activities. The media could further highlight the implications of food waste on the environment. One of the best tools to raise awareness would be through social networks, which are being used by about 95% of the country's population. As mentioned in a UNEP report 2021 that "life satisfaction is not the source of the owned goods, but rather the good social interactions that stimulate the exchange of feelings, such as the feeling of the other".

Acknowledgements This [chapter] is part of Qatar University Q-grant-H3P ["QPH3P-SESRI-2021-450"]. The findings achieved herein are solely the responsibility of the authors.

References

Abeliotis, K., Lasaridi, K., & Chroni, C. (2014). Attitudes and behaviour of Greek households regarding food waste prevention. *Waste Management & Research: The Journal for a Sustainable Circular Economy, 32*(3), 237–240. https://doi.org/10.1177/0734242X14521681

Abusin, S., Lari, N., Khaled, S., & Al Emadi, N. (2020). Effective policies to mitigate food waste in Qatar. *African Journal of Agricultural Research, 15*(3), 343–350. https://doi.org/10.5897/AJAR2019.14381

Adema, S. (2016). *Food Waste Woes in Qatar*. EcoMENA. https://www.ecomena.org/food-waste-in-qatar/

Alruqaie, I. M., & Alharbi, B. H. (2012). Environmental advantage assessment of recycling food waste in Riyadh, Saudi Arabia. *Research Journal of Environmental Sciences, 6*, 230–237.

Amery, H. A., & Series, G. I. (2019). Food security in Qatar: Threats and opportunities. *Gulf Insights, 7*, 1–6.
Arenas, D. J., Thomas, A., Wang, J., & DeLisser, H. M. (2019). A systematic review and meta-analysis of depression, anxiety, and sleep disorders in US adults with food insecurity. *Journal of General Internal Medicine, 34*(12), 2874–2882. https://doi.org/10.1007/s11606-019-05202-4
Awadh, M. A. (2020). *Analysis of the Agriculture Situation before and after the Blockade for the period 2016–2019*. Agricultural Affairs Department, Ministry of Municipality and Environment.
Cotugna, N., & Beebe, P. (2002). Food banking in the 21st century: Much more than a canned handout. *Journal of the American Dietetic Association, 102*, 1386–1388.
Davis, L. B., Sengul, I., Ivy, J. S., Brock, L. G. III., & Miles, L. (2014). Scheduling food bank collections and deliveries to ensure food safety and improve access. *Socio-Economic Planning Sciences, 48*(3), 175–188.
Edelstein, S. (2011). *Food, cuisine, and cultural competency for culinary, hospitality, and nutrition professionals*. Jones and Bartlett Publishers.
El Bilali, H., & Ben Hassen, T. (2020). Food waste in the countries of the gulf cooperation council: A systematic review. *Foods, 9*(4), 463. https://doi.org/10.3390/foods9040463
Evans, D. (2012). Beyond the throwaway society: Ordinary domestic practice and a sociological approach to household food waste. *Sociology, 46*(1), 41–56. https://doi.org/10.1177/003803851 1416150
Food and Agriculture Organization. (2011). *Global food losses and food waste. Extent, causes and prevention*. UN FAO.
Food and Agriculture Organization. (2015). *The State of Agricultural Commodity Markets IN DEPTH 2015–16*. https://www.fao.org/3/i5222e/i5222e.pdf. Accessed on line in 14 May 2022.
Food and Agriculture Organization. (2021). *The state of food security and nutrition in the world*. https://www.fao.org/state-of-food-security-nutrition. Accessed on 14 May 2022.
Global Organic Trade Guide for Qatar. (2020). *Qualitative Analysis*. https://globalorganictrade.com/country/qatar
Graham-Rowe, E., Jessop, D. C., & Sparks, P. (2014). Identifying motivations and barriers to minimising household food waste. *Resources, Conservation and Recycling, 84*, 15–23. https://doi.org/10.1016/j.resconrec.2013.12.005
Gunes, C., van Hoeve, W.-J., & Tayur, S. (2010). Vehicle routing for food rescue programs: A comparison of different approaches. In *Integration of AI and OR techniques in constraint programming for combinatorial optimization problems* (pp. 176–180). Springer.
Hifz Al-Naema. (2018). *Number & Statistics*. HIFZ ALNAEMA. http://hifzalnaema.com/en/index.php?page=chart. Accessed 9 April 2022.
Huddleston-Casas, C., Charnigo, R., & Simmons, L. A. (2009). Food insecurity and maternal depression in rural, low-income families: A longitudinal investigation. *Public Health Nutrition, 12*(8), 1133–1140. https://doi.org/10.1017/S1368980008003650
Karanisa, T., Amato, A., Richer, R., Abdul Majid, S., Skelhorn, C., & Sayadi, S. (2021). Agricultural production in Qatar's hot arid climate. *Sustainability, 13*(7), 4059.
Karina Enikeeva, E., & Carol Khadra, C. (2020). *Agrico Qafco Yara Trial and Demonstration Center*. Report on Trials Conducted on December 2019–December 2020.
Kirang, K., & Frongillo, E. A. (2007). Participation in food assistance programs modifies the relation of food insecurity with weight and depression in elders. *The Journal of Nutrition, 137*(4), 1005–1010. https://doi.org/10.1093/jn/137.4.1005
Koivupuro, H.-K., Hartikainen, H., Silvennoinen, K., Katajajuuri, J.-M., Heikintalo, N., Reinikainen, A., & Jalkanen, L. (2012). Influence of socio-demographical, behavioural and attitudinal factors on the amount of avoidable food waste generated in Finnish households: Factors influencing household food waste. *International Journal of Consumer Studies, 36*(2), 183–191. https://doi.org/10.1111/j.1470-6431.2011.01080.x
Kumar, A., & Samadder, S. R. (2017). A review on technological options of waste to energy for effective management of municipal solid waste. *Waste Management, 69*, 407–422.

Leung, C. W., Epel, E. S., Willett, W. C., Rimm, E. B., & Laraia, B. A. (2015). Household food insecurity is positively associated with depression among low-income supplemental nutrition assistance program participants and income-eligible nonparticipants. *The Journal of Nutrition, 145*(3), 622–627.

Luo, N., Olsen, T. L., & Liu, Y. (2021). A conceptual framework to analyze food loss and waste within food supply chains: An operations management perspective. *Sustainability, 13*(2), 927.

Martin, R., Sunley, P., Gardiner, B., & Tyler, P. (2016). How regions react to recessions: Resilience and the role of economic structure. *Regional Studies, 50*(4), 561–585.

Ministry of Municipality and Environment. (2020). *An analytical study to compare the agricultural situation before and after the blockade for the period from 2016 to 2019.* Report in Arabic, Doha, Qatar.

Myers, C. A., Mire, E. F., & Katzmarzyk, P. T. (2020). Trends in adiposity and food insecurity among US adults. *JAMA Network Open, 3*(8), e2012767.

Nair, D. J., Rashidi, T. H., & Dixit, V. V. (2017). Estimating surplus food supply for food rescue and delivery operations. *Socio-Economic Planning Sciences, 57*, 73–83.

Parizeau, K., von Massow, M., & Martin, R. (2015). Household-level dynamics of food waste production and related beliefs, attitudes, and behaviours in Guelph, Ontario. *Waste Management, 35*, 207–217.

Planning and Statistics Authority. (2018). *Qatar second national development strategy—psa.gov.qa.* The Qatar National Vision 2030. https://www.psa.gov.qa/en/knowledge/Documents/NDS2Final.pdf. Accessed 7 April 2022.

Pradhan, P., Lüdeke, M. K., Reusser, D. E., & Kropp, J. P. (2014). Food self-sufficiency across scales: How local can we go? *Environmental Science & Technology, 48*(16), 9463–9470.

Qatar Guide. (2020). *ILOVEQATAR Qatar Guide.* https://www.iloveqatar.net/guide/living/organic-vegetable-farms-and-marketsin-qatar

Reynolds, C. J., Piantadosi, J., & Boland, J. (2015). Rescuing food from the organics waste stream to feed the food insecure: An economic and environmental assessment of Australian food rescue operations using environmentally extended waste input-output analysis. *Sustainability (Basel, Switzerland), 7*(4), 4707–4726. https://doi.org/10.3390/su7044707

Sarkozi, A., & Rukikaire, K. (2020). *Food loss and waste must be reduced for greater food security and environmental sustainability.* FAO. https://www.fao.org/news/story/en/item/1310271/icode/

Scherhaufer, S., Moates, G., Hartikainen, H., Waldron, K., & Obersteiner, G. (2018). Environmental impacts of food waste in Europe. *Waste Management, 77*, 98–113.

Sillitoe, P., & Al-Misnad, S. (2014). *Sustainable Development: An appraisal focusing on the Gulf Region.* Berghan Books.

Solak, S., Scherrer, C., & Ghoniem, A. (2014). The stop-and-drop problem in nonprofit food distribution networks. *Annals of Operations Research, 221*, 407–426.

The Peninsula. (2021, January 13). Hifz Al Naema humanitarian projects benefit over 167,000 in 2020. *Peninsula.* https://thepeninsulaqatar.com/article/13/01/2021/Hifz-Al-Naema-humanitarian-projects-benefit-over-167,000-in-2020. Accessed 9 April 2022.

Van Garde, S. J., & Woodburn, M. J. (1987). Food discard practices of householders. *Journal of the American Dietetic Association, 87*(3), 322–329.

Wahab. (2021). https://www.wahab.qa/news-1. Accessed 9 April 2022.

Watson, M., & Meah, A. (2012). Food, waste and safety: Negotiating conflicting social anxieties into the practices of domestic provisioning. *The Sociological Review, 60*, 102–120.

Williams, H., Wikström, F., Otterbring, T., Löfgren, M., & Gustafsson, A. (2012). Reasons for household food waste with special attention to packaging. *Journal of Cleaner Production, 24*, 141–148.

Yang, T. C., & Matthews, S. A. (2010). The role of social and built environments in predicting self-rated stress: A multilevel analysis in Philadelphia. *Health & Place, 16*(5), 803–810.

Youn, A., Ollinger, M., & Kantor, L. S. (1999). Characteristics of mid-Atlantic food banks and food rescue organizations. *Food Review, 22*, 45–51.

Sana Abusin joined Qatar University, Social and Economic Survey Research Institute (SESRI), in 2019 as Senior Policy Analyst/Research Associate. She is an Environmental and Resource Economist in addressing environmental challenges at the international, national, sub-regional, and continent-wide levels. She is interested in policy-oriented research and analysis in sustainable development as general, focusing on issues of environmental economics-energy, renewal resources, regulatory economics, climate change, food waste management, water security and food security, sustainable agricultural production systems "Aquaponics," and air and water pollution. She has also excellent background on regulatory economics, crime, and punishment in natural resources. Her journal publications are under natural resources in addition to SESRI publications. She participated in various sustainability international conferences worldwide. As a sustainable development desk lead at SESRI, Qatar University, she completed many survey research projects including food waste management, fish consumption, and fish processing quality and safety, and currently working on corruption in education, social, economic and environmental impacts of camping in Qatar, indoor air quality and health impacts and other socioeconomic research. Currently, she is leading a grant project of innovative mobile application to sustain food security in Qatar.

Ebaidalla Mahjoub Ebaidalla is Assistant Research Professor of Economics at Ibn Khaldon Center for Humanities and Social Sciences, Qatar University. He engaged in many research projects related to issues of food security, inequality, and development economics; and he received grants from a number of regional and international organizations including the Economic Research Forum (ERF), African Economic Consortium (AERC), the Global Development Network (GDN). He worked as a part-time consultant for many international organizations and NGOs, such as the United Nations Development Program (UNDP), the World Bank, the International Labor Organization (ILO), and the Food and Agriculture Organization (FAO). He has an extensive background in econometrics and quantitative analysis, and he published several papers in international journals such as Journal of Economic Cooperation and Development (JECD), Middle East Development Journal (MEDJ), International Journal of Sustainable Economy, and the Journal of Developing Areas. He was a Visiting Research Fellow at the Humphrey School of Public Affairs, University of Minnesota (2018–2019), USA, and a Visiting Scholar at the United Nations University, World Institute for Development Economics Research (UNU-WIDER), Helsinki, Finland (April–June 2019). Before joining Qatar University in 2021, he has worked at University of Khartoum, Sudan, for more than 12 years and he was the Head of Economics Department.

Maryam F. Al-Thani is a Senior Research Assistant at the Social and Economic Research Institute at Qatar University, with an MA in Intercultural Communication and Education from Durham University. She has been engaged in many research projects and has many publications.

Chapter 13
Terrestrial Biodiversity in Arid Environments: One Global Component of Climate Crisis Resilience

April Torres Conkey, Cromwell Purchase, Renee Richer, and Nobuyuki Yamaguchi

Abstract The biological wealth of hyper-arid environments has traditionally been underestimated. However, with recent advancements in technology and understanding, desert biodiversity has become an important component of climate change resilience and will become increasingly important. Extreme weather events, such as drought and heat events, are becoming more common and the salinization of soils is increasing. The genetic and metabolic components of desert and saline-adapted organisms from microbes to plants to animals can be used to ensure agricultural resiliency, improve human health through unique bioactive components, and serve as the blueprint for biomimetic designs.

Keywords Arid environment · Terrestrial biodiversity · Climate change · Environment · Qatar

13.1 Introduction

The Convention on Biological Diversity (CBD), to which Qatar is a signatory, charges countries with preventing the loss of their national biological resources whilst also providing a framework for countries to benefit from their biological resources (Glowka et al., 1994). The CBD emphasizes the protection and promotion of biological resources native and unique to the country. Qatar is blessed with an, as yet largely, unrecognized plant, animal, and microbial-based store of biological diversity that can serve as a global resource for climate crisis resilience.

A. T. Conkey
Texas A&M University, College Station, TX, USA

C. Purchase
Doha, Qatar

R. Richer (✉)
Duke Kunshan University, Kunshan, China
e-mail: rricher@richerenvironments.com; renee.richer@dukekunshan.edu.cn

N. Yamaguchi
Qatar University, Doha, Qatar

L. Cochrane and R. Al-Hababi (eds.), *Sustainable Qatar*, Gulf Studies 9,
https://doi.org/10.1007/978-981-19-7398-7_13

Climate change is driving an increase in extreme weather events, such as drought and flooding, extreme temperatures (hot and cold), wind events, and even changes in salinity or pH. Thus, most organisms are likely to be exposed to environmental conditions outside of the normal historical range of tolerance. Over the last 30 years in Qatar, there has been a distinct warming trend with increasing daytime and nighttime temperatures (Cheng et al., 2017). Notably, there has been nighttime average and maximum temperature increases. Such nighttime increases in temperatures have been associated with decreased plant and animal success (Bechelet et al., 2016).

Biodiversity can be defined in multiple ways and at different scales from genetic and species diversity to functional and ecosystem diversity (Hamilton, 2005). Biodiversity is traditionally thought of as species richness and biodiversity hotspots were first defined based on plant species richness (Myers, 1988) with tropical forests being the focus. Today, tropical forest areas are still central to conservation efforts and global biodiversity hotspots (Lele et al., 2010). However, with an emphasis shifting to ensuring genetic (Swarup et al., 2021) diversity and functioning ecosystem processes (Bennett et al., 2009), biodiversity protection and conservation efforts are incorporating a new and expanded view.

The role that hot, arid, and saline ecosystems play in current global processes, ecosystem services assessment, and biodiversity is only starting to be fully appreciated (Khan & Weber, 2006). This may be due to conditions which limited earlier expeditions as well as aesthetic notions that may not have valued deserts as highly as other regions. With generally lower productivity and lower biodiversity values (based on current assessment techniques), desert resources are likely to have been simply underestimated (Razgour et al., 2018). Technological advancements, such as remote sensing and robotics, have improved desert access and research into even the most extreme environments (Wettergreen et al., 1999). Advancements in the field of molecular biology, similarly, have brought the adaptations of organisms in extreme environments to the forefront as a genetic reservoir to build resilience to the climate crisis.

13.2 Plant and Microbial Diversity in Qatar

The plants of Qatar exhibit a wide range of adaptations to hot and dry conditions. Species migration from Africa, Asia, and Europe is reflected in the unique plant community assemblage. More than 400 species have been recorded in Qatar including native, naturalized, invasive, and those restricted to urban or irrigated areas (Abdel Bary, 2012a, b; Norton et al., 2009) with new species records for Qatar continuing to be published (Richer et al., 2022). However, continued documentation of all species present in Qatar will require regular fieldwork during multiple seasons and multiple years. Unusually abundant rainfall in 2014 resulted in new plant species records for Qatar, most of which are thought to be native (Richer et al., 2022). Additional species are likely to be found in restricted areas or border areas that are protected from direct human stressors. Given the wide variability in rainfall, regular monitoring is necessary to maintain an accurate understanding of species present, their abundance, and population status.

Terrestrial species exhibit a wide range of morphological, physiological, and symbiotic adaptations that confer advantages in extreme conditions. Features such as excreting salt, succulence, photosynthetic adaptations, and resilient protein structures are attributes that can be exploited. Many of the species are adapted to the saline, dry, and hot conditions and display microbial associations that facilitate life in this extreme environment (Abdel Bary et al., 2006; Al-Thani & Yasseen, 2018). Halophytes in particular, those plants adapted to saline conditions, support a range of salt tolerant bacteria associated with plant roots. The bacteria themselves can promote salt tolerance in non-salt-tolerant agricultural species, promoting sustainable food production despite increasing salinization of soils (Etesami & Beattie, 2018). Halophtyic plants are now being utilized as animal feed with increasing global applications (Panta et al., 2014).

In deserts, the biological soil crust can play a dominant role in larger scale ecosystem processes (Al-Thani & Yasseen, 2018), facilitate soil water retention and seed germination, prevent soil erosion, and provide important symbionts for plant species. Thus, the hyper-saline sabkha, a coastal mudflat characterized by high evaporation rates (Barth et al., 2002; Chatziefthimiou et al., 2018) and inland biological soil crust are unique in their beauty as well as in the untapped genetic resources available (Oren, 1998). Cyanobacterial based biological soil crust can cover up to 87% of the land surface in northern Qatar, with a gradient in coverage from north to south (Richer et al., 2012). In addition to conferring improved tolerance to extreme conditions, cyanobacteria in Qatar produce a wide range of bioactive compounds (Metcalf et al., 2015; Richer et al., 2015). Although studies on species present and the compounds they produce are in the primary stages, investment in the study of these microbial communities is sure to produce biological and biomimicry solutions ranging from health care to climate change resilience.

The biodiversity of extreme environments in Qatar provides for ecosystem services, functional diversity, and importantly genetic diversity, which is only beginning to be understood. The marine and terrestrial environments provide examples. The Arabian Gulf supports reef building corals at the extremes of temperature and temperature variability (Howells et al., 2016), and the more we understand the relationship between the host and symbiotic algae, the greater the potential for applying that knowledge to problem solving. Recent work with Arabian Gulf corals has shown that these extreme temperature-adapted corals, when bred with non-adapted corals, can transfer heat tolerance, increasing survival up to 84% (Howells et al., 2021). The same may ultimately be true for fish species as well (Vaughan et al., 2021). The transfer of resilience to environmental extremes has also been illustrated in plants. In recent work, a key enzyme isolated from *Salsola laricifolia*, when transformed and expressed into *Arabidopsis*, conferred salt and drought resistance (Wen et al., 2021), thus, illustrating the role that the genetic resources from hot, arid, and saline ecosystems of terrestrial Qatar can provide for global resilience to climate change, particularly for crop species (Swarup et al., 2021).

To fully develop the biological resources of the Qatar desert, investment in protection and research is necessary. Select sites across the country support unusually high diversity for plants in the country (Richer, unpublished data) as well as important sabkha and soil crust areas. The designation of national Important Plant Areas or Important Resource Areas would be an important step to protect those most vulnerable areas that hold disproportionate numbers of species. Additionally, research stations in the field would provide opportunities for long-term studies and intensive resource development programmes.

13.3 Terrestrial Mammal Diversity in Qatar

As a group, mammals are often an important focal group for conservation efforts in general because they play a key role in many ecosystem functions and monitoring the status of the mammalian fauna can provide an indication of the status of the general ecosystem as a whole (Hurlbert, 1997; Landres et al., 1988). However, not only there is a paucity of knowledge with respect to the status of Qatar's mammalian species, but also even the diversity of the species present, and their distributions, in the country are not fully understood (Richer, 2008). Surprisingly little information exists on the terrestrial mammalian fauna of Qatar in contrast to the country's avifauna on which a comparatively large amount of information has been compiled (e.g. Avibase the bird checklist of the world: https://avibase.bsc-eoc.org/checklist.jsp?region=QA). The last mammal survey in Qatar was carried out nearly 40 years ago by Kamel and Madkour (1984), which was not a systematic large-scale survey. A review of Zoology in the Middle East, a peer reviewed journal that specifically publishes zoological information from the Middle East, showed that out of 68 volumes published between 1986 and 2022 only two papers concern terrestrial mammals in Qatar, which are Al-Musfir and Yamaguchi (2008) and Mohedano et al. (2014). At the moment, the Qatar e-Nature lists 20 non-domesticated terrestrial mammals occurring in Qatar (Qatar e-Nature: https://www.enature.qa/kingdom/mammals/) whilst Wikipedia lists 24 non-domesticated terrestrial mammals (Wikipedia: https://en.wikipedia.org/wiki/List_of_mammals_of_Qatar), neither of which, however, provides complete, verified detailed information.

A detailed review of available data indicates that there are 150 species of terrestrial wild mammals recorded in the Arabian Peninsula (Harrison & Bates, 1991), of which 19 appear to have been recorded in Qatar (Table 13.1). However, the occurrence of some species, such as honey badger (*Mellivora capensis*) and golden jackal (*canis aureus*), may be due to sporadic stray individuals. In addition to those 19 species, six other mammal species may occur in Qatar assessing from the species distribution range and the recorded locations in the Arabian Peninsula, but none has so far been recorded in Qatar (Table 13.2).

Approximately 40% of all recorded mammal species are rodents, and hence not surprisingly, rodents have the largest number of species in the non-domesticated terrestrial mammals in Qatar (Table 13.1). The indigenous gerbils, jerboas, and jirds

Table 13.1 Terrestrial mammalian species that occur in Qatar

Species	Latin name	IUCN status	Reference
Lagomorpha			
Arabian Hare	*Lepus capensis*	Least Concern	Kamel and Madkour (1984) and Johnston et al. (2019)
Carnivora			
Sand Cat**	*Felis margarita*	Near Threatened	Sliwa et al. (2016)
Red Fox	*Vulpes*	Least Concern	Kamel and Madkour (1984)
Honey Badger***	*Mellivora capensis*	Least Concern	Do Linh San et al. (2016)
Striped Hyaena**	*Hyaena*	Near Threatened	AbiSaid and Dloniak (2015)
Golden Jackal***	*Canis aureus*	Least Concern	Hoffmann et al. (2018)
Eulipotyphla			
Desert Hedgehog	*Paraechinus aethiopicus*	Least Concern	Kamel and Madkour (1984)
Chiroptera			
Desert Long-Eared Bat	*Otonycteris hemprichii*	Least Concern	Abdulrahman et al. (2021)
Geoffroy's Trident Leaf-Nosed Bat	*Asellia tridens*	Least Concern	Abdulrahman et al. (2021)
Kuhl's Pipistrelle	*Pipistrellus kuhlii*	Least Concern	Abdulrahman et al. (2021)
Rodentia			
Cheeseman's Gerbil	*Gerbillus cheesmani*	Least Concern	Kamel and Madkour (1984) and Shenbrot and Amr (2016)
Dwarf Gerbil	*Gerbillus nanus*	Least Concern	Kamel and Madkour (1984), Nader (1984), and Shenbrot (2016)
Lesser Egyptian Jerboa	*Jaculus*	Least Concern	Nader (1984) and Amori et al. (2021b)
Sundevall's Jird	*Meriones crassus*	Least Concern	Nader (1984) and Granjon (2016a)
Brown Rat	*Rattus norvegicus*	Least Concern	Nader (1984)
Black Rat	*Rattus*	Least Concern	Kryštufek et al. (2021)
House Mouse	*Mus musculus*	Least Concern	Kamel and Madkour (1984)
Cetartiodactyla			
Arabian Oryx*	*Oryx leucoryx*	Vulnerable	IUCN SSC Antelope Specialist Group (2017b)

(continued)

Table 13.1 (continued)

Species	Latin name	IUCN status	Reference
Arabian Sand Gazelle	*Gazella marica*	Vulnerable	IUCN SSC Antelope Specialist Group (2017a)

The primary source of data is Harrison and Bates (1991), and only additional references are shown. *The Arabian oryx has not been (re-)introduced into the wild in Qatar. **Possibly extinct. ***Questionable

Table 13.2 Terrestrial mammalian species that may occur in Qatar

Species	Latin name	Status	Reference
Carnivora			
Ruppell's Fox	*Vulpes rueppellii*	Least Concern	Mallon et al. (2015)
Arabian Wolf	*Canis lupus*	Least Concern	Boitani et al. (2018)
Chiroptera			
Naked-Rumped Tomb Bat	*Taphozous nudiventris*	Least Concern	Monadjem et al. (2017)
Sind Serotine Bat	*Eptesicus nasutus*	Least Concern	Benda et al. (2019)
Rodentia			
Fat Sand Rat	*Psammomys obesus*	Least Concern	Granjon (2016b)
Wagner's Gerbil	*Gerbillus dasyurus*	Least Concern	Amori et al. (2021a)

The primary source of data is Harrison and Bates (1991), and only additional references are shown

are all well adapted to the hyper arid environment in Qatar where there is no natural permanent freshwater source on the surface. However, it is almost certain that house mice (*Mus musculus*), black rats (*Rattus rattus*), and brown rats (*R. norvegicus*) have been unintentionally introduced into Qatar by human activities and survive in the country due to the recent changes of land use and increased availability of freshwater resources. When one of the authors (NY) started fieldwork in the northern Qatar in 2010, it was not common to come across a black rat. However, by 2017, they became very common at least around some human settlements and irrigated farms. On the other hand, during the same period, it was relatively common to come across lesser Egyptian jerboas (*Jaculus jaculus*) at the beginning, and yet, became very difficult towards the later part of the period. Once, the author observed a free-ranging cat hunting a small rodent right in front of his car. As human settlements expand, land-use changes and the population of human-associated animals, such as free-ranging cats, feral dogs, black rats, and brown rats, would increase, the indigenous rodents may likely suffer negative impacts from predation and competition. Although it's not a rodent, Arabian hares (*Lepus capensis*) were common in 2007 in the Qatar University campus in Doha. By 2019 they disappeared from the campus.

A recent survey by Qatar University researchers suggests that bats may be more common in the less disturbed southern part of the country (Abdulrahman et al., 2021). Less disturbed roosting sites in caves, sink holes, and rock crevices are essential for the survival of bats in Qatar. Key bats roosting sites need appropriate protection for the continuing survival of bats in Qatar. Urban environments can be designed to support bat populations with appropriately managed roosting sites (Lewanzik et al., 2022).

Although we highlighted the paucity of even the basic information of indigenous terrestrial mammals in Qatar, there is one exception. It's the Desert hedgehog (*Paraechinus aethiopicus*). In fact, Qatar is one of the world's centres of the research on various aspects of the biology of the desert hedgehog. Studies conducted by researchers from the Qatar University found desert hedgehogs go into torpor in spite of the relatively mild winter in Qatar (Al-Musfir & Yamaguchi, 2008), and they often bask in winter (Abu Baker, Reeve, et al., 2016). They have two breeding peaks a year between winter and summer (Yamaguchi et al., 2013), a female gives birth to, on average, 3.3 babies in captivity (Abu Baker, Mohedano, et al., 2016), and captive-bred animals appear to be able to survive in the wild well when they were released (Abu Baker et al., 2018). Desert hedgehogs have much larger home ranges than do hedgehogs in Europe likely due to the lower productivity of the hyper arid environment of Qatar (Pettett et al., 2018). The desert hedgehog does not show a strong sexual size dimorphism, and yet, males tend to have much larger home ranges than do females and appear to keep the large range all year round including the non-breeding season (Pettett et al., 2018). The population density of desert hedgehogs in Qatar is approximately seven per km^2 in the north of the country where there are many irrigated farms (Pettett et al., 2020). Understandably, desert hedgehogs are heavier, and their home ranges are smaller in an area with many irrigated farms in comparison to those in more "natural" desert environment likely because the available resource level is higher in the former (Abu Baker et al., 2017). In this context, the agricultural land-use change due to the economic miracle of Qatar in the last 50 years may have been helping desert hedgehogs to increase their population, and yet, urbanisation (e.g. building cities and roads) likely increases hedgehog mortality, including traffic accidents, resulting in local extinctions. Wildlife-friendly designs of cities and roads may make human hedgehog coexistence easier in Qatar. Although hedgehogs do not look like long-distance travellers, a genetic study suggests that there appears to be no population division, or clear population structure, amongst desert hedgehogs in Qatar (O'Meara et al., 2021).

Some mammals, including the Arabian sand gazelle (*Gazella marica*), have likely become extinct in Qatar (Al Hamar & Almutai, 2001; Vine & Casey, 1992). However, recently some species have been bred in captivity in Qatar with re-introduction into the wild as one of its goals. The Arabian sand gazelle has been re-introduced into the Al-Reem Biosphere Reserve, one of the two largest terrestrial protected areas in Qatar, and the Arabian Oryx (*Oryx leucoryx*) may be released as well in the future (Al Hamar & Almutai, 2001; IUCN SSC Antelope Specialist Group, 2017a, 2017b; Sillitoe et al., 2010). Ultimately, successful re-introduction requires appropriate habitat recovery and protection.

Such re-introduction likely plays an important role in re-establishing key ecosystem functions including seed dispersal, which sadly has been threatened by the declines in animal populations, limiting the potential for plants to adapt to climate change by shifting their ranges (Del-Claro & Dirzo, 2021; Fricke et al., 2022). Different ungulates tend to disperse a different assemblage of plant species with little overlap in a Middle Eastern desert (Polak et al., 2014). Especially, the Arabian oryx are the key dispersers of seeds of *Vachellia tortilis* (which is commonly found in Qatar), and oryx-ingested seeds planted with oryx pellets had a germination success 250 times higher than non-ingested seeds planted without pellets (Polak et al., 2014). It is important to re-introduce all the species lost (e.g. Arabian oryx) as a restorative process of the ecosystem functions lost even when other species from the same guild may still exist in the wild.

Domesticated or feral mammals commonly found in Qatar are dromedary camel (*Camelus dromedarius*), horse (*Equus caballus*), donkey (*Equus asinus*), sheep (*Ovis aries*), goat (*Capra hircus*), dog (*C. familiaris*), and cat (*F. catus*). Amongst them, feral dogs and seemingly a large number of feral cats are likely causing negative impacts on indigenous wildlife population due to predation. The Qatar Animal Welfare Society (http://www.qaws.org/newsite/) has been working on improving dog welfare and reducing feral dog population in the country. Domesticated grazers and browsers have placed particular stress on important native plant species including *Panicum turgidum* and *Rhanterium eppaposum.*

13.4 Birds in Qatar and Their Role in Maintaining Biodiversity and Culture

Birds and humans have been interconnected for thousands of years. Birds bring inspiration, entertainment, and food and even clothe humans. They have strongly associated with humans throughout the cultural development of our societies. Birds provide both cultural and provisioning ecosystem services. For example, in the form of bird art and eggs, more so falconry in the Middle East, as an entertainment and lucrative hobby, these are commodities that can be bought and sold. There are indirect services associated with birds too. For instance, they help maintain components of ecosystems that we depend on for food, shelter, and services like disease management (vultures) and pest control (owls, insectivores). These indirect services facilitate other ecosystem services in promoting biodiversity.

Birds play an important role in the stability and functioning of the environment. This can be seen in pollination and seed dispersal, which is crucial to any environment, and many plant species are dependent on birds for their dispersal both locally and beyond. Loss of these seed dispersers can seriously impact larger environment services (Del-Claro & Dirzo, 2021; Fricke et al., 2022). If seeds are not dispersed, germinating seedlings will crowd the parent plant and compete for light, space, water, and nutrients. Since plants cannot migrate, they need to rely on other means of seed dispersal. Due to the ability of birds to traverse large distances in a relatively short time, they are a terrestrial plant's most reliable seed disperser. The health and well-being of habitats rely heavily on a good strong biodiversity of creatures to maintain the environment, birds play a substantial role in this maintenance. Declines in avian pollinators and seed dispersers would indirectly affect many other human uses of plants and the habitats in which they occur.

Birds and birding in Qatar have grown remarkably in the last decade. Interest from locals and tourists in seeing Qatar's birding hotspots is an important development for biodiversity in Qatar. It is quite remarkable that 297 bird species are recorded in Qatar, which is a small state of 11,521 sq km, with no rivers, lakes, or natural surface water. Most of the bird species found in Qatar are visitors on their migratory routes, hosting some brightly coloured birds in the form of Bee-eaters (*Merops apiaster, Merops persicus and Merops orientalis*) and Rollers (*Coracias garrulus*), to mention but a few. Although Qatar has no endemic birds, it does host some special Arabian endemics such as the Arabian Scops owl (*Otus brucei*).

Qataris have a long history of falconry, and the country is the natural habitat of a few falcon species (e.g. Saker). This natural inclination of Qataris to falcons suggests that with more education on the broader needs for avian biodiversity and wild avian species richness, they should be willing and in fact be driven to protecting the Important Birding Areas. The importance of which is so great, that international organisations, such as Birdlife International, have come to this small country to help secure Qatar's avian biodiversity. These organizations understand the importance of bird population health as an indicator of the health of regional biodiversity, reflecting species' richness. Changes in bird populations can provide a useful indication for broader environmental change. Hence the avian component used in environmental impact assessments is crucially important.

For some international and national background to biodiversity conservation: Qatar joined the world to recognize the need to maintain biodiversity at the UN conference on environment and development in Brazil in 1992. The Earth summit continued the concerns about the global loss of biological diversity and the need for sustainable use of the components of biodiversity prompted 150 nations including Qatar to sign the Convention on Biological Diversity. The State of Qatar ratified the convention in 1996 through Act No 90. Qatar's Environment and Natural Reserves (SCENR) was responsible to conserve biodiversity under its jurisdiction. However, now the responsibility of biodiversity conservation is now held with the Ministry of Environment and Climate Change with 3 aims; conserve biodiversity, sustainable use of the components of biodiversity, and sharing the benefits arising from commercial and other use of genetic resources in a fair and equitable way.

With the awarding of the 2022 FIFA World Cup, Qatar has ventured out to pursue stronger environmental protection and growth in its state (see Chap. 4). In pursuit of this biodiversity conservation need, Qatar has established grants to conserve its Important Birding Areas. Some of the best birding areas with the highest species sightings included: Irakhiya farm, Abu Nakla, Al Khor, Al Ruwais, Al Sailiya Ponds, Al Karanah treatment Lagoons, Al Wakra beach, Al Thakhira, and Simaisma. These locations are all available to visit except for Abu Nakla, a large water treatment plant, which was closed down in 2016/2017 to a massive outcry from the local birding community. But for infrastructurally strategic reasons, the plant was closed down and water was no longer pumped into the man-made lake. However, the expansion of Al Karanah water treatment lagoons now compensates to some extent for the loss of Abu Nakla.

For people interested in birding and research on bird sightings and locations in Qatar, there is a website where this information can be accessed (ebird.org). This site is actively updated with the latest citizen scientists (birders) who log their sightings on the eBird app. This shows that to date 297 species have been observed within the borders of Qatar. Another important source of local birding information is the book *Common Birds of Qatar* (Eriksen et al., 2010).

Despite its harsh, arid environment, Qatar can provide excellent birding from August to May, with the right wind conditions in spring and autumn and even winter can bring thousands of migrants flooding into the country. A great time to be in Qatar for birding is in spring, when a regionally important bird migration takes place, as part of the West Asia-East Africa Flyway (Kirby et al., 2008). The peak periods to visit Qatar for birding are reported to be from September–November and February–April. Qatar's coastal habitats including mangroves are the most important and equally the habitats needing the most protection. Human encroachment on the coastal habitats needs to be closely monitored and more legislated protections would be helpful to meet the biodiversity conservation needs of the country.

Urban areas also host an array of invasive and introduced Arabian-Asian species including White-eared (*Pycnonotus leucotis*) and red vented bulbuls (*Pycnonotus cafer*), and even parrot species like Alexandrine (*Psittacula eupatria*) and Ring-necked parakeets (*Psittacula krameria*). From the parks and gardens of the city of Doha, many species can be seen and heard. However, it's the rural areas, beaches, and farmlands that bring birders the joys of variety and the satisfaction of finding some of those more elusive species like the Namaqua doves, Lilith owls. and the notoriously unpredictable and sort after Hypocolius (*Hypocolius ampelinus*).

Qatar has grown its Protected Area (PA) designations, which will go a long way to securing a stable and growing biodiversity. However, a publication in 2016 (Zogaris & Kallimanis, 2016), doing rapid assessment surveys of bird species richness and abundance during the spring migration, highlighted the need for further protection of the coastal habitats. There is abundant evidence that anthropogenic pressures and threats to natural habitats concentrate on coastal zones. The important sites for biodiversity, like coastal lagoons and mangroves, are under imminent threat (Al-Khayat & Balakrishnan, 2014).

The link between biodiversity conservation, avian functional families, and ecosystem services is becoming more and more evident. Some standard bird functions, like scavenging by vultures, nutrient deposition by seabirds, and vertebrate predation by raptors, are declining rapidly due to the decline of bird species in these groups. Affording better protection to the bird species will help the natural healing of the environments in which they live, and fundamentally allow for a stronger biodiversity and enrich our lives.

How can the people of Qatar help with protecting the biodiversity of the country and its ecosystems? There are many small farms located throughout Qatar, these little green "oasis" locations scattered around are the house of many species and give birds the opportunity to reside, breed, feed, and shelter in the harsh Qatar environment, they create in a way mini-protected areas. The same applies to the gardens in the urban sprawl of the cities, with the catching term "Urban Rewilding". The planting bird friendly trees and bushes, with fruits, berries, and seeds help create an increased biodiversity within the Urban environment. This helps to secure species that would otherwise be in decline and/or regionally extinct. If we all work together and play even the smallest part in the restoration of biodiversity conservation at all levels, our lives, and the lives of all the species we share our country and environment with will be better for it.

13.5 Toads

Biological invasions are mostly caused by human activities deliberately (or accidentally) moving non-native "alien" species to areas outside their "natural" ranges where they flourish, which is one of the main environmental problems today. Qatar is one of the only few countries with no permanent surface freshwater source where no indigenous amphibian likely occurs. However, the recent massive increase of freshwater resources and associated land-use changes in Qatar have likely provided comfortable new homes for African common toads (*Sclerophrys regularis*), which may have been introduced from Egypt (Yamaguchi et al., 2019). Interestingly, toads were present in sites where water was supplied from outside sources by irrigation pipes and/or water transportation vehicles more frequently than in sites where water was provided exclusively from inside sources (wells) (Abdulkarim & Yamaguchi, 2021). The distribution pattern of toads in Qatar may suggest that toads were initially introduced to eastern parts of the country and have dispersed from there through water transportation networks (Abdulkarim & Yamaguchi, 2021).

13.6 Terrestrial Reptiles

Reptiles are surprisingly diverse in desert ecosystems. With adaptations that allow them to withstand arid conditions, lizards and snakes are thought to be widely dispersed across Qatar and are important components in maintaining ecosystem balance. In addition, they provide benefits by helping control "pest" species (insects and rodents) and maintaining biodiversity and habitat quality by dispersing plant seeds. Habitat requirements include microenvironments where they can seek refuge from temperature extremes (shade from plants or rock formations, crevices between and under rocks, and soil requirements to be able to dig underground burrows). Habitat loss to land development and agricultural use, including over-grazing of livestock, is considered the greatest threat to reptiles across the Arabian Peninsula (Cox et al., 2012). Over-harvesting of reptiles for consumption (Spiny-tailed Agama) and pet trade and the persecution and killing of snakes and large lizards is also a threat (Cox et al., 2012), although the extent of these issues in Qatar remains unknown.

Lizards are the most common reptile group in Qatar, with 21 species documented in the most recent and comprehensive survey by Cogălniceanu et al. (2014). Seven different families of lizards were documented from this study (Gekkonidae, Lacertidae, Agamidae, Scincidae, Varanidae, Trogonophiidae, and Sphaerodactylidae), with geckos being the most diverse group. Most lizards are carnivorous, feeding on insects and arthropods, and help keep invertebrate populations under control. In contrast, the Spiny-tailed Agama, or Dhub (*Uromastyx aegyptia microlepus*) is primarily herbivorous and an important seed disperser of desert plants with the seeds of nearly 40 species recorded in the faeces (Conkey et al., 2010). Large and easily identified and captured, Dhubs are well studied and were a supplemental food source for people (Monchot et al., 2014). Dhubs are also popular in the exotic pet trade and for the production of medicinal oils in Malaysia, which has led to over-harvesting in some regions (Ching & Chng, 2016).

Little documentation exists on the snake species of Qatar. A review of the book *Snakes of Arabia* (Egan, 2007) and an 18 March 2022 web-search on Reptile Database (www.reptile-database.org) and Google Scholar (https://scholar.google.com/) revealed no published scientific literature for snakes in Qatar, other than two articles referencing venomous snakebites (Elmoheen et al., 2020; Haidar & Deitch, 2015) and a survey of citizens, asking if they had seen sea snakes in the waters of Qatar (Castilla et al., 2017). Qatar eNature lists 11 terrestrial snake species, two of which are venomous, but no citations are given for species documentation within Qatar (www.enature.qa, accessed 18 March 2022).

The biological wealth in the reptiles of Qatar remains largely unknown. Options to promote protection and knowledge-based practices include conducting and publishing standardized and periodic reptile surveys and habitat assessments (including terrestrial snakes) and developing and implementing best practice guidelines for land development and agricultural/grazing practices to minimize impacts on habitat.

13.7 Terrestrial Arthropods (Insects, Arachnids, and Crustaceans)

Invertebrates represent a large and important portion of most ecosystems, yet they are often overlooked and lack study. Since 1979, a few insect surveys in Qatar have been published. Abdu and Shawmar (1985) documented 170 insect species across Qatar in various habitats using several different trap types over a 34-month period. Abushama has conducted several surveys (1997, 1999, 2002, 2006), with the 2006 survey lasting an entire year of both daytime and nighttime sampling of small flying insects in the urban city of Doha. Eleven different insect orders (Hemiptera, Coleoptera, Diptera, Hymenoptera, Collembola, Odonata, Psocoptera, Thysanoptera, Mecoptera, Lepidoptera, and Neuroptera) were documented, and on average, less than 10 species were collected per month. Most recently, AlHajri (2013) documented 110 different insect species from June to December 2012 using pit-fall ground traps in four different habitat types (no vegetation, vegetated, hedge row, and farm) in northern Qatar.

Because of the ease of sampling and their important place in the food web, biodiversity surveys of insects and other arthropods are often used as indicators of environmental health. When their abundance or diversity changes, it is often a signal that impacts to other flora and fauna maybe seen in the ecosystem. Thus, standardized arthropod surveys are good options for monitoring biodiversity and ecosystem health. In addition, monitoring invertebrates is critical for human health. Mosquitoes, flies, and ticks are the most common vectors that transmit diseases from animals to humans. Qatar's National Malaria Surveillance System is a good example and can be used as a model for other zoonotic diseases (Chehab et al., 2018; Schaffner et al., 2021). Domestic livestock, pets, exotic animals, and wild animals could all be disease hosts (Barradas et al., 2020; Chavshin & Seyyed-Zadeh, 2021; Lima et al., 2019; Schaffner et al., 2021; Seimenis, 2008; Wernery, 2014), thus, a "One Health" approach to include human, animal, and environmental factors in a national zoonotic disease management programme would help keep Qatar healthy.

Options to promote protection and knowledge-based practices include developing standardized periodic arthropod surveys to monitor ecosystem health and a National zoonotic disease surveillance, planning, and management programme. Harvesting the biological wealth of arthropods includes pursuing biomimicry in the development of desert-adapted infrastructure.

13.8 Summary

The range of communities and organisms from microbial to plant and animal in Qatar remains with few detailed studies. However, these organisms are uniquely adapted to extreme conditions of temperature, aridity, and salinity. At all levels of the biological hierarchy, these adaptations provide a wealth of biological resources that can be used directly to build climate change resilience through genetic engineering and breeding, promote human health with novel bioactive compounds developed from plants, animals, and microbes, and serve as a biomimetic resource. The last century has seen the discovery of the fossil biological resources (fossil fuel energy resources) in the region and the technology to develop those resources has brought great riches. The next century will surely see the next wave of wealth comes from the living biological resources of the amazing desert.

Appendix

Photographs

Arabian oryx

Arabian sand gazelle

Cheeseman's gerbil

Desert hedgehog

Bats (*Asellia tridens*)

Red fox

Others—Geckos toads invertebrates

References

Abdel Bary, E. M. M. (2012a). *The Flora of Qatar:Volume 1:The Dicotyledons*. Environmental Studies Center, Qatar University.

Abdel Bary, E. M. M. (2012b). *The Flora of Qatar:Volume 2:The Monocotyledons*. Environmental Studies Center, Qatar University.

Abdel Bari, E. M. M.,Yasseen, B.T., & Al-Thani, R. F. (2006). *Halophytes in the State of Qatar*. Environmental Studies Center, Qatar University.

Abdulkarim, A., & Yamaguchi, N. (2021). How does an alien amphibian expand its distribution in an hyper-arid environment?: African common toad (*Sclerophrys regularis*) in Qatar. *Journal of Arid Environments*, 184. https://doi.org/10.1016/j.jaridenv.2020.104250

Abdulrahman, M., Gardner, A., & Yamaguchi, N. (2021). The occurrence and distribution of bats in Qatar. *Journal of Arid Environments*, 185. https://doi.org/10.1016/j.jaridenv.2020.104379

Abdu, R. M., & Shawmar, N. T. F. (1985). A preliminary list of insect fauna of Qatar. *Qatar University Science Bulletin, 5*, 215–232.

AbiSaid, M., & Dloniak, S. M. D. (2015). *Hyaena hyaena. The IUCN Red List of Threatened Species* 2015: e.T10274A45195080. https://doi.org/10.2305/IUCN.UK.2015-2.RLTS.T10274A45195080.en. Accessed on 6 March 2022.

Abu Baker, M. A., Mohedano, I., Reeve, N., & Yamaguchi, N. (2016). Observations on the postnatal growth and development of captive Ethiopian hedgehogs, *Paraechinus aethiopicus*, in Qatar. *Mammalia, 80*, 515–520.

Abu Baker, M. A., Reeve, N., Conkey, A. A. T., Macdonald, D. W., & Yamaguchi, N. (2017). Hedgehogs on the move: Testing the effects of land use change on home range size and movement patterns of free-ranging Ethiopian hedgehogs. *PLoS ONE, 12*, e0180826.

Abu Baker, M. A., Reeve, N., Mohedano, I., Conkey, A. A. T., Macdonald, D. W., & Yamaguchi, N. (2016). Caught basking in the winter sun: Preliminary data on winter thermoregulation in the Ethiopian hedgehog, *Paraechinus aethiopicus*, in Qatar. *Journal of Arid Environments, 125*, 52–55.

Abu Baker, M. A., Villamil, L. A., Reeve, N., Karanassos, C., Mahtab, H., & Yamaguchi, N. (2018). Into the wild: Survival, movement patterns, and weight changes in captive Ethiopian hedgehogs, *Paraechinus aethiopicus* following their release. *Journal of Arid Environments, 158*, 43–46.

Abushama, F. T. (1997). Density and diversity of the desert Arthropoda of Qatar. *Qatar University Science Journal, 17*(2), 359–375.

Abushama, F. T. (1999). Dominance value and community production of desert Arthropoda in Qatar. *Qatar University Science Journal, 18*, 137–153.

Abushama, F. T. (2002). Desert arthropods of Qatar: A study to spark conservation assessment. *Antenna-Bulletin Royal Entomological Society London, 26*(2), 106–111.

Abushama, F. T. (2006). Night and early morning flying insects in a residence backyard in Doha city, Qatar. *Qatar University Science Journal, 26*, 83–90.

AlHajri, S. H. (2013). *Seasonal changes in biodiversity and abundance of invertebrates in different ecological environments* (Master's thesis). Qatar University.

Al Hamar, Y., & Almutai, A. (2001). Qatar. In D. P. Mallon & S. C. Kingswood (Eds.), *Antelopes, global survey and regional action plans. Part 4: North Africa, the Middle East and Asia* (pp. 82–83). IUCN.

Al-Khayat, J., & Balakrishnan, P. (2014). *Avicenna marina* around Qatar: Tree, seedling and pneumatophore densities in natural and planted mangroves using remote sensing. *International Journal of Sciences, 2*, 18–27.

Al-Musfir, H. M., & Yamaguchi, N. (2008). Timings of hibernation and breeding of Ethiopian Hedgehogs, *Paraechinus aethiopicus*, in Qatar. *Zoology in the Middle East, 44*, 67–74.

Al-Thani, R. F., & Yasseen, B. T. (2018). Biological soil crusts and extremophiles adjacent to native plants at Sabkhas and Rawdahs, Qatar: The possible roles. *Frontiers in Environmental Microbiology, 4*(2), 55–70.

Amori, G., Hutterer, R., Kryštufek, B., Yigit, N., Mitsainas, G., Palomo, L., & Aulagnier, S. (2021a). *Gerbillus dasyurus* (amended version of 2016 assessment). *The IUCN Red List of Threatened Species* 2021: e.T9116A197509063. https://doi.org/10.2305/IUCN.UK.2021-1.RLTS.T9116A197509063.en. Accessed 6 March 2022.

Amori, G., Hutterer, R., Kryštufek, B., Yigit, N., Mitsainas, G., Palomo, L,. & Aulagnier, S. (2021b). *Jaculus jaculus* (amended version of 2016 assessment). *The IUCN Red List of Threatened Species* 2021: e.T10912A197517244. https://doi.org/10.2305/IUCN.UK.2021-1.RLTS.T10912A197517244.en. Accessed 6 March 2022.

Bachelet, D., Ferschweiler, K., Sheehan, T., & Strittholt, J. (2016). Climate change effects on southern California deserts. *Journal of Arid Environments, 127*, 17–29.

Barradas, P. F., Mesquita, J. R., Lima, C., Cardoso, L., Alho, A. M., Ferreira, P., Amorim, I., de Sousa, R., & Gärtner, F. (2020). Pathogenic Rickettsia in ticks of spur-thighed tortoise (*Testudo graeca*) sold in a Qatar live animal market. *Transboundary and Emerging Diseases, 67*(1), 461–465.

Barth, H. J., Boer, B., & Böer, B. (Eds.). (2002). *Sabkha Ecosystems: Volume I: The Arabian Peninsula and Adjacent Countries* (Vol. 1). Springer Science & Business Media.

Benda, P., Srinivasulu, C., & Srinivasulu, B. (2019). *Eptesicus nasutus. The IUCN Red List of Threatened Species* 2019: e.T7935A22117147. https://doi.org/10.2305/IUCN.UK.2019-3.RLTS.T7935A22117147.en. Accessed 6 March 2022.

Bennett, A. F., Haslem, A., Cheal, D. C., Clarke, M. F., Jones, R. N., Koehn, J. D., Lake, P. S., Lumsden, L. F., Lunt, I. D., Mackey, B. G., & Nally, R. M. (2009). Ecological processes: A key element in strategies for nature conservation. *Ecological Management & Restoration, 10*(3), 192–199.

Boitani, L., Phillips, M., & Jhala, Y. (2018). *Canis lupus* (errata version published in 2020). *The IUCN Red List of Threatened Species* 2018: e.T3746A163508960. https://doi.org/10.2305/IUCN.UK.2018-2.RLTS.T3746A163508960.en. Accessed 6 March 2022.

Castilla, A. M., Riera, R., Humaid, M. A., Garland, T., Jr., Alkuwari, A., Muzaffar, S., Naser, H. A., Al-Mohannadi, S., Al-Ajmi, D., Chikhi, A., & Wessels, J. (2017). Contribution of citizen science to improve knowledge on marine biodiversity in the Gulf Region. *Journal of the Association of Arab Universities for Basic and Applied Sciences, 24*, 126–135.

Chatziefthimiou, A. D., Deitch, E. J., Glover, W. B., Powell, J. T., Banack, S. A., Richer, R. A., Cox, P. A., & Metcalf, J. S. (2018). Analysis of neurotoxic amino acids from marine waters, microbial mats, and seafood destined for human consumption in the Arabian Gulf. *Neurotoxicity Research, 33*(1), 143–152.

Chavshin, A. R., & Seyyed-Zadeh, S. J. (2021). Ticks and tick-borne diseases. *Climate, Ticks and Disease, 18*, 469.

Chehab, M. A., Bala, M. O., Al-Dahshan, A., Selim, N. A., Al-Romaihi, H. E., Al-Thani, M., & Farag, E. A. (2018). Evaluation of the completeness and timeliness of National Malaria Surveillance System in Qatar, 2016. *Cureus, 10*(6).

Cheng, W. L., Saleem, A., & Sadr, R. (2017). Recent warming trend in the coastal region of Qatar. *Theoretical and Applied Climatology, 128*(1), 193–205.

Ching, O. O., & Chng, S. C. (2016). The use of spiny-tailed lizards *Uromastyx spp.* for medicinal purposes in Peninsular Malaysia. *TRAFFIC Bulletin, 28*(1), 35.

Cogălniceanu, D., Castilla, A. M., Valdeon, A., Gosa, A., Al-Jaidah, N., Alkuwary, A., Saifelnasr, E. O., Mas-Peinado, P., Richer, R., & Al-Hemaidi, A. A. M. (2014). A preliminary report on the distribution of lizards in Qatar. *ZooKeys, 373*, 67.

Conkey, A. A. T., Richer, R., Castilla, A. M., Tribuna, J., & Chan, R. (2010). Non-invasive method to examine the diet of the spiny-tailed lizard, *Uromastyx aegyptia microlepis*, in Qatar. *Qatar Foundation Annual Research Forum, 2010*(1), EEP18.

Cox, N. A., Mallon, D., Bowles, P., Els, J., & Tognelli, M. F. (2012). *The conservation status and distribution of reptiles of the Arabian Peninsula.* IUCN and Environment and Protected Areas Authority.

Del-Claro, K., & Dirzo, R. (2021). Impacts of Anthropocene defaunation on plant-animal interactions. In *Plant-animal interactions* (pp. 333–345). Springer.

Do Linh San, E., Begg, C., Begg, K., & Abramov, A. V. (2016). *Mellivora capensis. The IUCN Red List of Threatened Species* 2016: e.T41629A45210107. https://doi.org/10.2305/IUCN.UK.2016-1.RLTS.T41629A45210107.en. Accessed on 6 March 2022.

Elmoheen, A., Salem, W. A., Haddad, M., Bashir, K., & Thomas, S. H. (2020). Experience of snakebite envenomation by a desert viper in Qatar. *Journal of Toxicology, 2020.*

Egan, D. (2007). *Snakes of Arabia: A field guide to the snakes of the Arabian Peninsula and its shores.* Motivate Pub., Dubai Media City.

Eriksen, H., Erisksen, J., & Gillespie, F. (2010). *Common birds of Qatar*. Published privately by the authors.

Etesami, H., & Beattie, G. A. (2018). Mining halophytes for plant growth-promoting halotolerant bacteria to enhance the salinity tolerance of non-halophytic crops. *Frontiers in Microbiology, 9*, 148.

Fricke, E. C., Ordonez, A., Rogers, H. S., & Svenning, J. C. (2022). The effects of defaunation on plants' capacity to track climate change. *Science, 375*(6577), 210–214.

Glowka, L., Synge, H., & Burhenne-Guilmin, F. (1994). *A guide to the Convention on Biological Diversity.* IUCN—The World Conservation Union.

Granjon, L. (2016a). *Meriones crassus. The IUCN Red List of Threatened Species* 2016: e.T13161A22433449. https://doi.org/10.2305/IUCN.UK.2016-3.RLTS.T13161A22433449.en. Accessed 6 March 2022.

Granjon, L. (2016b). *Psammomys obesus* (errata version published in 2017). *The IUCN Red List of Threatened Species* 2016: e.T18418A115143282. https://doi.org/10.2305/IUCN.UK.2016-3.RLTS.T18418A22416093.en. Accessed on 6 March 2022.

Haidar, N. A., & Deitch, E. (2015). Snake bites in the Arabian Peninsula, a review article. *Journal of Arid Environments, 112*, 159–164.

Hamilton, A. J. (2005). Species diversity or biodiversity? *Journal of Environmental Management, 75*(1), 89–92.

Harrison, D. L., & Bates, P. J. J. (1991). *The Mammals of Arabia* (Harrison Zoological Museum, 354).

Hoffmann, M., Arnold, J., Duckworth, J. W., Jhala, Y., Kamler, J. F., & Krofel, M. (2018). *Canis aureus* (errata version published in 2020). *The IUCN Red List of Threatened Species* 2018: e.T118264161A163507876. https://doi.org/10.2305/IUCN.UK.2018-2.RLTS.T118264161A163507876.en. Accessed 6 March 2022.

Howells, E. J., Abrego, D., Liew, Y. J., Burt, J. A., Meyer, E., & Aranda, M. (2021). Enhancing the heat tolerance of reef-building corals to future warming. *Science Advances, 7*(34), eabg6070.

Howells, E. J., Abrego, D., Meyer, E., Kirk, N. L., & Burt, J. A. (2016). Host adaptation and unexpected symbiont partners enable reef-building corals to tolerate extreme temperatures. *Global Change Biology, 22*, 2702–2714.

Hurlbert, S. H. (1997). Functional importance vs. keystoneness: Reformulating some questions in theoretical biocenology. *Australian Journal of Ecology, 22*, 369–382.

IUCN SSC Antelope Specialist Group. (2017a). *Gazella marica. The IUCN Red List of Threatened Species* 2017: e.T8977A50187738. https://doi.org/10.2305/IUCN.UK.2017-2.RLTS.T8977A50187738.en. Accessed 6 March 2022.

IUCN SSC Antelope Specialist Group. (2017b). *Oryx leucoryx. The IUCN Red List of Threatened Species* 2017: e.T15569A50191626. https://doi.org/10.2305/IUCN.UK.2017-2.RLTS.T15569A50191626.en. Accessed 2 March 2022.

Johnston, C. H., Robinson, T. J., Child, M. F., & Relton, C. (2019). *Lepus capensis. The IUCN Red List of Threatened Species* 2019: e.T41277A45186750. https://doi.org/10.2305/IUCN.UK.2019-1.RLTS.T41277A45186750.en. Accessed 6 March 2022.

Kamel, A., & Madkour, G. (1984). New records of some mammals from Qatar: Insectivora. Lagomorpha, and Rodentia. *Qatar University Scientific Bulletin, 4*, 125–128.

Khan, M. A., & Weber, D. J. (2006). *Ecophysiology of high salinity tolerant plants* (1st ed.). Springer Netherlands. https://doi.org/10.1007/1-4020-4018-0

Kirby, J. S., Stattersfield, A. J., Butchart, S. H. M., Evans, M. I., Grimmett, R. F. A., Jones, V. R., O'Sullivan, J., Tucker, G. M., & Newton, I. (2008). Key conservation issues for migratory land- and waterbird species on the world's major flyways. *Bird Conservation International, 18*, S49–S73.

Kryštufek, B., Palomo, L., Hutterer, R., Mitsainas, G., & Yigit, N. (2021). *Rattus rattus* (amended version of 2016 assessment). *The IUCN Red List of Threatened Species* 2021: e.T19360A192565917. https://doi.org/10.2305/IUCN.UK.2021-1.RLTS.T19360A192565917.en. Accessed 6 March 2022.

Landres, P. B., Verner, J., & Thomas, J. W. (1988). Ecological uses of vertebrate indicator species: A critique. *Conservation Biology, 2*, 316–328.

Lele, S., Wilshusen, P., Brockington, D., Seidler, R., & Bawa, K. (2010). Beyond exclusion: Alternative approaches to biodiversity conservation in the developing tropics. *Current Opinion in Environmental Sustainability, 2*(1–2), 94–100.

Lewanzik, D., Straka, T. M., Lorenz, J., Marggraf, L., Voigt-Heucke, S., Schumann, A., Brandt, M., & Voigt, C. C. (2022). Evaluating the potential of urban areas for bat conservation with citizen science data. *Environmental Pollution, 297*, 118785.

Lima, C., Colella, V., Latrofa, M. S., Cardoso, L., Otranto, D., & Alho, A. M. (2019). Molecular detection of *Leishmania* spp. in dogs and a cat from Doha, Qatar. *Parasites & Vectors, 12*(1), 1–4.

Mallon, D., Murdoch, J. D., & Wacher, T. (2015). *Vulpes rueppellii. The IUCN Red List of Threatened Species* 2015: e.T23053A46197483. https://doi.org/10.2305/IUCN.UK.2015-4.RLTS.T23053A46197483.en. Accessed 6 March 2022.

Metcalf, J. S., Banack, S., Richer, R., & Cox, P. (2015). Neurotoxic amino acids and their isomers in desert environments. *Journal of Arid Environments, 112*, 140–144.

Mohedano, I., Abu Baker, M. A., Hunter, B., Buchan, J., Michaels, C. J., & Yamaguchi, N. (2014). On the diet of the Pharaoh Eagle Owl, *Bubo ascalaphus* (Savigny, 1809), in Qatar, with an overview of its feeding habits. *Zoology in the Middle East, 60*, 111–119.

Monadjem, A., Racey, P. A., Amr, Z. S. S., Fahr, J., Palmeirim, J., Benda, P., Bates, P., Aulagnier, S., Bergmans, W., Hutson, A. M., & Kock, D. (2017). *Taphozous nudiventris. The IUCN Red List of Threatened Species* 2017: e.T21462A22109884. https://doi.org/10.2305/IUCN.UK.2017-2.RLTS.T21462A22109884.en. Accessed 6 March 2022.

Monchot, H., Bailon, S., & Schiettecatte, J. (2014). Archaeozoological evidence for traditional consumption of spiny-tailed lizard (Uromastyx aegyptia) in Saudi Arabia. *Journal of Archaeological Science, 45*, 96–102.

Myers, N. (1988). Threatened biotas: Hotspots in tropical forests. *The Environmentalist, 8*, 178–208.

Nader, I. A. (1984). First records of rodents from the state of Qatar (Mammalia: Rodentia) with a checklist of the mammals of the State. *Zeitschrift fur Saeugetierkunde, 49*, 117–121.

Norton, J., Abdul Majid, S., Allan, D., Al Safran, M., Böer, B., & Richer, R. (2009). *An Illustrated Checklist of the Flora of Qatar*. Browndown Publications.

O'Meara, D., O'Reilly, C., Abdullahi, A. A., Abu Baker, M., & Yamaguchi, N. (2021). Phylogeography of desert hedgehogs (*Paraechinus aethiopicus*) in Qatar: Implications for its intra-specific phylogeny and taxonomy. *Journal of Arid Environments*. https://doi.org/10.1016/j.jaridenv.2021.104584

Oren, A. (1998). *Microbiology and biogeochemistry of hypersaline environments* (Vol. 5). CRC Press.

Panta, S., Flowers, T., Lane, P., Doyle, R., Haros, G., & Shabala, S. (2014). Halophyte agriculture: Success stories. *Environmental and Experimental Botany, 107*, 71–83.

Pettett, C. E., Al-Hajri, A., Al-Jabiry, H., Macdonald, D. W., & Yamaguchi, N. (2018). A comparison of the Ranging behaviour and habitat use of the Ethiopian hedgehog (*Paraechinus aethiopicus*) in Qatar with hedgehog taxa from temperate environments. *Scientific Reports, 8*, 17783. https://doi.org/10.1038/a41598-018-36117-5

Pettett, C., Macdonald, D. W., Hajiri, A. A., Jabiry, H. A., & Yamaguchi, N. (2020). Characteristics and demography of a free-ranging Ethiopian Hedgehog, Paraechinus aethiopicus, population in Qatar. *Animals (Basel), 10*(6), 951. https://doi.org/10.3390/ani10060951

Polak, T., Gutterman, Y., Hoffman, I., & Saltz, D. (2014). Redundancy in seed dispersal by three sympatric ungulates: A reintroduction perspective. *Animal Conservation, 17*, 6, 565–572.

Qatar eNature. (2022). *Snakes*. https://www.enature.qa/kingdom/snakes/. Accessed 18 March 2022.

Razgour, O., Persey, M., Shamir, U., Korine, C., & Syphard, A. (2018). The role of climate, water and biotic interactions in shaping biodiversity patterns in arid environments across spatial scales. *Diversity and Distributions, 24*(10), 1440–1452.

Richer, R. (2008). *Conservation in Qatar: Impacts of increasing industrialisation* (p. 23). Centre for International and Regional Studies, Georgetown University.

Richer, R., Anchassi, D., El-Assaad, I., El-Matbouly, M., Ali, F., Makki, I., & Metcalf, J. (2012). Variation in the coverage of biological soil crusts in the State of Qatar. *Journal of Arid Environments, 78*, 187–190.

Richer, R., Banack, S. A., Metcalf, J. S., & Cox, P. A. (2015). The persistence of cyanobacterial toxins in desert soils. *Journal of Arid Environments, 112*, 134–139.

Richer, R., Knees, S., Norton, J., Sergeev, A. (2022). Hidden Beauty: and Exploration of Qatar's native and naturalized Flora. Akkadia Press.

Schaffner, F., Bansal, D., Mardini, K., Al-Marri, S. A., Al-Thani, M. H. J., Al-Romaihi, H., Sultan, A. A., Al-Hajri, M., & Farag, E. A. B. A. (2021). Vectors and vector-borne diseases in Qatar: Current status, key challenges and future prospects. *Journal of the European Mosquito Control Association, 39*(1), 3–13.

Seimenis, A. M. (2008). The spread of zoonoses and other infectious diseases through the international trade of animals and animal products. *Veterinaria Italiana, 44*(4), 591–599.

Shenbrot, G. (2016). *Gerbillus nanus* (errata version published in 2018). *The IUCN Red List of Threatened Species* 2016: e.T9136A135436781. https://doi.org/10.2305/IUCN.UK.2016-2.RLTS.T9136A22463551.en. Accessed 6 March 2022.

Shenbrot, G., & Amr, Z. (2016). *Gerbillus cheesmani* (errata version published in 2017). *The IUCN Red List of Threatened Species* 2016: e.T9113A115089958. https://doi.org/10.2305/IUCN.UK.2016-3.RLTS.T9113A22464809.en. Accessed 6 March 2022.

Sillitoe, P., Alshawi, A. A., & Hassan, A. K. A. A. (2010). Challenges to conservation: Land use change and local participation in the Al Reem Biosphere Reserve, West Qatar. *Journal of Ethnobiology and Ethnomedicine, 6*. https://doi.org/10.1186/1746-4269-6-28

Sliwa, A., Ghadirian, T., Appel, A., Banfield, L., Sher Shah, M., & Wacher, T. (2016). *Felis margarita. The IUCN Red List of Threatened Species* 2016: e.T8541A50651884. https://doi.org/10.2305/IUCN.UK.2016-2.RLTS.T8541A50651884.en. Accessed 6 March 2022.

Swarup, S., Cargill, E. J., Crosby, K., Flagel, L., Kniskern, J., & Glenn, K. C. (2021). Genetic diversity is indispensable for plant breeding to improve crops. *Crop Science, 61*(2), 839–852.

Vaughan, G. O., Shiels, H. A., & Burt, J. A. (2021). Seasonal variation in reef fish assemblages in the environmentally extreme southern Persian/Arabian Gulf. *Coral Reefs, 40*(2), 405–416.

Vine, P., & Casey, P. (1992). *The heritage of Qatar*. Immel.

Wen, Z., Wang, Y., Xia, C., Zhang, Y., & Zhang, H. (2021). Chloroplastic SaNADP-ME4 of C3–C4 woody desert species Salsola laricifolia confers drought and salt stress resistance to Arabidopsis. *Plants, 10*(9), 1827.

Wernery, U. (2014). Zoonoses in the Arabian Peninsula. *Saudi Medical Journal, 35*(12), 1455–1462.

Wettergreen, D., Bapna, D., Maimone, M., & Thomas, G. (1999). Developing Nomad for robotic exploration of the Atacama Desert. *Robotics and Autonomous Systems, 26*(2–3), 127–148.

Yamaguchi, N., Al-Hajri, A., & Al-Jabiry, H. (2013). Timing of breeding in free-ranging Ethiopian hedgehogs, *Paraechinus aethiopicus*, from Qatar. *Journal of Arid Environments, 99*, 1–4.

Yamaguchi, N., Vasconcelos, R., Mohamed, T., Ahmed, T., Alatalo, J. M., & Barnett, R. (2019). Toads in Qatar: The species present and their probable original source. *Journal of Arid Environments, 160*, 91–94.

Zogaris, S., & Kallimanis, A. (2016). Coastal zone habitat-use by birds in Qatar: Insights from a rapid assessment method during spring migration. *Tropical Conservation Science, 9*(2), 658–676.

April Torres Conkey is a wildlife biologist with a Ph.D. in Wildlife and Fisheries Sciences from Texas A&M University. From 2009 to 2012, she lived and worked in Qatar and was an Adjunct Assistant Professor of Biology at Carnegie Mellon University in Qatar. While in Qatar, she participated in various research projects including work on the desert hedgehog and spiny-tailed lizard. She currently serves as the Vice-Chair Board Member of the Texas Children in Nature Network and teaches as an adjunct instructor in the Department of Biological and Health Sciences at Texas A&M University-Kingsville.

Cromwell Purchase has from a young child enjoyed wildlife and the natural world, with a special interest in birds. Having grown up on the North East coast of South Africa, he was spoilt with incredible wildlife and biodiversity rich habitats. Following his interests, he pursued an education in zoology and ornithology, becoming a conservation biologist. In 2010, after completing his Ph.D., he moved to Qatar to take on the Blue Macaw Captive coordinator position at the esteemed Al Wabra Wildlife Preservation, where he took on the Spix's macaw de-extinction project among others, and within a few years ended up as Director of the Preservation. In 2018, it was time to move forward with the flagship Spix's macaw project and he moved to Germany and now resides in Brazil, preparing for the first release of an extinct parrot, back into its former habitat.

Renee Richer received her BA in Biology from the University of Chicago and her Ph.D. in Biology from Harvard University addressing the impact of atmospheric CO_2 increases on southern African savanna tree species. She worked for Weill Cornell Medicine in Qatar from 2007 to 2014 and has recently published a book on the flora of Qatar. She is currently employed by Duke Kunshan University. Her conservation work at the American University of Armenia was recognized by the Whitley Award, the UK's largest conservation award. Her current work focuses on the intersection of biological processes and sustainable development in arid and hyper-arid regions, with a focus on the physiological ecology of plants and cyanobacteria.

Nobuyuki Yamaguchi was born and brought up in Japan where he studied biology for his BSc and endocrinology for his MSc at the Waseda University, Tokyo. Then, he moved to the UK and obtained his DPhil from the University of Oxford, Oxford, on reproductive biology and behavioral ecology of feral American mink in the lowland England. Later, he moved to the Qatar University, Doha, Qatar, and worked on fauna of Qatar, especially the desert hedgehog. His main research interests are evolution, biogeography, and conservation of terrestrial carnivores, using research methods linked with molecular systematics, morphology, and behavioral ecology, and he has been working on those subjects on big cats (especially the lion) for more than 20 years. Currently, he is based in the University of Malaysia Terengganu, Kuala Nerus, Malaysia.

Chapter 14
Doha as a 15-Minute City: An Urban Fereej

Velina Mirincheva, Jason Twill, and Nihal Al-Saleh

Abstract The 15-Min City is a proposed new way of re-assessing and upgrading the liveability of our cities. Put forward by Professor Carlos Moreno, it is becoming a prevalent discussion among policymakers and urbanists across many cities. At its core is a belief that cities should be planned for walking and biking and all daily essentials should be located within fifteen minutes of one's residence. In this chapter, we use the 15-Min concept as a lens to discuss Doha's potential of adapting itself to the new notion of chrono-urbanism, poised by recent shifts in people' mobility patterns, their proximity needs, and their own perceptions and experiences in the city, and considering post-pandemic reflections on strengths and weaknesses of cities to remain malleable, less car-dependent and to continuously synchronize with the needs of their residents.

Keywords Chrono-urbanism · Density · Digitization · Liveability · Spatial Planning · Proximity · Ubiquity

14.1 Life Per Minute, or How Much Is Our Time Worth?

Imagine that you escort your kids to their school as part of your early morning neighborhood run, ride, or walk. Imagine then, that your kids can remain on the school grounds, which by early afternoon are publicly open and flocked by neighborhood kids and a few parents, passersby, and dog walkers. The latter watch their impromptu games with what Jane Jacobs describes as the 'eyes of the proprietors' as they subconsciously and naturally police the neighborhood streets and open spaces

V. Mirincheva
University College London, Greater London, UK
e-mail: ucftvmi@ucl.ac.uk

J. Twill
Urban Apostles, Minnesota, USA

N. Al-Saleh (✉)
Environmental Management, University of London, SOAS, London, UK
e-mail: nmmohamed@qf.org.qa

L. Cochrane and R. Al-Hababi (eds.), *Sustainable Qatar*, Gulf Studies 9,
https://doi.org/10.1007/978-981-19-7398-7_14

(Jacobs, 1961). As early afternoon approaches, a few mobile kiosks have moved onto the now empty school parking lots to offer snacks, ice cream, and juices. Crowds spill out from the adjacent mosque and majlises and converge on the shared grounds during prayer time. In the late afternoon, children run in and out of neighboring households to visit friends and acquaintances and have a snack. Outside, perhaps a few more kiosks with local produce appear offering fruits and vegetables or a hot meal for the evening. As you pick up a meal and head home with your kids, you walk along familiar neighborhood streets, buzzing with other kids telling their parents about their day at school.

In a few hours the city has transformed, catering to one's needs within a space-time frame that diminishes passive experiences and expands on self-enriching activity and essential provisions—places for education, worship, access to open space, recreational diversity, pop-up food and beverage, daily essentials, and community and family connections. Within such a scenario of the *malleable city* (Gwiazdzinski, 2007), displacement car-based time is contracted or eliminated, and traversable spaces shrunk; the social and economic offer at play fits a spatio-temporal frame, which is optimized around one's carless reach. The city itself becomes the active agent that depreciates the car and adapts itself to a human-centric needs bounded by space and time. Our life per minute is not only increased but multiplied exponentially via the convergence of otherwise singular experiences and destinations.

This convergence or hybridization is what is currently underpinning the notion of chrono-urbanism, which not only recognizes the agency of time in the planning and shaping of urban activity but further argues that time-based planning needs to disassociate from the linear notion of distance to a destination and the speed of travel and instead needs to warp into space-time proximity module, where time, in and of itself, becomes almost insignificant or inconsequential as a singular decisive metric. The unraveling of the concept is similar to our dismissal of the archaic view that cities exist in some kind of an equilibrium, a system of destinations and infrastructure that only accommodate planned or otherwise formally recognized change. We now understand that all cities are social products and as such they have rhythms, chaos, informalities, and reciprocities—qualities that ignite our fascination with discovering them, identifying with them, and testing them. Our relationship with them is much like our relationship with the universe, redefined only in 1915 by the general *theory of relativity*, which dismissed the idea that space and time were a fixed stage onto which events simply took place without a reciprocal effect on the whole system (Hawking, 1998). Since time and space are not fixed, but relational, then neither should our cities be. Spaces within the cities should be flexible, convertible, and reusable—at different times, for varied duration and with assorted rhythms. This would allow for the convergence of our *lifetimes and our urban space*—lifetimes being the sum of our life per minute measure—thus, seeding time across a multitude of simultaneous or nearby experiences to the point that a singular measure of any event becomes inconsequential.

The concept of chronotopia as an arm of chrono-urbanism, relates to the alternating uses of spaces or buildings through different time periods. The weekend market is a prime example of chronotopian space. Campo dei Fiori in Rome transforms completely throughout a single day on the weekend—morning markets densely occupy the piazza while the restaurants and bars are closed and outside chairs put away to make space; in the afternoon, the market goers recede, stalls are taken away and after the cleaning crew leaves, a new wave of visitors arrives, and bars and restaurant re-activate the perimeter while the central space transforms from service space to leisure one in just few hours.

The chronotopian urban planning is the way to achieve the 'malleable city'—a sustainable city that is able to bend, reshape, and respond to our space-time frames without breaking (Gwiazdzinski, 2007). Such an approach to city shaping can be an answer to our deepening urban crisis, rooted in menacing climate change, environmental degradation, overconsumption, hustling and anxiety from an ever-accelerating urban lifestyle, traffic fatigue, unpredictable pandemic, and other aspects of daily life that have blurred our vision and have cornered us to think there must be a global solution to these issues. On the contrary, the solutions are local and often in front of our eyes. If we follow principles of cultural norms and habits, proximity, multilevel scalability, horizontal and vertical alternating usage, versatility, convertibility, adaptability, and diversity—we can begin to optimize and thus limit the passive consumption of space and time, achieve urban sustainability, address spatial and social justice, mitigate pandemic restrictions, and improve on mental health. If we only ask how much life per minute is our city giving us, we can begin to understand and better address our urban environments.

14.2 What Is the 15-Minute City?—A New Lens for Sustainable Urbanism

Using time as a metric to measure qualitative urban planning and livability within our cities and towns is not a new concept. In the predominantly urbanized world we live in today, time-distance relationships between home, work, school, daily amenity, and leisure activities have become an increasingly critical factor in measuring urban livability and quality of life. There have been many iterations of time and space scales for measuring cities. Most notably the 30-min city concept used in many world cities today or the 20-min neighborhood concept made famous by Portland, Oregon's city planning team in the early 2000s. Most recently, however, Carlos Moreno's 15-Min City (Moreno et al., 2021) has drawn the international attention of politicians, planners, and the media largely owing to Paris Mayor Anne Hidalgo's public endorsement of the concept in her 2020 re-election campaign.

While earlier versions of time-bound urban planning measured the distance one could travel between various destinations using multiple-modes of mobility (e.g., walking, cycling, mass transit, car), the most current iteration of the 15-Min City concept conceived by Professor Carlos Moreno has more rigorous parameters for modes of mobility. Moreno's definition excludes more energy-intensive modes of transport such as mass transit or automobiles to a planning framework where 'residents will be able to enjoy a higher quality of life where they will be able to effectively fulfill six essential urban social functions to sustain a decent urban life' all within a 15-min walk or bike ride. These functions include (a) living, (b) working, (c) commerce, (d) healthcare, (e) education, and (f) entertainment (Moreno et al., 2021). In their 2021 paper *Introducing the '15-min City': Sustainability, Resilience and Place Identity in Future Post-Pandemic*, Moreno and co-authors introduce a modified version of the original 15-Min City concept which incorporates four critical urban dimensions premised on observations to the urban challenges presented by the COVID-19 global pandemic. These are (a) density, (b) proximity, (c) diversity, and (d) digitalization (Fig. 14.1; Moreno et al., 2021). These four dimensions are explored in sub-sections that follow.

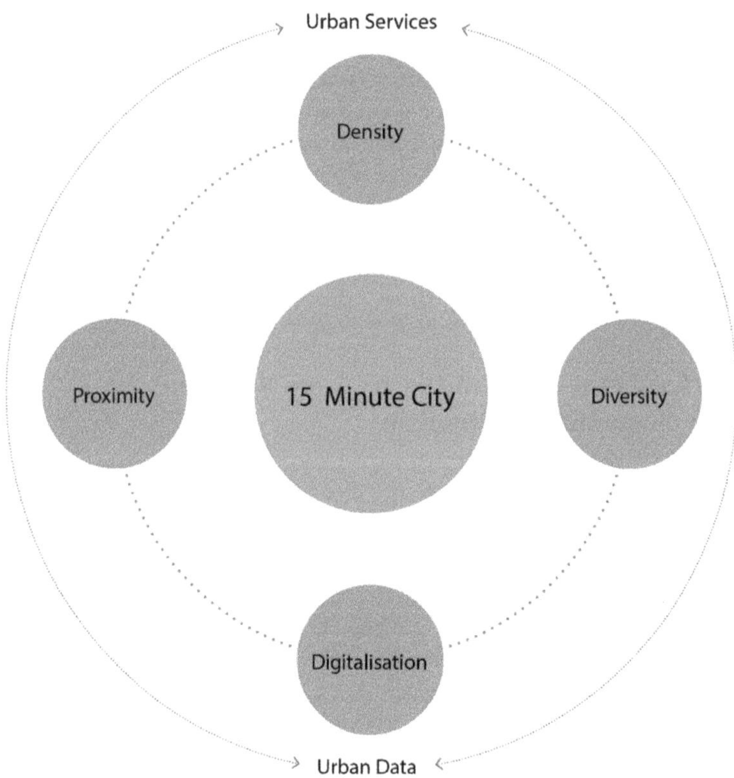

Fig. 14.1 The 15-Minute City framework (Moreno et al., 2021)

14.2.1 Density

Urban density is a critical aspect of sustainable urbanism and livability. It is a concept used in city planning to describe the concentration of people, jobs, housing units, total floor area of buildings, or some other measures of human occupation, activity, and development across a defined unit of area (Hess, 2014). In general terms, urban density describes the degree of concentration or compactness of people or development in a city. The density of cities is relevant to a broad range of issues related to the quality of urban life including 'environmental quality, transportation systems, physical infrastructure and urban form, social factors, and economic factors' (Churchman, 1999, p. 398).

It is accepted that the lower the urban density, the greater the environmental burdens and the more dramatic the societal impact. There are countless scholarly articles about the ills of suburban sprawl and low-density development (see Habibi & Asadi, 2011; Polidoro et al., 2012; Russell-Evans & Hacker, 2011). When people are spread out, public transportation systems lose their effectiveness, private vehicle use rises, pollution increases, and the citizens' interconnectedness and cultural connections suffer (Al-Kodmany, 2018; McLennan, 2009). This is very evident in the recent development of Qatar's urban landscape, where families that would have otherwise been co-located in the same area are now living in different regions of the country for a variety of reasons such as work proximity, school access, and cost of living.

In conventional planning, density is viewed in terms of building heights, urban form, block dimensions, and compactness, but in the 15-Min City concept, density is viewed more from a human perspective in terms of people per kilometer squared (Moreno et al., 2021). The authors' rationale for measuring density in terms of human-centric metrics in lieu of more traditional density measures is that by focusing on people you can better analyze the balance between demand on resources in the form of urban amenity (i.e., energy, water, food, services, etc.) that support a minimum standard quality of life against the biological carrying capacity of the land and region upon which the urban population resides.

By combining urban morphology with population density, Moreno and others further argue that planners may better solve for 'optimal density' whereby it is possible to 'effectively plan the available space such that all the essentials could be accessible to residents without the need for time- and energy-consuming automobiles' (Moreno et al., 2021; Salingaros, 2006).

In our 15-Min City assessment of Doha, we link density and urban form and space and showcase what their relationship means within the built landscape of the city.

14.2.2 Proximity

Proximity is another core dimension of the 15-Min City concept as it considers time-distance relationships between one's home and place of work as well as other basic services and amenities such as entertainment and recreational offerings. The scale, speed, and spatial distribution of urban growth are typically major concerns for urban policymakers. Indeed, urban accessibility, neighborhood scale, ability for social interactions, and spatial-driven policies are argued to be the most influential factors on contemporary land-use change (Verburg et al., 2004).

In Moreno's 15-Min City context, the only two modes of mobility are again walking and biking as these modes not only help cities reduce the amount of time lost in commuting but have the add-on benefits of reduced environmental and economic impacts and increased social interaction among urban residents (Alexander et al., 1977; Duany et al., 2000; Jacobs, 1961). Moreno further argues that good proximity allows residents to transition from residential areas, work, commercial areas, education centers, health facilities, and other basic institutions in a reduced timespan (Moreno et al., 2021).

More recently, and where available, indicators such as Walk Score™ and Bike Score™ have become useful tools for urban planners and policymakers to measure and improve upon urban proximity. These tools provide a numerical index score to define how good the access to a neighborhood is on foot or by bike.

In the discussion on Doha that follows, we elaborate on the concept of proximity in the sense of carless reach around the city and the urban templates that promote it.

14.2.3 Diversity

The third dimension of Moreno's 15-Min City model considers diversity. Many urban theorists and practitioners have long argued that mixed-use, socio-economic, and culturally diverse urban communities are foundational elements that underpin healthy and vibrant cities as they have a positive bearing on economic competitiveness, employment, tourism, heritage offerings, and place identity (see Rodríguez-Pose, 2019; Rose, 2016; Whyte, 2012). Another essential component to urban diversity is ensuring that housing is affordable and accessible for all ages, abilities, and income levels and located in close proximity to workplaces and other basic services.

Diversity in the context of Moreno's 15-Min City model is considered from two critical perspectives: (i) the need for mixed-use neighborhoods which are primary in providing a healthy mix of residential, commercial, and entertainment components and (ii) diversity in culture and people. Moreno and his co-authors further emphasize the inter-dependency of the various dimensions by stating that in the pursuit of a 15-min City model, the adoption of mixed-use neighborhoods is paramount in ensuring that an optimal density and proximity of essential amenities are achieved, while also providing for development of walkable streets and bicycle lanes. This approach ensures that residents can benefit from essentials within their residential areas, thus reducing the need for them to travel to access them (Moreno et al., 2021).

Building scale, particularly at the street level, also plays a vital role in ensuring mixed-use and diverse urban communities. Active street frontages and spatial diversity of urban retail floor spaces all work together to support affordable rents, services, and products for local business owners and consumers alike (Jacobs, 1961).

Coupling diversity with convertibility makes a good pair to discuss in relation to Doha's built environment and its dynamics. Our subsequent discussion on Doha emplaces these with relevant examples.

14.2.4 Digitalization

The final dimension of Moreno's 15-Min City model focuses on the role of digitalization in cities and how technology can enable and support the actualization of the three other dimensions. Largely drawing on elements from the Smart Cities' movement, Moreno and colleagues discuss various ways that technology and digitalization can improve livability, inclusiveness, and sustainability objectives for urban planners, government, and citizens alike.

From a livability perspective, the digital age in information technology, communications, and mobile applications has had a profound impact on modern urban life in our cities (see Menon, 2017; McKinsey & Company, 2018). From on-demand mobility and delivery services to healthcare management and remote working capabilities, our lives have been transformed by technological developments over the past several decades. This too has coincided with the rise of the sharing economy and digital platforms and tools to improve urban life such as bike sharing, car sharing, and platform cooperatives that not only provide citizens with greater ease and access to urban amenity but also support more equitable access to these services and products through peer-to-peer enabled technologies (Enochsson et al., 2021).

In support of the proximity dimensions, digitalization has proven effective where services such as online shopping, cashless transactions, and virtual communications and interactions among others are implemented and promoted (Cohen et al., 2020; Han et al., 2019; Reinartz et al., 2019). From a diversity and inclusivity perspective, technology use has demonstrated an increase in citizen participation. Recent studies have shown that digital tools such as crowd sourcing platforms properly designed and used by government can empower citizens, create legitimacy for the government with the people, and enhance the effectiveness of public services and goods (Liu, 2017). Lastly, through a sustainability lens, the amalgamation of digital solutions within our cities and effective use of big data is leading to greater efficacy in urban decarbonization programs through reduced automobile use and more optimal consumption patterns of materials and resources (Enochsson et al., 2021).

There is much Doha can acquire from cities like Toronto, Amsterdam, Milan, and Seoul on how smart city solutions and digitization strategies can be applied and adapted within the dry arid climate and resource constraints within the Gulf region. Here, in the subsequent discussion on Doha, we emphasize that digitalization needs to be ubiquitous to support equal access to services among societal and income strata.

14.3 Doha in 15 Minutes

The premise of the 15-Min City concept is that it can work for any city, that the concept is adjustable to the unique conditions of the city at play, and that the 15-min radius can be a 10- or a 20-min one, even a 30-min one in relation to density or location (edge cities or suburbs would be more suitable to the bigger spatio-temporal scale). The 15-Min City concept also posits itself as a ubiquitous solution to the failures of modernist planning, of which Gulf Cities have consistently fallen victims. In the past century, the latter has transformed at an alarming pace, evolving to cause a rapid socio-economic rupture of the traditional fabric in the *khaleej*. Large corporations—in order to attract the talent and labor required to build those oil states—have had to move swiftly and create multinational microcosms to accommodate foreigners into the desert. The result was a quick shift from the traditional way of life in the Gulf to a more fragmented, destination-based planning—cities within cities, dependent on expansive infrastructure. This phenomenon has had deep implications for the distribution of density, proximity, and accessibility, environmental sustainability, infrastructure loads, and social cohesion.

It is time, as Moreno says, to 'repair urban and social fragments, largely fueled by modernist approaches' (Moreno et al., 2021). The way out is to look inward and adapt the resulted spatial and physical capital into a new timeframe, tailored to the individual carless reach and rights to equal access to services. Can Doha become a 15-Min city? Yes, although many would argue that one cannot walk for fifteen minutes in the scorching summer sun and that this number needs to be adjusted for the climatic reality of Qatar. But in fact, this number is only a starting point, a north star to adjust a planning trajectory toward a human-centric urban design. Each city needs to assess its own needs and create its own measures, according to its own rhythms and demographics. The frequency of prayer far exceeds the frequency of grocery shopping, for instance, and so these two services necessitate a varied temporal urban location—the former perhaps within a 5-min reach, while the latter within a 10- or a 15-min reach. Whatever the case might be, building upon the principles of chrono-urbanism requires establishing some core relationships, whose spatio-temporal balance ameliorates our urban experience, diminishes our environmental impact, fosters our social and economic relations, and brings about physical and mental wellbeing.

14.3.1 Density and Urban Space and Form in Doha

Optimizing urban density optimizes the vitality of a place. Density is not height; it is not the image of Al Dafna and its high rises. In a people-centric urbanism, it is people per unit area. The densest areas of Doha are Fereej Abdel Aziz, Doha Al Jadeed, and Old Al Ghanem (according to the author's calculations based on GIS data and the 2015 mini census). The Qatar National Master Plan and Municipal

Spatial Development Plans rightfully set targets for people-based densities across the country. Centering density on people centers the service delivery and resource consumption on people as well. Linking then the density of people to their mobility patterns measured in time for walking or biking—not driving or distance traveled—would begin to adjust urban spaces and forms toward a more people-centric scale.

Let us consider a comparison between Al Dafna's high-rise density and the mid-rise one of Old Al Ghanem (Zone 16). With mostly zero front setbacks, greater plot coverage and continuous ground floor retail on many streets, along the 300-m-long Tariq Bin Ziyad Street in Old Al Ghanem one can reach 78 establishments, or 1 in every 4 m—restaurants, mini markets, corner bakeries with earth ovens, hotels, numerous daily essential services, as well as a park and a library. In Al Dafna, in 4 m one has not overcome a setback distance or moved past the lonely standing high rise. Buildings that line up Tariq Bin Ziyad Street have residential tops and retail ground floors, offering great opportunities for live-work spaces. The density of people is clearly noticeable as the streets in Zone 16 are always buzzing with people, many on their bikes, many walking around, including children walking to nearby schools. With the right kind of intervention to repair the degrading urban fabric in downtown areas, the car can start being expunged and surface car parking space recuperated, appealing to the walker and biker to occupy these lively areas with easy 5-min access to all daily essentials, not ignoring the tremendous role of the metro in this equation as a way to transcend the 15-min scale, dispersing people out, and bringing people into the area in a carless, more sustainable way (Fig. 14.2).

To work toward chrono-urbanism based on walking and biking times, we must recognize the relationship between people density, urban space, and form, as seen in our example. Cervero and Kockelman (1997) find that density, diversity, and pedestrian-oriented designs are in fact linked to reduced trip rates and do in fact encourage non-auto travel, in statistically significant ways. Salingaros (2006) even pushes the idea of replacing both the high-rise, ultra-high-density megacity model and the low-rise suburban sprawling model with a *compact city* model, following New Urbanist and Smart Growth ideas. The compact city mode, to borrow Lehmann's definition (2016), is 'a mixed-use spatial urban form characterized by *compactness*,

Fig. 14.2 (Left) Density of active frontages along Tariq bin Ziyad Street; (Right) An illustrative study by Makower Architects for the Ministry of Municipality (Makower Architects)

which defines a relatively dense urban area linked by easy access to public transport systems and designed to have minimal environmental impact by supporting walking and cycling…The compact city with four- to eight-story urban perimeter blocks represents the optimum use of space' (3). This would put Old Al Ghanem exactly in line with the compact mid-rise city (of the same exact density in fact as the Eixample district in Barcelona—360 pph), and with far greater accessibility to services and walkability than Al Dafna.

The greatest advantage of the balanced compact city is its more intimate relationships with urban spaces and the ability to localize solutions, in tune with local daily rhythms that prioritize walking and biking. In 2016, a winning entry to the Doha Unlimited Design Prize was a vision for Doha Al Jadeeda and Old Al Ghanem as flexible child-centric neighborhoods, where streets are converted to safe play areas at certain times; where spaces adopted multiple functions; and where essentially new chronotopes were introduced, tailored to the nature of the neighborhoods and its demographic. More than anything, this project was praised for recognizing the true nature of these downtown neighborhoods, their spirits, their demographic, and their multi-layered nature, and for proposing a humanizing solution which is centered around the most vulnerable demographic—children. Having the flexibility to adapt a neighborhood to the rhythm of a child during specific periods means having the flexibility to alter the spatio-temporal frames around children and bend the city, make it more flexible, adaptable to change, and resilient (Fig. 14.3).

Although each city tailors its density differently according to their demographic—Asian cities are on the upper scales of density, followed by European cities and last American cities—there is a desired balance to be achieved between density of people, urban space, and form and the provision of services, to avoid overconsumption of resources and infrastructure. Proponents of compact cities do warn of over-densifying where the latter can contribute to reduction in the access to renewable energy (i.e., sunlight and wind). Too much clustering can reduce daylight, access to solar energy, and increase air pollution. Hence, the benefits of the compact city need to be balanced according to 'climate, land use type, culture and latitude' (Edwards, 2014, p. 144).

14.3.2 On Proximity and Carless Reach in Doha

The recent COVID-19 pandemic has motivated an overdue debate on proximity, access to services, and social equality. It showed us that while some people had access to basic services and products within a kilometer of their house (an imposed COVID-19 movement restriction radius in some cities), others did not and had to rely on a car to supply daily essentials. Many cities—Berlin, Vienna, Rotterdam, Turin, Edinburgh, Denver, Dallas, Bogota, and Vancouver—reacted to this disparity by launching temporary and permanent urban initiatives, such as bike lanes, hyperlocal markers, shipping container hospitals, or pop-up stores (Moreno et al., 2021). The most obvious measure introduced in many cities as a pandemic response was new or temporary bike lanes and the promotion of shared rides, such as electric scooters.

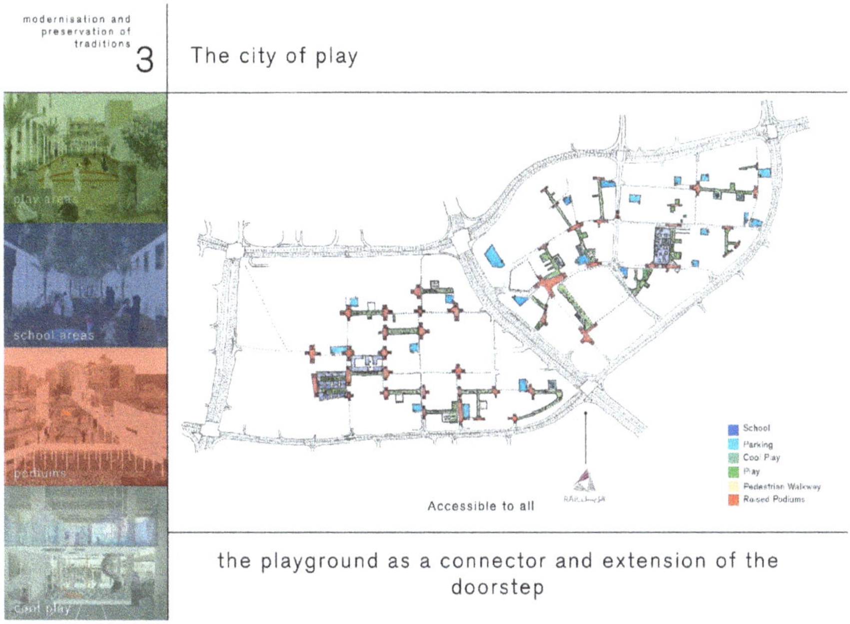

Fig. 14.3 The winning entry from the Unlimited Doha Design Prize on regenerative strategies for Doha Al Jadeeda (Zone 15) and Old Al Ghanem (Zone 16) (Alex Scott-Whitby and Cherng-Min Teong from the UK; and Omer Can Aksoy, Gizem Kahraman and Deena Al Terawi)

Doha was among the many cities that saw a tremendous spike in bike ownership, with people of all ages and demographic taking to the streets to ride a bike, along the newly built bike lanes or overtaking existing roads and sidewalks. Suddenly, a whole new culture was born, a bike demographic that not only used the bike to get some fresh air, but also wanted to now reach a destination—light shopping at the local store, meet a friend at a local coffee shop, and organize a bike ride on a weekend to a further off destination. Whoever was not on their bikes, walked and looked for the same nearby services, reachable via the relative safety of the outdoor fresh air during the pandemic.

Another important legacy of the pandemic is that people appreciated the time saved in working from home and many have not gone back to working in an office. Companies like Airbnb, Zillow, and PwC have moved to a 100% remote work scenario (*Airbnb Tells Employees They Can Work Remotely Forever—The New York Times*, n.d.), and there are now numerous vacated office buildings being converted into apartments across a myriad of cities (Liu, 2021; Mekouar, n.d.). Working from home allowed people to save the time in driving or traffic and instead invest that same time in outdoor activities—they have the time to walk or bike to the store; they experimented with reaching a friend's house via a free scooter rental, rather than driving; and they liberated their minds from the default choice of the car and became experimental, invigorated, and curious by what else does the city have to offer. Recognizing the health benefits of the outdoor, people connected with it on a daily level, they changed their rhythm, and that changed the rhythm of their city.

Doha felt these new rhythms and now, more than any other time, it is fundamentally important to carry this momentum and introduce the right mechanisms to alter our urban landscapes toward a proximity-based dynamic.

The Pearl Qatar offers several urban proximity-based templates that are quite successful. Firstly, the way Porto Arabia is configured allows for the car to be quickly tucked away in resident or visitor multi-story parking—nestled behind the commercial clusters and within the residential towers; people are quickly liberated from it and channeled along the pedestrian-only commercially led promenade connecting all towers. Residents, once home and parked away, walk to the grocery stores or barber shops or ice cream shop 'downstairs' and do not generally resort to using their cars once inside the Pearl. Many residents choose to walk even to distances beyond a comfortable 15 min to reach a movie theater or a further-away restaurant.

Qanat Quartier also hosts a spatial template that tucks away the car in multi-story car parks and structures most movement along pedestrian-only waterfronts. Different in that area is the vertical land-use mix, with many non-residential units located on top of podium parking. The mix of commercial land use allows for small businesses, such as real estate offices, business centers, or travel bureaus, to co-exist among fitness centers, cafes, yoga studios, pet stores, and pharmacies. In this proximity-based urban template, both residents and visitors have an immediate access to a host of daily essentials in a spatial proximity that also fosters a sense of community, belonging, and at the same time intrigue in the random and diverse encounters.

Finally, one must mention Medina Centrale (at the Pearl Qatar) as a third urban template, where both cars and pedestrians co-exist on ground level; where on-street parking and multi-story parking have found space; where streets are always buzzing with people who are more than willing to park their car and join the urban dynamic—be it for leisure or while running an errand (Fig. 14.4).

All these proximity-based templates create spatio-temporal frames, centered on the pedestrian experience. To a large extent, they are very successful and have offered several models to separate cars from pedestrians and from back-of-house servicing, so that priority is given to the walker or biker.

In downtown Doha, Msheireb has also achieved a proximity-based urbanism, where the car is forced underground, and pedestrian movement is encouraged along narrow shaded streets with zero setback and active ground floors. Residents within this 35-ha development have a 15-min access to daily essentials, including places of employment, school, cultural facilities, food and beverage services (F&B), and retail outlets. What Msheireb is doing in a dense downtown setting is what proximity-based urbanism was in the past in the *khaleej*—climatically sensitive urbanism that was pedestrian oriented and community building via opportunities for random encounters and easy access to daily essentials; it is a model of the compact neighborhood with presumably the right kind of density to allow for optimum resource consumption per capita and a more sustainable way of living (Fig. 14.5).

Fig. 14.4 Pedestrian-friendly environment at Medina Central, The Pearl Qatar (Authors)

Fig. 14.5 Pedestrian-friendly environments at Msheireb, Doha (Deepthi John)

14.3.3 Diversity and Convertibility in Doha

Moreno speaks of a twofold diversity: that of the mixed-use neighborhood—with a healthy mix of residential, commercial, and entertainment components, as well as that of cultures and people. The vitality, resilience, and sustainability of places have a strong dependence on the right mix of uses that residents have proximal access to. For Moreno's 15-Min urban framework, this mix needs to include residential, employment, commerce, health care, education, and entertainment (Moreno et al., 2021). These need to exist in spatio-temporal dynamic that elevates the quality of life and makes life per minute 'inversely proportional to the amount invested in transportation' (100). Gehl (2010) further posits that compact urbanism, based on walking and biking will inspire the creation of more public spaces—most importantly neighborhood parks—thus creating an equal opportunity among all demographics to access green and public space, regardless of car ownership, which has been a marginalizing force of modernist car-led planning.

This renewed interest in the compact, mixed-use walkable neighborhood as an integral part of the sustainable resilient, socially equitable city is being echoed in current regeneration and activation schemes around Doha. A recent study by NeighbourlyticsTM for Qatar Foundation builds an evidence-based analytical data pond around five main ingredients for placemaking—volume of activity, variety of activity, vitality, relevance, and character (Neighbourlytics, 2022).

If we take vitality as an example of improving the proximity dynamics, we can use the data analytics images to pinpoint the heavy activity clusters shown in red in the visual below. These clusters indicate the critical mass of activity in the precinct and can support addressing spatial distribution opportunities to connect these clusters and enrich the precinct experience through activities, programs, services provided, and events. By strengthening the connectivity of these clusters, we also strengthen the walkability, diversity, and convertibility of the area (Fig. 14.6).

The compact city template in Doha—where all five factors are at play—seems to exist only in the historic downtown districts of Doha. On a global scale, however, it seems that the pendulum of chrono-urbanism swings between the very dense, 5-min downtown city dynamic, with activity at every few meters, and the low density, low

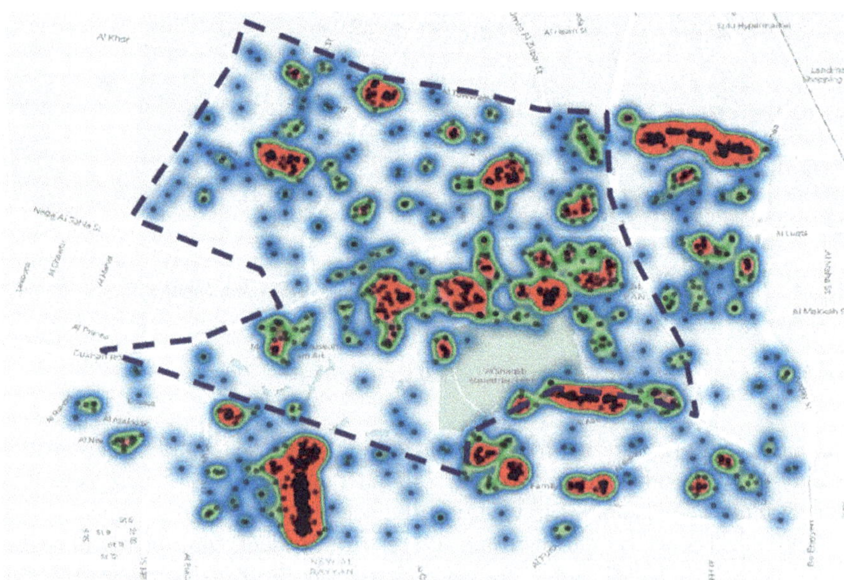

Fig. 14.6 Education City: Vitality—linkages between place activity (Neighbourlytics, 2022)

activity, overly spaced out, car-dependent destinations, which depend on vast parking lots and only activate after work hour or on weekends.

But the reborn post-pandemic consciousness is ready to borrow the best of both worlds and establish its compact, chrono-based neighborhoods with balanced density and range of goods and services along the six main areas of daily essentials by Moreno. At the core of this initiative needs to be the residential community—the one that echoes the past where rich and poor lived side by side in Najma, for instance, with access to the same services and opportunities, where equality, family, and community were the building bricks of the physical and non-physical landscape. Building *social circularity* rather than fueling speculative real estate would ease the risks of sudden gentrification and ground the communities in long-term commitments and generational resilience. Increasing density in both people and culture would cross-pollinate mono-cultured developments and start bringing up those placemaking indicators as tested in Education City by Neighbourlytics.

The Pearl Qatar can very well exist as a compact, proximity-based neighborhood. Since it is mostly active at night and on weekends, an opportunity exists to add vitality and variety through diversification of its economic ecosystems and the addition of a metro line extension to allow additional footfall. An added employment component will enrich the daily critical mass and add day-time footfall to the restaurants in Medina Central. Al Dafna, on the other hand, has a large employment component, but lacks dense activation via small F&B ground outlets. It also requires additional variety of office sizes to welcome small, medium, and large businesses side by side and foster a corporate urban buzz. And most importantly, the Al Dafna area can improve its pedestrian permeability and become even a 5-Min district. Education City, through the data analytics findings, shows that it rates high on the community pillar, but requires critical residential mass to enliven its premises and engender more activation, easier linkages, and greater density of people and assets. Finally, downtown Doha aligns best with the compact city model, having most ingredients in place, but still not able to expunge the car significantly and regain its walking and biking proximity dynamic. Whenever this happens, sidewalk space will be liberated, restaurants and shops will spill out, and frontages will be crafted so neighbors greet each other and have a chat on a *datcha* seat and watch over the kids walking back from school.

This is our city, our vision for a slower-pace, compact neighborhood living, where our reach to daily essentials is not measured in car ownership and parking capacity, but in the enriching experience of walking, biking, and engaging with the outdoors. To help us get there, it is essential that we allow flexibility and convertibility play its role in adapting our environment from the car dynamic to the people dynamic. Many spaces in our neighborhoods are left unused, unplanned, and empty. These need to be lent to informal community use, which can help shape their future. Onaiza, for instance, hosts an incredible amount of SLOAPS (Spaces Left Over After Planning). The same can be activated in variety of ways throughout the day or the year—improvised kids' playgrounds, private functions areas, weekend outdoor markets, mobile kiosks spaces, board games areas for the elderly, Ramadan tents, charity location points, recycling points, community fruit and veggie gardens, planting tree

Fig. 14.7 The linear park in Al-Hilal is a brilliant example of turning left-over spaces into green areas with rest areas, for the recreational enjoyment of residents (Deepthi John)

grounds as part of the million tree Doha initiative, etc. The Onaiza community boasts a favorable mixed-use offer peppered around its premises—the Furjan Markets are an essential recent ingredient, the various schools and nurseries that are within a walking reach, embassies provide employment opportunities, and Katara across the highway offers a variety of leisure activities. To strengthen even further this community, residents need to be given flexibility to shape their community spaces, craft their frontages, engage with public land in their vicinity, and tighten up their spatio-temporal dynamics around their places of residence and work (Fig. 14.7).

People tend to indicate their spatial needs in various ways. The informal cricket plays on vacant lots around F- and G-Ring Roads in the Al Thumama areas, or even in Al Dafna are begging the city for space, telling it there is a communal bond that needs its space-time bubble to thrive. Can more open space be shared between schools and the community, so that play (football, cricket, etc.) spills between the generations and within the same space? Communal open space is the most easily accessible leisure space which does not marginalize the carless residents, yet it is currently still scarce in Doha. Even the Pearl Qatar, who boasts a communal living and recently introduced a school and a hospital, is largely deprived of small communal flexi spaces where random play occurs, or a birthday party or a yoga session, without having to book in advance formal and fenced spaces or ask permission. We have learned the importance of such spaces, especially having lived through a pandemic which made us acutely aware of our urban pre-existing condition—a temporal crisis and preoccupation with speed, wide streets for cars, and remote destinations. It is time to look next to our homes and shape the spaces within our 5- to 15-min reach by foot or bike (Fig. 14.8).

That flexibility needs to extend to both spaces and buildings. Historic Doha is full of derelict buildings of great value—for their heritage character, but also because they exist at the right place and are deeply relevant. The recent conversion by Qatar Museums of the historic building of the first officially recognized

Fig. 14.8 Informal cricket play on empty grounds between F- and G-Ring Road (Subash Kumar)

school for girls—now called the Liwan Design Studios and Labs, 'a space dedicated to supporting and connecting design professionals who are integral to Qatar's burgeoning creative community' (*Qatar Tribune*)—is a brilliant example of adaptive reuse where it is most relevant, at the heart of Doha. A place of creative employment in the vicinity of other opportunities to convert courtyard houses to living quarters for staff and creative professionals, with easy access to all daily essentials within a short walking distance—a regeneration toward chrono-urbanism that is founded on flexibility, malleability, and creative solutions (Fig. 14.9).

The compact downtown living is by no means a thing from the past in Doha, and the walking and biking culture exists in historic Doha by all means. Daily, residents cycle or walk around while running errands or while taking a stroll. It is the car that seems at odds with the context, not the other way around. There are local families that have not moved away to new typologies in the outskirts. The Al Naama family and house in Old Al Ghanem is one of many to stand its ground and keep its doors open to its neighbors to remind us that there is a place for both a courtyard house, as well as a mid-rise residential block or a higher-rise corner block. The gradation of density, activation, and convertibility can all manifest themselves simultaneously without victimizing or gentrifying, but with a common desire to make the neighborhood more walkable, more breathable, more resilient, and where residents actively participate in its livelihood and are given the right incentive to preserve and cherish its nature. We cannot speak of equal rights to the city, unless we make our cities malleable, bendable, and adaptable—via an intense and proliferous mixing of uses—both horizontally and vertically—and via the flexibility of new chronotopes that ensure a more continuous, more sustainable use of buildings and spaces.

Fig. 14.9 Liwan building, by Qatar Museums (Qatar Museums)

14.3.4 Ubiquity and Digitalization in Doha

Another offspring of our pandemic life has been an intensified technological innovation in work-from-home solutions and service and product delivery. Digital solutions for online shopping and delivery, online communication platforms, and online social media have contributed greatly to reduction of car usage and hence, emissions. Working from home has also taken the car off the street and has recuperated that commuting time to serve other, more socially enriching experiences. Hence, our temporal frames have been warped to our advantage, helped by the rise in digital solutions and the imposed distancing restrictions. This new post-pandemic paradigm prompts a new urban digital consciousness that looks to optimize the Smart City model to work for the purposes of the proximity-based urbanism, and away from its negative impacts, such as centralization, hyper-urbanization, and resource consumption (Allam, 2020). The 'malleable' 15-Min City would be a digital city, which knows and follows its human-centric evolution—collectively and intelligently—so that building uses, spaces, furnishing, and signage are amendable and even experimental, coordinated with temporal scales that go from the day to the year (Gwiazdzinski,

2015). Then we can subserve the Smart City concept to the needs of chrono-urbanism where 'the ensemble of plans, schedules and agendas…coherently act upon space and time, enabling the optimal organization of technical, social and aesthetic functions in the city, in an attempt to create a more human, more accessible, welcoming city' (Gwiazdzinski, 2015, p. 1).

The recent introduction of Falcon e-scouters is a fitting example of Smart City technologies working to improve proximity-based urbanism. Residents download an app and upon registering with their Qatar identification card, are able to unlock and hop on a scouter to travel around the island. Such micro-mobility solutions create opportunities to replace short car trips significantly within the 15-min temporal frame. A study in Cardiff, Wales, finds that half of daily trips for residents were less than 3 miles long; hence, taking into account personal habits and constraints, active travel was found to be able to realistically substitute 41% of short car trips, saving nearly 5% of CO_2 emissions, in addition to the 5% of 'avoided' emissions from cars due to the presence of residents who already walk and cycle regularly (Neves & Brand, 2019) (Fig. 14.10).

Elsewhere in Doha, the Loop e-scooters cover West Bay, Corniche, Al Waab, Lusail, Legtaifiya, QNCC (Qatar National Convention Centre), and Al Bayt Stadium, working on the same principle. But are these mostly leisure-oriented luxuries for the paying customer? Do they have the ability to replace car trips, where it is most

Fig. 14.10 Scooters can be a way to reduce congestion in downtown areas of Doha. This photo is taken at the Mansoura intersection in Najma (Deepthi John)

needed—in dense urban areas with air pollution and congestion problems? The cycling and walking culture in Doha is certainly begging for more affordable options, better connectivity, and safety. The e-scooter and cycle usage needs to become ubiquitous if Doha is to become a cycling and scooter capital and raise up its livability standard. The expansion of the cycling lanes brings us one step closer to satisfying the need for more active mobility corridors, but the inability to bring a bike on the metro, for instance, draws us one step back.

Beyond the ride-sharing ability to bring the city closer to us via sustainable active mobility, the power of the Smart City concept needs to be harnessed for measuring the overall performance of the built environment across all sectors—energy consumption in buildings and infrastructure, emissions, air quality, waste disposal, etc. Measuring performability against benchmarks of proximity-based urbanism with time-saving mechanisms that increase our life per minute will ensure the economic, social, and environmental resilience of our cities. Smart Technologies need to be deployed with a great amount of social responsibility, so they do not become exclusive vehicles of revenue pursuit but are used to bridge the social inequalities that exist too ubiquitously in our cities. Smart Cities, for example, are often a marketing tool to drive up real estate values, thus marginalizing whole communities. Thus, local governments have a responsibility to recognize that 'tech-driven solutions are as important to the poor as they are to the affluent' (Remes & Kharas, 2018).

For Moreno's proximity urbanism, smart technologies can serve to collect necessary data and deploy innovative solutions that provide equal access to daily essentials for a diverse demographic, combating traffic congestion, poor air quality, overconsumption of non-renewable energy, and passive time, providing equal access to green spaces, close by places of employment, education, health care, and entertainment activities—all accessible via walking and biking and assisted with the calibrating force of smart cities technology.

14.4 Qatar's Opportunity for Positive Change

The compact, proximity-based versions of our cities are already at play in many cities and many parts of cities. When we think of the densities of London, New York, and Tokyo, we do not imagine ourselves driving through their cores, but rather walking, biking, or taking public transport. Congestion charges minimize car travel to essential and motivate residents to leave their cars at home or use them only on weekends to reach further away destinations. Most do not even own a car. But more than a congestion charge, the reason for minimizing car usage is the urban form and configuration themselves—shaped to balance off pedestrian space and car space—continuous sidewalks, active frontages, traffic-calming measures and frequent crossings, and bike lanes as part of shared car space, to name a few. The residents—on foot, by car, or bike—have been given a more balanced opportunity to use the urban space without necessarily competing for it. Having said that however, we have also become aware that these same cities have become too dense and have begun to cause environmental problems—air and noise pollution from traffic congestion, daylight deficiency from buildings too close to each other, and high non-renewable energy consumption. New Urbanism and Smart Growth proponents are now arguing that the most efficient city model is the compact one, where the built environment is balanced off with nearby agricultural land and where there is a controlled consumption of resources and increased number of ways to replenish them (Salingaros, 2006). This is not to say that ultra-high densities or ultra-low densities cannot exist anymore, but rather than these need to be more balanced and graduated, to avoid both the overpowering and exhausting draw on resources that both hyper-dense downtowns and sprawling suburbia cause, because of their sizes.

Doha is a very suitable candidate for the compact, density-balanced city template, which seems to be topping the Global Liveability Index charts with medium-size cities like Vienna, Melbourne, Sydney, Calgary, Vancouver, Toronto, Copenhagen, and Adelaide taking 8 out of the 10 top spots in pre-COVID-19 rankings (*The Global Liveability Index 2019*, 2019). Salingaros (2006) argues that the balance in size and density can be achieved with mechanisms such as transect zoning which implies a gradient of height and density from an urban 6-story core toward suburban single-house layout with mechanisms (narrow streets and sharp curb radii) that eliminate the fast car on large roads, or the extensive parking in front of buildings or the large parking lots.

This model can be well suited for Doha outside of the high-rise districts (which are not seen as an urban evil, but rather as opportunities for positive change), as a mechanism to mediate drastic density swings, such as the one in Onaiza 63 between the high rise sitting next to the single-family house.

Transect zoning and form-based codes have already been suggested in documents such as the Qatar Urban Design Compendium and the Regeneration Framework for Zones 15 and 16. The organizing principles of these mechanisms underpin proximity-based urbanism and would be essential for elevating Doha's liveability standards by connecting, integrating, and bringing parts to a whole via a people-centric set of principles.

Without a doubt, the recent cycle lanes' additions and beautification works by Ashghal are core engines for positive change. With more people biking in shaded and safe streets, more active streets will surface to offer easy access to daily essentials for bikers. The neighborhoods will shrink in time and space and car journeys will start to be replaced by bike journeys.

Neighborhood commercial streets (high streets) are essential in this transformation. Great examples are already at play—Al Kinana Street in Al Sadd, Al Jazeera Street in Fereej bin Mahmoud, and Environment Street in Al Duhail South, to name just three, are becoming the hearts of their neighborhoods, offering easy access to all daily essentials from Moreno's framework. Of the same spatial and transformative power are streets, such as Tariq bin Ziyad (mentioned earlier) in Old Al Ghanem, or Abdul Aziz bin Ahmed in Al Asmakh—two neighborhoods that historically have been among the most walkable in Doha and their conservation (both of spatial and physical nature) should be at the forefront of regenerative agendas in the city (Boussaa, 2014, 2021).

It is also worth mentioning Qatar National Open Space Strategy by the Ministry of Municipality, which envisions greater connectivity, quality, and reach to open spaces in Doha and throughout the country. It re-looks at the Green Belt to potentially transform it into agricultural land, which tallies with the concept of the compact city next to natural resources that balances consumption with replenishment. It also equips the recent achievement of Doha and Al Rayyan municipalities as Healthy Cities and Qatar Foundation as Healthy Education City with a spatial framework that improves on access to open space and the health, social, and mental-wellbeing benefits this carries in line with WHO (World Health Organization) criteria.

Qatar's future in proximity urbanism is visible and attainable. Doha is well suited to become a 15-Min City if a people-centric approach is at play as a core mechanism for urban transformation. The recent pandemic's silver lining has been a reflection upon pasts, current, and future strategies that better align with people's shifting perception of what they need from the cities, how they want to maneuver daily and what brings about happiness and wellbeing in their urban lives. Ultimately, it is the people that will drive change and policymakers should be receptive to change, adaptability, temporality, flexibility, and cultural nuances that allow us to experience our cities outside of our cars—in all their sensorial and wonderfully complex natures.

References

Alexander, C., Ishikawa, S., & Silverstein, M. (1977). *A Pattern language*. Oxford University Press.

Al-Kodmany, K. (2018). New Suburbanism: Sustainable spatial patterns of tall buildings. *Buildings, 8*(9), 127. https://doi.org/10.3390/buildings8090127

Allam, Z. (2020). *Cities and the digital revolution: Aligning technology and humanity*. Springer International Publishing.

Boussaa, D. (2014). Al Asmakh historic district in Doha, Qatar: From an urban slum to living heritage. *Journal of Architectural Conservation, 20*(1), 2–15. https://doi.org/10.1080/13556207.2014.888815

Boussaa, D. (2021). The past as a catalyst for cultural sustainability in historic cities; the case of Doha Qatar. *International Journal of Heritage Studies, 27*(5), 470–486. https://doi.org/10.1080/13527258.2020.1806098

Cervero, R., & Kockelman, K. (1997). Travel demand and the 3Ds: Density, diversity, and design. *Transportation Research Part D: Transport and Environment, 2*(3), 199–219. https://doi.org/10.1016/S1361-9209(97)00009-6

Churchman, A. (1999). Disentangling the concept of density. *Journal of Planning Literature, 13*, 389–411. https://doi.org/10.1177/08854129922092478

Cohen, N., Rubinchik, A., & Shami, L. (2020). Towards a cashless economy: Economic and socio-political implications. *European Journal of Political Economy, 61*, 101820. https://doi.org/10.1016/j.ejpoleco.2019.101820

Duany, A., Plater-Zyberk, E., & Speck, J. (2000). *Suburban nation: The rise of Sprawl and the decline of the American dream*. North Point Press.

Economist Intelligence Unit. (2019). *The Global Liveability Index 2019*. https://www.eiu.com/n/the-global-liveability-index-2019/. Accessed 17 May 2022.

Edwards, B. (2014). *Rough guide to sustainability: A design primer*. Routledge. https://www.routledge.com/Rough-Guide-to-Sustainability-A-Design-Primer/Edwards/p/book/9781859465073. Accessed 3 April 2022.

Enochsson, L., Palgan, Y. V., Plepys, A., & Mont, O. (2021). Impacts of the sharing economy on urban sustainability: The perceptions of municipal governments and sharing organisations. *Sustainability, 13*(8), 4213. https://doi.org/10.3390/su13084213

Gehl, J. (2010). *Cities for people.* Island Press.

Gwiazdzinski, L. (2007). Redistribution des cartes dans la ville malléable. *Espace, Population, Sociétés, 2007.* https://doi.org/10.4000/eps.2270

Gwiazdzinski, L. (2015). Space-time design for chrono-urbanism in Smart City. *Smart Sustainable City and Big Data, The 3rd International Conference.* https://halshs.archives-ouvertes.fr/halshs-01248865. Accessed 15 February 2022.

Habibi, S., & Asadi, N. (2011). Causes, results and methods of controlling urban Sprawl. *Procedia Engineering, 21,* 133–141. https://doi.org/10.1016/j.proeng.2011.11.1996

Han, C., Wu, Y., Chen, Z., & Wang, X.. (2019). Terahertz communications (TeraCom): Challenges and impact on 6G wireless systems. arXiv:1912.06040 *[cs, eess, math]* [Preprint]. http://arxiv.org/abs/1912.06040. Accessed 10 April 2022.

Hawking, S. (1998). *A brief history of time.* Bantam Books.

Hess, P. (2014). Density, urban. In A. C. Michalos (Ed.), *Encyclopedia of quality of life and well-being research* (pp. 1554–1557). Springer Netherlands. https://doi.org/10.1007/978-94-007-0753-5_698.

Jacobs, J. (1961). *The death and life of Great American cities.* Random House.

Lehmann, S. (2016). Sustainable urbanism: Towards a framework for quality and optimal density? *Future Cities and Environment, 2,* 8. https://doi.org/10.1186/s40984-016-0021-3

Liu, H. K. (2017). Crowdsourcing government: Lessons from multiple disciplines. *Public Administration Review, 77*(5), 656–667. https://doi.org/10.1111/puar.12808

Liu, J. (2021). The top 10 cities turning old office buildings into apartments—Take a look inside. *CNBC.* https://www.cnbc.com/2021/11/26/the-top-10-cities-turning-old-office-buildings-into-apartments.html. Accessed 18 May 2022.

McKinsey and Company. (2018). *Smart city solutions: What drives citizen adoption around the globe?* McKinsey Center for Government. https://www.mckinsey.com/industries/public-and-social-sector/our-insights/smart-city-solutions-what-drives-citizen-adoption-around-the-globe. Accessed 10 April 2022.

McLennan, J. (2009). *Density and sustainability: A radical perspective.* Second Quarter. Trim Lab.

Mekouar, D. (n.d.). *As pandemic empties offices, record number of buildings converted to apartments.* VOA. https://www.voanews.com/a/as-pandemic-empties-offices-record-number-of-buildings-converted-to-apartments-/6281999.html. Accessed 18 May 2022.

Menon, A. (2017). Smart cities, livable cities. *Gfk Marketing Intelligence Review, 9.* https://doi.org/10.1515/gfkmir-2017-0008

Moreno, C., Alam, Z., Chabaud, D., Gall, C., & Pratlong, F. (2021). Introducing the "15-Minute City": Sustainability, resilience and place identity in future post-pandemic cities. *Smart Cities, 4*(1), 93–111. https://doi.org/10.3390/smartcities4010006

Neighbourlytics. (2022). *Qatar Foundation: Education City as an Innovation District.*

Neves, A., & Brand, C. (2019). Assessing the potential for carbon emissions savings from replacing short car trips with walking and cycling using a mixed GPS-travel diary approach. *Transportation Research Part A: Policy and Practice, 123,* 130–146. https://doi.org/10.1016/j.tra.2018.08.022

New York Times. (n.d.). *Airbnb tells employees they can work remotely forever.* https://www.nytimes.com/2022/04/28/business/airbnb-employees-remote-work.html. Accessed 18 May 2022.

Polidoro, M., de Lollo, J. A., & Barros, M. V. F. (2012). Urban Sprawl and the challenges for urban planning. *Journal of Environmental Protection, 3.* https://doi.org/10.4236/jep.2012.39117

Qatar-Tribune. (n.d.). *Qatar Museums opens Liwan design studios & labs to support creative industry professionals.* http://www.qatar-tribune.com/news-details/id/232089/qatar-museums-opens-liwan-design-studios-labs-to-support-creative-industry-professionals. Accessed 5 April 2022.

Rodríguez-Pose, A., & von Berlepsch, V. (2019). Does population diversity matter for economic development in the very long term? Historic migration, diversity and county wealth in the US. *European Journal of Population, 35*(5), 873–911.
Reinartz, W., Wiegand, N., & Imschloss, M. (2019). The impact of digital transformation on the retailing value chain. *International Journal of Research in Marketing*, 350–366.
Remes, J., & Kharas, H. (2018, June 11). Can smart cities be equitable? *Brookings*. https://www.brookings.edu/opinions/can-smart-cities-be-equitable/. Accessed 6 April 2022.
Rose, J. F. (2016). *The well-tempered city: what modern science, ancient civilizations, and human nature teach us about the future of urban life.* Harper Collins
Russell-Evans, V., & Hacker, C. S. (2011). Expanding waistlines and expanding cities: Urban Sprawl and its impact on obesity, how the adoption of smart growth statutes can build healthier and more active communities. *Virginia Environmental Law Journal, 29*(1), 63–113. https://www.jstor.org/stable/24789325. Accessed 10 April 2022.
Salingaros, N. A. (2006). *Compact city replaces Sprawl.* Crossover: Architecture, Urbanism, Technology.
Verburg, P. H., Ritsema van Eck, J. R., & Schot, P. (2004). Determinants of land-use change patterns in the Netherlands. *Environment and Planning B: Planning and Design, 31*(1), 125–150. https://doi.org/10.1068/b307
Whyte, W. H. (2012). *City: Rediscovering the center*. University of Pennsylvania Press.

Velina Mirincheva is an urban design and planning specialist with a stronghold in both industry and academia. She believes in and promotes a sustainable people-based urbanism, contextualized and attuned architectures that belong to their sites and spaces, and a deep responsibility for making a difference for the better. Her professional experience started in the USA, from where she holds a professional license and membership with the AIA. After the completion of her first master's degree, she moved to Doha, where for the past thirteen years she has built not only a formidable portfolio of large-scale building and urban projects, but also a deep knowledge of the country's history, urban traditions, social practices, places, and spaces that are constantly undergoing transformations. Currently, a founder of a consulting practice (ARRUS International) with offices in both Chicago and Doha, her most engaging and prominent projects include Qatar's National Open Space and Recreational Facilities Strategy, Qatar's Urban Design Compendium, and previously, the Regeneration Framework for Doha Al Jaeeda and Old Al Ghanem. In parallel, she is also a Researcher and Frequent Lecturer at Qatar University, participating in several grants as an industry partner and having lectured extensively on subject of sustainable urban design and placemaking, urban economics, vernacular architecture, heritage preservation, and spatial and movement analytics.

Jason Twill has a career spanning over 22 years in climate change, education, sustainability, and urban regeneration. He has been at the forefront of social, ecological, and economic transformation. A globally recognized pioneer in regenerative development and design, his work is advancing next generation solutions to reverse global warming and restore planetary health. His career experience includes managing award-winning urban regeneration projects throughout North America and Asia Pacific, serving on the world change advisory team for Paul G. Allen, heading sustainability and innovation for Lendlease, Australia, and consulting multiple local and national governments on climate action and social equity strategies. In Doha, Qatar, he served as Director of the Qatar Foundation's World Cup Program helping shape the national legacy of Qatar's hosting of the 2022 FIFA World Cup™. A regular lecturer at multiple universities across the world, he previously served as a Runstad Fellow at the University of Washington and a Research Innovation Fellow at the University of Technology, Sydney, where he led research into regenerative urbanism, housing affordability, and Indigenous Knowledge Systems. He was designated a LEED Fellow by the United States Green Building Council in 2014. He is a co-founder of both the International Living Future Institute and the Green Sports Alliance.

Nihal Al-Saleh has an extensive career in the oil and gas industry starting with Occidental Petroleum Oman and later moving to Qatar Shell GTL. Her focus has been new business development, negotiations and systems implementation projects in Qatar, Oman, Malaysia, UK, and the Netherlands. As a Commercial Lead and Business Opportunity Manager in Shell, she has managed sustainable industrial solutions projects related to water efficiency, energy contracts, plant optimization, and volatile organic compounds capture in Ras Laffan. Following her departure from oil and gas, she has worked on consulting for private and government entities in Qatar. She holds a Bachelor of Science degree in Business Administration in Finance and Management Information Systems from Boston University. She also holds master's degree in Sustainable Development with a specialism in Environmental Management from University of London—SOAS. She is now a Program Manager for Qatar Foundation's new sustainability center (Earthna, Center for a Sustainable Future). In her current role, she is managing programs and projects related to innovative sustainability pilots testbeds that support policy advocacy and development as well as creating sustainable cities solutions in hot and arid climates.

Chapter 15
From Planning Cities to Sustaining Communities: Smart and Sustainable Urban Strategies for the Post-carbon Transition in the Gulf States

Esmat Zaidan

Abstract The achievement of smart urbanism and sustainable urbanism is one of the more prominent themes of the research on urban planning worldwide. Smart sustainable cities (SSCs) are rapidly impacting global debates about the future prospects of urban development. Despite several instances of the characteristics of a smart or sustainable city, a precise definition and a theoretical framework remain lacking. An analysis of regional development plans and city layout reveals that large-scale development will continue in an unsustainable manner. The current trends indicate that infrastructure planning in the cities of the Arabian Gulf must be reoriented towards compact, linked and people-centric characteristics as well as low-carbon infrastructure (UN Habitat in Smart sustainable cities and smart digital solutions for urban resilience in the Arab region: Lessons from the pandemic, 2021). The governments of the Gulf states are confronted with major obstacles due to a projected increase in population and significant reliance on hydrocarbon energy sources. Gulf cities are at the forefront of increasing demand and pressure on critical resources, such as electricity, food and water, which results in a substantial carbon footprint on the environment. In the Gulf region, SSCs emerge as a potential answer to concerns associated with resilience and sustainability, which also emerge due to the unprecedented rate of urbanisation. SSCs are designed to deliver a succession of intelligent solutions for mitigating the economic, social and environmental effects of urban issues (UN Habitant, 2021). This chapter presents a framework for sustainable urbanism by integrating concepts derived from several disciplines. This knowledge-based approach integrates the multidimensional domain knowledge of urban sustainability to produce a comprehensive road map. Given that smart and sustainable urbanism incorporates elements derived from society, technology, policy and the environment, this integration is relatively complex. The proposed framework can contribute to an informed and well-defined understanding of problems faced by cities in the Gulf nations as they strive to achieve sustainable urbanism with a low-carbon footprint.

E. Zaidan (✉)
College of Public Policy, Hamad Bin Khalifa University, Doha, Qatar
e-mail: ezaidan@qu.edu.qa

L. Cochrane and R. Al-Hababi (eds.), *Sustainable Qatar*, Gulf Studies 9,
https://doi.org/10.1007/978-981-19-7398-7_15

Keywords Smart and sustainable cities · Smart communities · Urban development · Sustainability · Gulf cities · Qatar

15.1 Introduction

Cities, which house more than 50% of the global population, are key economic hubs. Cities account for 55.72% of the global population in 2019, and this figure is expected to increase to 68% by 2050. In the Arab region, the proportion of urban residents to the total population in 2019 was 59.2%, which is higher than the global average. Rapid urbanisation makes cities more susceptible to issues associated with growth and development, competitiveness, performance and livelihood, as well as potential strains on services and infrastructure (UN Habitat, 2021). From now until 2050, the developing world is predicted to account for 93% of the total growth with the majority of the growth in urban population occurring in medium-sized cities in Asia and Africa. More than 800 million people currently live in poverty-stricken localities with the majority residing in Africa, Asia and South America. Real estate ideas and values dominate the prevalent urbanisation model, where automobile service is the primary source of transportation. These claims are supported by innovative methodologies for ecological and system-based thinking in architectural, urban and landscape design as well as bottom-up models of urbanisation. Emerging ideas, such as citizen science, co-creation and community-led design, are propelling the social sciences to the forefront of research in terms of sustainable urban design (Grichting, 2018).

In addition, unplanned or poorly managed development coupled with unanticipated shocks and pressure (Abulibdeh & Zaidan, 2020) may place metropolitan systems under significant stress, which could result in long-term strains on fundamental infrastructure (UN Habitat, 2021). In general, the subsequent urban form is characterised by high degrees of zoning, tall buildings or superblocks, freeways, decorative landscapes (in contrast to functional and ecological landscapes) and a stark difference between formal (blocks and superblocks or gated residential areas) and informal living spaces (poverty-stricken localities, Grichting, 2018).

Alternatively, all member states of the UN accepted the 2030 Agenda for Sustainable Development in 2015. Since then, the majority of Arab nations have begun to connect national plans and objectives with the 17 Sustainable Development Goals of the Agenda (Katramiz et al., 2020). In addition, the Arab region is one of the world's most urbanised regions, which places cities in the region at risk of natural and man-made hazards as well as of various regional issues such as shrinking living spaces; conflicts and rentier economies; poverty and water scarcity; gender inequality; and unsustainable extraction patterns of resources and youth unemployment (UN ESCWA, 2020). To address these issues, numerous countries in the region pursue significant measures towards urban resilience and sustainable development, albeit at a slower rate than those of other regions. In addition, a number of nations, including Algeria, Egypt, Jordan, Morocco, the State of Palestine, Qatar, Saudi

Arabia and the United Arab Emirates, are undertaking local projects to develop SSCs by redeveloping the existing cities or constructing new ones (UN Habitat, 2021).

Furthermore, the worldwide demographic weight is shifting inexorably towards cities as the predominant organising force of human coexistence. The administration of large and small cities is the driving force of the modern time, which is an aspect of urban planning that inefficiently stifles human development or that can be exploited to create prosperity and distribute advantages in the search for social tranquillity and individual wellbeing. Along with economies of scale, smart infrastructure underpinned by new finance mechanisms and integrated planning can offer optimum outcomes for beneficial cohabitation in cities, according to the State of Arab Cities 2020 Executive Summary (UN Habitat, 2020). The shift in the urbanisation model of the twenty-first century presents a unique set of possibilities for the development of smart cities. Moreover, advancements in the twenty-first century support the transition from sustainability evaluation to smart city objectives. Discussions on SSCs increasingly influence global debates on the future of urban development. However, very little is known about the concept of *smart sustainable urbanism* despite the abundance of examples of the characteristics that comprise a *smart* or *sustainable* city. The reason underlying this notion is the lack of definitional specificity and corresponding predisposition towards self-congratulation. In the face of ever-increasing urbanisation, new technological advancements fundamentally alter the manner in which cities are structured, planned, designed, operated, managed and governed using smart and sustainable methods (Al-Khalifa, 2021). The concept of the smart city and its related technological initiatives serve as a catalyst for enhancing the living standards of urban residents. These initiatives strengthen the digital intelligence of the economy, urban infrastructure, environment, social capital and other facets of the city, which, thus, enhances the interaction between citizens and their urban environment (Al-Khalifa, 2021).

The achievement of smart and sustainable urbanism is one of the more prominent themes of the research on urban planning worldwide (Al-Khalifa, 2021). SSCs are rapidly impacting global debates about future prospects of urban development. Various terms, such as digital, ecological, information, eco-, resilient, intelligent, green, sustainable, smart, knowledge, liveable and zero- and low-carbon cities and even combinations of these terms are used in today's urban policy, design and planning discourses. Despite the fact that *sustainable city* is the most commonly used word in the literature, it is also the largest and most interconnected (De Jong et al., 2015). The term *smart city* refers to the use of advanced technologies to elevate living standards and the economic development of a city. Cities combine clever technologies and inventive design to provide inhabitants with high-quality services and to revamp urban spaces to enhance the standards of living. Over a thousand smart cities have been established globally, primarily in Europe, North America and Asia (Ghofrani et al., 2021; Zaidan et al., 2022). However, this term lacks a clear-cut definition in academics and industry.

The purpose of the smart city exceeds the concepts of intelligent, digital and informed cities, as it examines technologies as a better approach for assisting people live better lives and make cities function efficiently. The objective of urban sustainability is to achieve economic, social, environmental, cultural and political effectiveness. As part of the smart city agenda, this concept emphasises the use of locally accessible resources and digital advances to develop appropriate, resilient and self-sufficient cities. In terms of urban sustainability, a clear connection is observed between smart and sustainable urbanism (Al-Khalifa, 2021). However, the literature demonstrates a gap between these two concepts, which highlights the need to develop a regional understanding of smart sustainable urbanism that considers regional circumstances and value judgements of communities.

15.2 Beyond Oil: The Inevitability of Smart and Sustainable Urbanism

Smart and sustainable urban development strategies are of key importance in rentier states given the distinct issues present in the urban landscape of various oil-rich rentier states. Rentier states are characterised by generating a significant portion of their income from the rent of local resources to foreign clients. Typically, the revenue is obtained through the extraction and sale of precious natural resources that are completely under the authority of the government, which is primarily directed by a ruling elite group (Kaya et al., 2019). Revenue from oil and gas resources is utilised to fund governmental budgets as well as infrastructure projects in order to improve the quality of life for citizens, and act as a tool for wealth redistribution. These investments in infrastructure include road, electricity supply, transportation, school, hospital and telecommunications projects. Owing to the substantial revenues generated from oil and gas resources, rentier states in the GCC region have witnessed an unparalleled level of economic development, accompanied with rapid urbanisation and motorisation (Azzali & Tomba, 2018). This has resulted in new social and business opportunities for the local populace, as well as significant problems for the local governments, particularly in terms of sustainable urban planning.

These major issues pertain to significant conurbations surrounded by urban sprawl, deep fragmentation of the urban landscape and an overall lack of a comprehensive master plan for urban development (Zaidan & Abulibdeh, 2021). The urban planning process of the GCC rentier states has been largely based on western patterns with little focus on the unplanned urban evolution and social complexity of the city. In addition, land speculation has contributed to a pronounced distortion of the planned organisational structure of the urban environment. This has further led to poor land zoning and prevalence of private residential land use ahead of recreational and public land use which are key aspects of urban living spaces (Furlan & Faggion, 2017). In the light of the major issues present in the urban living spaces of GCC rentier states and owing to an overall global shift towards sustainability, the governments and relevant policymakers of the GCC region have adopted smart and sustainable urban development strategies. In order to guide the transformation of their urban living spaces towards smart and sustainable cities, the GCC countries have formulated comprehensive visions (e.g. Qatar National Vision 2030, Abu Dhabi Policy Agenda 2030 and Dubai Plan 2021) accompanied by specific and technical documents (e.g. Qatar National Development Strategy 2018–2022, Qatar National Urban Master Plan, Plan Abu Dhabi 2030/Urban Structure Framework and Dubai 2020 Urban Master Plan) in order to achieve the aims of the visions through various initiatives (De Jong et al., 2019).

Major investments have been made in the development of new industrial trade zones that extend beyond oil and gas exploration, research and education/knowledge cities where high-tech innovation is cultivated for future generations to lean on, and liveable and smart urban living spaces where residents can reside in a comfortable and environmentally friendly manner. In addition, conglomerate spaces which serve multiple functions have emerged throughout the GCC region. For instance, Abu Dhabi's Masdar City serves three functions: free economic zone, houses a world-class research institute and a world-renown sustainable city (De Jong et al., 2019). To adopt sustainable urbanism in the region, the GCC rentier states need a comprehensive and innovative urban planning strategy. This strategy should combine land use and strategic planning based on building national capacities, as well as a complete understanding of the social, cultural, and oil-dominated economies and community engagement (Zaidan & Abulibdeh, 2021).

Furthermore, rapid urbanisation has resulted in increased social and economic inequality (a widening of the wealth gap), an increase in informal urbanisation, poor standards of safety and the lack of appropriate governance in many rapidly urbanising countries. Increased urbanisation has exerted negative impact on the provision of fundamental infrastructure and services (e.g. water, sanitation and energy), which leads to a low proportion and quality of public spaces in low-density urban forms; overcrowding and traffic congestion; high rates of unemployment (more than 50% among young people); and an informal economy that can occasionally reach 70%. This scenario is also a reflection of a poor public sector, which does not stimulate agglomeration or urbanisation economies and results in the loss of thc city of its primary role as a multiplier of wealth (Grichting, 2018).

Although cities and urbanisation in the twenty-first century face numerous obstacles, potential exists for the construction of sustainable cities and urban areas. Among societal and technological developments with the potential to positively affect change in an urbanised and urbanising world are access to technologies for communicating; gathering and sharing data; networks and citizen-driven initiatives; online learning and training; increase in the availability of micro-scale infrastructure to produce green energy (e.g. solar); clean water or food production; and new modes of financing (e.g. crowd funding, micro-credits and public–private partnerships, Grichting, 2018).

Efforts towards smart and sustainable urbanism are required to avert impending urban ecocide and advance towards post-anthropocentric urbanism, according to the research (Yigitcanlar, 2018). In the twenty-first century, the focus of sustainable city planning is shifting to smart city objectives. Conversely, a concrete definition of a *smart* or sustainable urban environment is lacking (Höjer & Wangel, 2015). Sustainability and smartness benefit from international debates on their respective interpretations and definitions (Maclaren, 1996). The literature reveals that diverse societies should possess slightly or significantly varied understandings of smart and sustainable urban forms based on their economic, social, political, environmental and cultural characteristics as well as the value judgements of their society (Al-Khalifa, 2021).

Projects and initiatives towards SSCs will only succeed if all stakeholders, including citizens, engage with one another and actively participate in realising the principles and goals of SSCs. Thus, maintaining a balance amongst the diverse needs of community stakeholders should be prioritised for the ideologies of SSCs. Therefore, efforts towards SSCs should not be restricted to the comprehension of local legislators, enthusiasts and governance populists. If city planners make smart and sustainable decisions, then the living standards of residents could significantly improve (Al-Khalifa, 2021). As a result, smart and conscientious residents are an absolute necessity for SSCs. A smart city prioritises smart citizens over the adoption of smart technologies, and numerous efforts are ongoing to educate city residents on the utilisation of public information and actions and procedures that shape their community and support infrastructure (Roseland, 2012). Today, citizens are encouraged to learn more about participating in the design and planning of neighbourhoods and cities. Various data, scenarios and models, which are influenced by the urban environment and communication technologies, could be used to engage citizens on-site or remotely (Al-Khalifa, 2021).

15.3 Human-Driven Urban Planning: Economy, Society, Technology and Policy

Since the discovery of oil and gas in the Gulf region, strategies and policies towards energy management have focused on the exploitation of fossil fuel to meet local energy needs and to export this valuable and highly profitable resource to the rest of the globe. In the light of the worldwide push to cut carbon emissions, scholars have anticipated that energy policies in the Gulf will undergo drastic changes (Gritchting, 2018). Recently, Gulf countries have issued pledges towards carbon neutrality in line with its projected benefits and global inclination towards it. The majority of these stances have been adopted shortly before the 26th UN Climate Change Conference (COP26) and varied in level of commitment (Gritchting, 2018). Such commitments have been considered hindered by several obstacles that lack further examination in the literature. In terms of achieving a zero-carbon city or community, the deployment of solely technology and technology-driven solutions is insufficient. Thus, engaging society in sustainability goals is vital for the success of these solutions and, ultimately, for improving the performance of urban sustainability (Al-Saidi & Zaidan, 2020).

Although carbon dioxide emissions, which are energy related, may have decreased in 2020 as a result of the COVID-19 pandemic and the downturn in oil prices, a comeback could restore the long-term trend. Consequently, governments worldwide are addressing major issues related to the sustainability of numerous sectors such as construction, transportation, industry, health care and education. Moreover, a growing strain is observed on key resources, such as energy, water and food (Abulibdeh et al., 2019), with the increase in the global population, which could result in a substantial carbon footprint on the natural environment. With their anticipated population growth as one of the highest in the world and extreme reliance on hydrocarbon energy sources, GCC countries are at the forefront of these increasing demands (Gritchting, 2018). Transitioning to zero-carbon emission is a particularly complex topic, because it involves the interaction of many factors that impact the demand and supply sides of the energy market. Effective zero-carbon transformation involves the active participation of every community entity. Social considerations and acceptance may hinder the adoption, extension and use of clean energy systems, which may affect the transition to a zero-carbon society. Studies that evaluate the social acceptability of transitioning into zero-carbon communities are essential for comprehending the social perspective of communities towards energy efficiency and the use of renewable energy technologies. For its successful implementation, zero-carbon technology must be promoted and implemented through participatory mechanisms in which social actors are actively involved.

Post-carbon cities have recently emerged from a break in the carbon-dependent urban system, which has led to the emergence of considerable amounts of anthropogenic greenhouse gases. To address numerous issues, such as the degradation of ecosystems, climate change, economic strains and social equity, this threat necessitates a complete overhaul in cognition about and planning interactions amongst energy, climate change and the society. Sustainable urban design and management involve striking a balance amongst several competing priorities, including those related to energy, the environment and the society as a whole. Consequently, urban design strategies for post-carbon cities must consider not only the control, efficiency and speed of complex infrastructure systems but also the behaviour of residents and their response to building performance and available urban services. Moreover, the general public should be informed, involved and self-sufficient in terms of energy, which include not only individual consumers but also prosumers who generate their power for personal use when feasible.

Gulf nations incorporated climate change threats into resilience-building processes and urban planning (UNDP, 2018). For instance, the ruler of Dubai incorporated guidelines for green buildings in 2007 to improve water conservation and energy efficiency in newly constructed buildings in the Emirate. According to UN Habitat (2021), the Housing Minister of Bahrain announced in January 2021 that the construction of new towns will, henceforth, be green and intelligent. This initiative was proposed in accordance with the relevant goals of the United Nations 2030 Agenda, particularly SDG 11, by utilising eco-friendly materials, providing open green spaces and planting trees along streets. Another example is Masdar City, which is located near Abu Dhabi in the United Arab Emirates and holds the claim of being the first fully eco-friendly city in the world. Developed from the ground up, it is one of the newest SSCs in the region with zero emission, low waste, access to smart services and green energy use.

As the State of Qatar prepares for the 2022 FIFA World Cup and work towards the goals of Qatar National Vision 2030, the population and economy of the country have increased at unprecedented rates. However, hydrocarbons are the country's sole source of energy. This notion is problematic, because increased economic and environmental consequences are due to the significant reliance of the country on conventional energy sources (Charfeddine et al., 2018). As a result, environmental goals and the development of environmental stewardship and alternative energy sources figure prominently in Qatar's vision statement. The nation possesses an opportunity to reduce carbon emission and create techniques and technology that can play a significant role in attaining global emission-reduction goals without undergoing a fundamental economic restructuring. Improved renewable energy management systems, new technology adoption and a shift to zero-carbon energy systems are necessary conditions for achieving the goals of emission reduction. Moreover, energy efficiency and carbon emission reduction can be achieved at low overall costs by using zero-carbon technologies and systems in Qatar. In addition, the country can serve as a platform for the global development of zero-carbon energy technology. Despite an increase in energy consumption, Qatar intends to decrease its carbon emission (Zaidan et al., 2021, 2022). This goal gives the nation the motivation to adapt and develop new energy technologies and techniques that will serve as sources of sustainable economic growth. Zero-carbon energy systems and technologies are essential for Qatar to achieve its economic and environmental goals. Thus, developing policies that encourage their adoption and integration is imperative. Accordingly, strategies for smart energy have become a national priority in the State of Qatar, where increased government funding encourages a community of data scientists, technologists and key city stakeholders to take advantage of the growing digital revolutions, low-cost sensors and development partnerships to create novel and socially beneficial solutions. Despite the acknowledgement of this large community cooperation, such measures are industry- and government-driven and place extremely less emphasis on creating community demands that are all inclusive and that consider the capacity of the community to profit from and utilise smart energy technologies. Evidently, community institutions are largely side-lined due to the ongoing discourse on smart energy communities (Zaidan et al., 2022).

15.3.1 The Proposed Framework: Smart Sustainable Cities in the Gulf States

Two primary approaches to the smart city concept are suggested. The first is a technocentric strategy that bestows information and communication technology (ICT) with a central role and concentrates on cutting-edge technology and digital infrastructure. For instance, Harrison et al. (2010) defined smart cities as a concept that combines real-time data into a computing platform to provide services through the implementation of visualisation, modelling, optimisation and information processing. According to Hancke et al. (2012), smart cities operate in an environment that is sustainable and intelligent due to the use of sophisticated monitoring and control technologies. Furthermore, the infrastructure and public services of cities are seamlessly interwoven (Zaidan et al., 2022). The second strategy is a human-centric approach that highlights the importance of social and human capital in smart cities. Numerous authors emphasise the importance of human capital and education in urban development and sustainability. For example, Coe et al.'s (2001) definition of smart communities highlights the important role of social and environmental capital in urban development. Consequently, engagement in decision-making processes, social inclusion and community affairs are essential elements for achieving social and environmental goals. The relationship between a highly educated workforce and smart cities is the subject of a few studies (Zaidan et al., 2022), as growth in employment, economic progress and technological advancement are dependent on this relationship (Eger, 2009). Moreover, social innovation and the significance of human elements and communities are the subject of considerable discussion. Innovation in urban regeneration is heavily reliant on the involvement of local communities. Hence, the foundation of smart communities is built on the synergy of the use of infrastructure and technology that is technically possible and relevant to every region. Nonetheless, considering crucial characteristics, such as affordability, acceptance, privacy and coherence, is vital to the discussion on the concept of 'community'. In addition, the term 'smart community' refers to a concept that incorporates cutting-edge technologies and infrastructures geared towards achieving sustainable development (Zaidan et al., 2022). As a result, smart communities render possible the fostering of social innovation for governments, infrastructure and technological forces. Consequently, local stakeholders, including non-profit organisations, corporations and individuals, can be involved in the development of new approaches for addressing issues related to wellbeing, inclusiveness and growth.

SSCs are the result of the efforts of numerous stakeholders. The literature describes the utilisation of stakeholder perceptions to establish key performance indicators for the achievement of smart and sustainable urbanism. By establishing the necessary guidelines and procedures and by ensuring that mitigation plans, remedial measures and implementation activities are conducted in a comprehensive manner, statutory law-making authorities play an important role in the development of smart sustainable communities (Al-Khalifa, 2021). The study proposes a framework for SSCs, which is described in the following section.

Currently, smart cities are being implemented to improve urban life in a wide range of areas, including economy and society (e.g. cultural heritage preservation, digital education, entrepreneurship and human capital development), in addition to government (e.g. transparency, e-government and e-governance) and daily life (e.g. public space management, cultural preservation, social inclusion, public health, public safety and pollution management) and smart building (e.g. waste management, food and agriculture, renewable energies, intelligent street lighting and smart grids) (Zaidan et al., 2022). Thus, moving from connected communities to a connected world necessitates a shift from community targets and plans to those at the city level.

This scenario is extremely challenging due to the exponential increase in complexities, especially given the inevitable inclusion of social and policy factors on a large scale. To achieve the transformation of a connected network of smart communities into large-scale smart cities, the transition at the city level should be undertaken in a distributed and progressive manner. Public engagement, asset management and possession, available resources, varied and versatile energy demands and scale are all shared characteristics of communities considered capable of a smart and sustainable evolution. As a consequence, constructing a system that is resource-driven and incorporates its residents is conceivable. A good example of the economic and social benefits that can be seen at the community and city levels is the transition involving communities. In addition, big business centres, transportation hubs and ports exert a large environmental impact and are extremely vulnerable (Zaidan et al., 2021, 2022) (Fig. 15.1).

To achieve SSCs, formulating new and effective strategies for zero-carbon systems based on a comprehensive and thorough understanding of the components that can be utilised during the design is crucial. The initial phase of the investigation consists of conducting an exhaustive assessment of socioeconomic elements that exert an impact on the transformation to a zero-carbon community. Based on a preliminary analysis of the relevant literature, it is shown that the socioeconomic factors that are driving the shift towards zero-carbon communities may be broken down into the following categories.

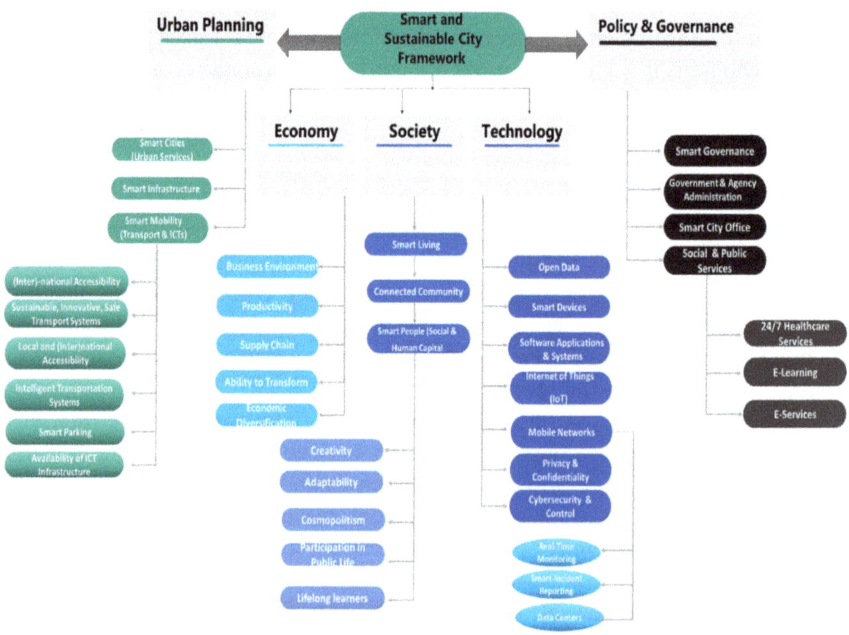

Fig. 15.1 Framework of smart sustainable cities

Demographic factors, such as gender, age, education, work status and income, influence the decision to implement pro-environmental policies (Ghofrani et al., 2021).

Education is one of the most crucial factors. For example, according to Mills and Schleich (2009), heads of households in Germany with post-secondary education are more likely to employ solar water heating for their homes. In a further study, Mills and Schleich (2012) stated that university-educated households are more likely to employ energy-efficient technologies. Sardianou and Genoudi (2013) proposed that educated Greeks are more likely to turn to renewable forms of energy. Moreover, Michelsen and Madlener (2012) suggested that German homeowners with a university degree are more inclined to use residential heating systems that are based on renewable energy. Niamir et al. (2020) indicated that highly educated households in two European locations are more likely to invest in home insulation and solar panels as well as to switch to green energy suppliers.

Furthermore, gender influences the decision to pursue environmentally friendly strategies. Although the literature remains inconclusive regarding the effect of gender on energy efficiency, a few studies indicate that females are more inclined to implement energy-conserving measures. According to a meta-analysis conducted by Zelezny et al. (2000), women are significantly more likely than men to engage in environmentally conscious activities. However, other studies reveal opposite findings. For example, evidence exists that men are more likely than women to instal solar panels, as reported by Niamir et al. (2020).

Additionally, age influences the level of reliance on green energy. Older families who are apprehensive about their return on investment in green energy are less likely than younger households to demonstrate pro-environmental attitudes (Nair et al., 2010). Nevertheless, certain studies (Barr et al., 2005; Long, 1993) indicate that elderly homeowners are typically more energy efficient than younger homeowners. In terms of income, studies argue that households with high incomes are more willing to invest in renewable energy. For example, Sardianou and Genoudi (2013) report that the probability of adopting renewable energy policies is strongly correlated with monthly income. Moreover, Sidiras and Koukios (2004) suggest that homeowners with high incomes are more inclined to invest in solar hot water technology. Conversely, Niamir et al. (2020) did not report a positive correlation between income and investment in home insulation and solar panel instalation.

Psychological motivators of the transition to zero-carbon emission: This category consists of two factors, namely, social norms and personal norms.

- Social norms. They are defined as ‘perceived social pressures from significant others and/or beliefs about how significant others expect one to act in a given situation’ (Chen et al., 2017) and influence the degree of the reliance of individuals on policies that protect the environment. For example, Allcott (2011) demonstrated that societal norms in the United States impact energy efficiency. For example, the author presented the case of ‘OPOWER’, is a corporation that sends letters to homeowners, compares their energy compensation to those of their neighbours and offers advice on the reduction of energy use. The study proposed that these letters contributed to low energy use. Wang et al. (2011) pointed to a correlation between societal standards in China and increased energy conservation in households. Similarly, Chen et al. (2016) found a positive correlation between societal norms and people's intentions in China to use solar water heating systems and automobiles that use non-fossil fuel energy (Ghofrani et al., 2021). According to Han et al. (2010), perceived social standards motivate consumer to select environmentally friendly hotels.

- Personal norms. In terms of energy efficiency, personal norms can be characterised as a sense of moral obligation to pursue pro-environmental conduct (Huijts et al., 2012). Werff et al. (2019) found that households in the Netherlands with stronger personal norms are more likely to demonstrate an energy conservation mindset. Moreover, Niamir et al. (2020) revealed that people in the Netherlands and Spain are more likely to spend money on energy-efficient appliances due to personal norms.
- Cultural barriers to low-carbon transport and buildings (e.g. masculinity, ethnic prejudice, social status, social exclusion, vandalism and thermal comfort (Sovacool & Griffiths, 2020). As an example, social status is a major cultural barrier to shared transportation in GCC nations. Indeed, car ownership in the GCC is a sign of wealth and social status. In the United States and Australia, aggressive driving and speeding are primarily attributable to masculinity and machismo, which result in substantial carbon dioxide emission. These cultural factors hinder the shift to a zero-carbon economy.
- Structural factors, such as the type, age and size of buildings, are key contributing factors for the transition towards a zero-carbon economy. For example, Mills and Schleich (2009) presented evidence that older homes are typically less energy efficient. Niamir et al. (2020) found that homeowners are more inclined to invest in renewable energy compared with apartment owners. New building owners are more (less) inclined to install solar panels (house insulation). In addition, the authors identified a positive relationship between the size of the home and investments in insulation and solar panels. This finding was consistent with the rationale that homeowners of large homes have high incomes and can, therefore, afford to make investments in renewable energy sources. Additionally, large homes typically have a larger rooftop area, which renders installing solar panel systems possible for homeowners.
- Financial policy factors include the cost of adopting zero-carbon electricity for consumers as well as carbon neutral products and buildings (Cheng et al., 2019; Zhang et al., 2011). These factors are considered to be among the most influential in determining the transition to zero-carbon energy. Lowering these costs and facilitating the transition to a zero-carbon economy are possible by reducing value-added tax on energy-saving appliances and technologies as well as programmes that provide customers with credit for the purchase of energy-efficient cooling and heating systems (e.g. Bulgaria). Multiple studies provide evidence supporting this viewpoint. For example, Sardianou and Genoudi (2013) find that tax deductions and subsidies exert a positive effect on the use of renewable energy by households. Li et al. (2016) demonstrate that government subsidies and advantageous tax

policies in China encourage consumers to use new energy cars. Furthermore, Wang et al. (2011) reveal that the majority of respondents are more inclined to select energy-saving options if the government provides financial assistance in the form of subsidies. Households with high incomes can afford to invest in green energy; thus, financial measures, such as subsidies, credits and tax breaks, will not exert a significant impact on investment in zero-carbon technologies for this group. Instead, financial measures should be channelled towards low-income households. Imposing a charge on fossil fuel may also encourage consumers to make the move towards cleaner forms of energy. This notion is further supported by Sardianou and Genoudi (2013), who suggested that people are more inclined to invest in green energy with the increase in the price of fossil fuel.

Alternatively, when the government plays an active role in building SSC projects, the influence of growth becomes significant in terms of the smart governance component of the framework. In addition, the literature reveals the necessity of viewing smart governance from a different angle. For a long time, specialists with backgrounds in information technology and computer science dominated the literature on e-governance. Nevertheless, the current issues associated with the creation of smart and sustainable communities demand a step back and the adoption of a holistic view of smart governance. E-governance must be evaluated from the perspective of urban domain specialists, such as experts in urban development, urban design and planning, climate change, carbon accounting, energy and water governance and the public realm (Al-Khalifa, 2021). As a result, the framework for SSCs may help to define smart and sustainable urbanism in the Gulf region and to determine the most important themes, factors and components required to better understand the level of smartness and sustainability at present in the urban environment with a focus on Qatar.

Reaping the benefits of innovative thinking, systematic stakeholder collaboration and bright scientific approaches (i.e. smart solutions) is possible for Qatar. Such benefits include the effective management of population surge, boost in economic development and improvement in the wellbeing of the people. A dire need also exists for scalable solutions that utilise many advantages offered through the appropriate use of ICT. Currently, Qatar possesses a unique opportunity to significantly improve its entire critical infrastructure, which includes the systematic development, integration and monitoring of its roads, railway network, airports, communication systems and buildings (MICT, 2014). This step will enable Qatar to effectively maximise the use of its resources and provide the highest possible level of service to residents while preserving the integrity of its security.

The most important benefit of smart cities is that they increase the quality of life of residents. Social, economic and environmental benefits can be gained if urban issues are addressed using smart approaches and technologically enabled solutions (e.g. safety and transportation). Furthermore, the current study suggests that officials must simultaneously identify and address the most pressing urban issues while formulating a futuristic vision for the nation. To illustrate, the transportation sector is a top concern in Qatar; therefore, the government may seek immediate and smart solutions to the increasing traffic difficulties. The development of smart cities can lead to the creation of new services and jobs. The overall economy will gain from the improved allocation of resources, utilisation, innovation and an enhanced entrepreneurship ecosystem, which offers additional opportunities. The smart and sustainable features of cities can also be promoted to improve competitive advantage in terms of location and attract new enterprises and talents. Currently, cities in Qatar can be developed in an environmentally sound manner with initiatives that enable waste minimisation, low-carbon footprint and the use of renewable forms of energy. Residents will enjoy better quality of life and a safer environment if cities are more environmentally friendly. Conversely, developing cities in Qatar with the consideration of cultural sensitivity is of equal importance. In non-western cities that experience rapid westernisation, recognising the proponents of sustainability is essential. Sustainable urbanism in Gulf cities emphasises the importance of achieving sustainability as a cultural asset by identifying members of society, organisations and businesses who possess a culture of resourcefulness and resilience as internalised values and the competencies required to advance these methods. In addition, exploring the potential contribution of *Old Doha* to this dialogue and whether or not all generations can be represented at the table is essential (Gritchting, 2018; Zaidan, 2019).

15.3.1.1 Challenges and Strategies of SSCs in Gulf States

Analysis of the proposed projects in the Gulf region reveals that large-scale developments will continue in a manner that cannot be maintained in a sustainable manner in the long run (UN Habitat, 2021; Zaidan et al. 2019). The current patterns illustrate that infrastructure planning in the cities of the Arabian Gulf needs to be reoriented towards a compact, linked, people-centric and low-carbon infrastructure. Furthermore, systemic and integrated infrastructure planning is required for future projects, such as highways, airports, mega-developments and new cities, to manage the interdependencies and links amongst various sectors, as well as to meet current and long-term needs. Such a vision would entail small-scale initiatives that cater to city inhabitants, improve quality of life and reduce entrenched socioeconomic differences (UN Habitat, 2021). The outcome is a clean and compact city in the Arab Gulf committed to providing equitable welfare and opportunities for all people, including future generations, while ensuring no one is left behind.

The complexity of Gulf cities is revealed in numerous aspects of the local government, including governance, policy and interlinkage and interaction among sectors. Conversely, other sectors may be unaware of the complex interplay between their activities and those of other sectors in the context of city development and operations. As a result, stakeholders may find that agreeing on implementation strategies for smart solutions is difficult. An overarching vision and a forum for collaboration among all stakeholders is required in the Gulf region, where various sectors may collaborate to discuss challenges, exchange ideas and develop solutions together. This forum will support all stakeholders in their efforts to become smarter. Collaboration is a crucial feature, because smart cities cannot be produced by a single entity.

Smart initiatives nearly always necessitate the participation of various stakeholders, as evidenced by international best practices. To facilitate the flow of information, a collaborative and collective command and control centre is required; currently, each organisation operates independently, which is an unproductive method. Furthermore, several nations, including those in the European Union, have put in place good governance procedures to foster public–private partnerships and citizen participation. Hence, allowing sectors and stakeholders to share experiences and success stories has endorsed innovation and the generation of new ideas.

Alternatively, concerns regarding confidentiality and privacy must be resolved for types of procedures related to smart cities to be implemented in Qatar. Formulating a government policy that promotes access and collaboration is critical; otherwise, organisations will struggle to become transparent. Concern also emerges about the potential exploitation of sensitive data if they are widely available and open, such as those on the energy consumption or water usage of the country. Nonetheless, information from one sector can exert a substantial impact on and provide value for other sectors. Consequently, the government must implement efficient privacy and security-related policies and methods to help mitigate these issues.

Another challenge is the divergent perspectives among stakeholders on the elements that constitute *smart*. In Qatar, for instance, the perspectives of an architect and an environmentalist may greatly differ. No consensus exists on the requirements for smart cities, and no consistency is in place with regard to policies and standards. This difficulty reaffirms the previously indicated need for a smart city organisation in Qatar to coordinate all related issues. During city planning, the concept of SSCs must be integrated. The transformation of an ordinary city into a smart one necessitates the incorporation of ICT, and its proponents must collaborate with government bodies and relevant political stakeholders, such as city councils, national institutions and urban planning authorities, which hold the potential to influence the outcome of ICT initiatives (MICT, 2014). The seamless implementation of smart city endeavours heavily relies on institutional readiness, which includes the lack of legal and regulatory obstacles. Interoperability between systems and entities is another corresponding issue. Good governance is required to meet the demands not only of entities but also of people and inhabitants. This aspect can be aided by marketing that targets users that promote and provide valuable information on the adoption of smart technologies. Below are certain challenges and counterstrategies for SSCs in the Gulf region (Fig. 15.2).

Scholars acknowledge that the efficient use of data will detect and remedy problems. Health care and energy are only two of the many areas in which Qatar possesses a wealth of information at its disposal. However, merely collecting data is insufficient if they are not effectively used and shared. In this regard, Qatar lags behind. Thus, mechanisms are required for the analysis and application of data. Initial investment in smart city solutions may result in a hike in expenses, which residents may not appreciate. However, although smart cities may initially result in a short-term increase in the cost of services for inhabitants, smart city solutions will eventually result in efficient and cost-effective techniques (MICT, 2014). Despite this notion, advancing the smart city agenda through governmental stakeholders and real estate

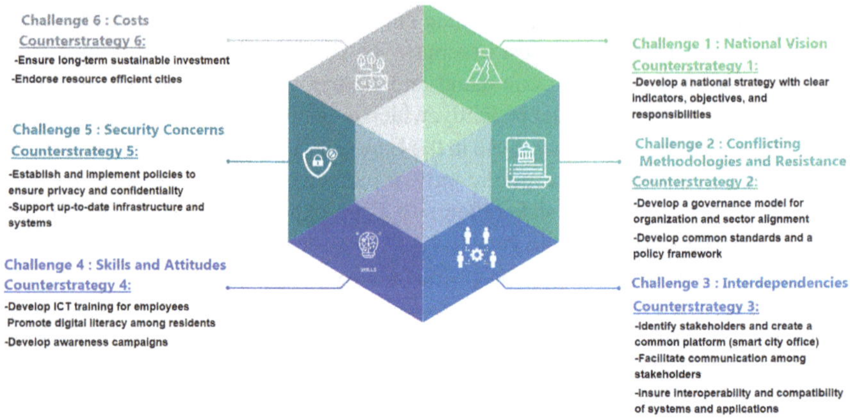

Fig. 15.2 Challenges and counterstrategies. Adopted from Zaidan et al. (2022) and MICT (2014)

developers in Qatar may be profitable given the need for investment. Qatar must establish a futuristic vision for the city that aims to improve the lives of residents, who will live in a hyper-connected, high-speed and zero-emission metropolis, which is similar to Barcelona, a leader in the smart city movement.

15.4 Conclusion

The examination of the manner in which governments and societies shape the urban future of the region now indicates that many Gulf countries are embarking on massive infrastructure and megaproject efforts in addition to the emergence of new smart cities. Despite the fact that this study focuses on all factors necessary to create smart cities that will improve the quality of life in cities of the Gulf, adopting a proactive role in implementing supportive policies and initiatives and coordinating efforts is imperative for governments in the Gulf, particularly in Qatar, to ensure that the necessary ingredients for success are in place, including collaboration, uniform standards, data sharing and a comprehensive national vision. Stakeholders seek the actions of the government to launch a strategy and facilitate the subsequent collaboration. In this manner, it gains attention and goodwill from all stakeholders to construct a governance model that guarantees adherence to a shared national vision. On the other hand, human-oriented urban planning is a requirement for post-oil urbanism in Gulf states and represent an unavoidable alternative for the development of SSCs. Uncertainties will increase when society, which is at the core of the smart transformation to zero-carbon community, is unengaged.

A bottom-up approach is required to acquire a holistic understanding of society, its constraints, driving forces, motivations, adaptability and acceptability as well as the role of human factors in the transition to smart communities and smart cities.

Simply put, no tangible and sustainable solutions will be observed without society engagement. Decision- and policymakers in Gulf states may endeavour to bridge gaps between the expected and actual outcomes of smart transformation policies, if the role of socioeconomic and behavioural dimensions in consumption patterns is considered mainly in regions that host high proportions of migrant communities with diverse cultures and ethnicities. Technology-driven solutions for SSCs must be supplemented by human-driven solutions to enable society to become aligned with the goals for national strategic sustainability, particularly in reducing the carbon footprint of Gulf nations by curbing the demand on the energy sector.

Lastly, effective regulations and an acceptable legal framework must be formulated for the successful use of ICT in stimulating progress in the creation of smart cities. Furthermore, Qatar may need to embrace an integrated strategy based on the premise that human and social capital in conjunction with technology create a suitable ecosystem, which fosters a continuous process of development and innovation. Knowledge-based activities, institutional structures for the development of innovation and social cooperation and digital infrastructure unite to form smart cities (i.e. ICT infrastructure, tools and applications). As a result, the concept of smart cities refers to a setting that holds the potential to combine elements of sustainable and smart urbanism and competitiveness through the incorporation of many aspects of urban development (i.e. social, economic, environment, governance, and people and mobility). The focus should not be limited to the impact of ICT drivers on urban development. The definitions of smart communities share three concepts, namely, processes (networking of multiple participants), methods of communication (ICT, technology and network architecture) and desired outcomes (public involvement or others).

Acknowledgements This publication was made possible by an NPRP award [NPRP13S-0206–200272] from the Qatar National Research Fund (a member of Qatar Foundation). The statements made herein are solely the responsibility of the authors.

References

Abulibdeh, A., & Zaidan, E. (2020). Managing the water-energy-food nexus on an integrated geographical scale. *Environmental Development, 33*, 100498. https://doi.org/10.1016/j.envdev.2020.100498

Abulibdeh, A., Zaidan, E., & Al-Saidi, M. (2019). Development drivers of the water-energy-food nexus in the Gulf Cooperation Council Region. *Development in Practice, 29*(5), 582–593

Al-Saidi, M., & Zaidan, E. (2020). Gulf futuristic cities beyond the headlines: Understanding the planned cities megatrend. *Energy Reports, 6*, 114–121.

Al-Khalifa, F. (2021). An approach to define smart sustainable urbanism locally through expert's perspective. *International Journal of Sustainable Building Technology and Urban Development, 12*(1), 14–26.

Allcott, H. (2011). Social norms and energy conservation. *Journal of Public Economics, 95*(9–10), 1082–1095. https://doi.org/10.1016/j.jpubeco.2011.03.003

Azzali, S., & Tomba, M. (2018). Urban development and planning practice in Doha. *Middle East Journal, 180.*

Barr, S., Gilg, A. W., & Ford, N. (2005). The household energy gap: Examining the divide between habitual- and purchase-related conservation behaviours. *Energy Policy, 33*(11), 1425–1444. https://doi.org/10.1016/j.enpol.2003.12.016

Charfeddine, L., Yousef Al-Malk, A., & Al Korbi, K. (2018). Is it possible to improve environmental quality without reducing economic growth: Evidence from the Qatar economy. *Renewable and Sustainable Energy Reviews, 82*, 25–39. https://doi.org/10.1016/j.rser.2017.09.001

Chen, C., Xu, X., & Day, J. K. (2017). Thermal comfort or money saving? Exploring intentions to conserve energy among low-income households in the United States. *Energy Research & Social Science, 26*, 61–71. https://doi.org/10.1016/j.erss.2017.01.009

Chen, C., Xu, X., & Frey, S. (2016). Who wants solar water heaters and alternative fuel vehicles? Assessing social–psychological predictors of adoption intention and policy support in China. *Energy Research & Social Science, 15*, 1–11. https://doi.org/10.1016/j.erss.2016.02.006

Cheng, X., Long, R., Chen, H., & Yang, J. (2019). Does social interaction have an impact on residents' sustainable lifestyle decisions? A multi-agent stimulation based on regret and game theory. *Applied Energy, 251*, 113366. https://doi.org/10.1016/j.apenergy.2019.113366

Coe, A., Paquet, G., & Roy, J. (2001). E-governance and smart communities. *Social Science Computer Review, 19*(1), 80–93. https://doi.org/10.1177/089443930101900107

De Jong, M., Hoppe, T., & Noori, N. (2019). City branding, sustainable Urban development and the rentier State. how do Qatar, Abu Dhabi and Dubai present themselves in the age of post oil and global warming?. *Energies, 12*(9), 1657. https://doi.org/10.3390/en12091657

De Jong, M., Joss, S., Schraven, D., Zhan, C., & Weijnen, M. (2015). Sustainable–smart–resilient–low carbon–eco–knowledge cities; Making sense of a multitude of concepts promoting sustainable urbanization. *Journal of Cleaner Production, 109*, 25–38. https://doi.org/10.1016/j.jclepro.2015.02.004

Eger, J. M. (2009). Smart growth, smart cities, and the crisis at the pump a worldwide phenomenon. *I-WAYS, Digest of Electronic Commerce Policy and Regulation, 32*(1), 47–53. https://doi.org/10.3233/iwa-2009-0164

Furlan, R., & Faggion, L. (2017). Urban regeneration of GCC cities: Preserving the urban fabric's cultural heritage and social complexity. *Journal of Historical Archaeology & Anthropological Sciences, 1*(1).https://doi.org/10.15406/jhaas.2017.01.00004

Ghofrani, A., Zaidan, E., & Jafari, M. (2021). Reshaping energy policy based on social and human dimensions: An analysis of human-building interactions among societies in transition in GCC Countries. *Humanities and Social Sciences Communications, 8*(1). https://doi.org/10.1057/s41599-021-00904-9

Grichting, A. (2018). Edges, interfaces, and Nexus: New paradigms for blue urban landscapes in the Gulf. *International Journal of Middle East Studies, 50*(3), 580–585. https://doi.org/10.1017/s0020743818000582

Han, H., Hsu, L., & Sheu, C. (2010). Application of the theory of planned behavior to green hotel choice: Testing the effect of environmental friendly activities. *Tourism Management, 31*(3), 325–334. https://doi.org/10.1016/j.tourman.2009.03.013

Hancke, G., Silva, B., & Hancke, G., Jr. (2012). The role of advanced sensing in smart cities. *Sensors, 13*(1), 393–425. https://doi.org/10.3390/s130100393

Harrison, C., Eckman, B., Hamilton, R., Hartswick, P., Kalagnanam, J., Paraszczak, J., & Williams, P. (2010). Foundations for smarter cities. *IBM Journal of Research and Development, 54*(4), 1–16. https://doi.org/10.1147/jrd.2010.2048257

Höjer, M., & Wangel, S. (2015). Smart sustainable cities: Definition and challenges. In L. Hilty, & B. Aebischer (Eds.), ICT *innovations for sustainability* (pp. 333–349). Springer-verlag.

Huijts, N., Molin, E., & Steg, L. (2012). Psychological factors influencing sustainable energy technology acceptance: A review-based comprehensive framework. *Renewable and Sustainable Energy Reviews, 16*(1), 525–531. https://doi.org/10.1016/j.rser.2011.08.018

Kaya, A., Tok, E., Koc, M., Mezher, T., & Tsai, I.-T. (2019). Economic diversification potential in the rentier states towards for a sustainable development: A theoretical model. *Sustainability, 11*(3), 911. https://doi.org/10.3390/su11030911

Katramiz, T., Okitasari, M., Masuda, H., Kanie, N., & Takemoto, K. (2020). *Local implementation of the 2030 agenda in the Arab world: Addressing constraints and maximising opportunities*. United Nations University Institute for the Advanced Study of Sustainability (UNU-IAS). https://collections.unu.edu/eserv/UNU:7577/UNU-IAS-PB-No19-2020.pdf. Accessed 20 May 2022.

Li, W., Long, R., & Chen, H. (2016). Consumers' evaluation of national new energy vehicle policy in China: An analysis based on a four paradigm model. *Energy Policy, 99*, 33–41. https://doi.org/10.1016/j.enpol.2016.09.050

Long, J. E. (1993). An econometric analysis of residential expenditures on energy conservation and renewable energy sources. *Energy Economics, 15*(4), 232–238. https://doi.org/10.1016/0140-9883(93)90012-g

Maclaren, V. W. (1996). Urban sustainability reporting. *Journal of the American Planning Association, 62*(2), 184–202. https://doi.org/10.1080/01944369608975684

Michelsen, C. C., & Madlener, R. (2012). Homeowners' preferences for adopting innovative residential heating systems: A discrete choice analysis for Germany. *Energy Economics, 34*(5), 1271–1283. https://doi.org/10.1016/j.eneco.2012.06.009

Mills, B. F., & Schleich, J. (2009). Profits or preferences? Assessing the adoption of residential solar thermal technologies. *Energy Policy, 37*(10), 4145–4154. https://doi.org/10.1016/j.enpol.2009.05.014

Mills, B., & Schleich, J. (2012). Residential energy-efficient technology adoption, energy conservation, knowledge, and attitudes: An analysis of European countries. *Energy Policy, 49*, 616–628. https://doi.org/10.1016/j.enpol.2012.07.008

Ministry of Information and Communications Technology. (2014). Sustainable smart cities: Improving quality of life.

Nair, G., Gustavsson, L., & Mahapatra, K. (2010). Factors influencing energy efficiency investments in existing Swedish residential buildings. *Energy Policy, 38*(6), 2956–2963. https://doi.org/10.1016/j.enpol.2010.01.033

Niamir, L., Ivanova, O., Filatova, T., Voinov, A., & Bressers, H. (2020). Demand-side solutions for climate mitigation: Bottom-up drivers of household energy behavior change in the Netherlands and Spain. *Energy Research & Social Science, 62*, 101356. https://doi.org/10.1016/j.erss.2019.101356

Sardianou, E., & Genoudi, P. (2013). Which factors affect the willingness of consumers to adopt renewable energies? *Renewable Energy, 57*, 1–4. https://doi.org/10.1016/j.renene.2013.01.031

Sidiras, D. K., & Koukios, E. G. (2004). Solar systems diffusion in local markets. *Energy Policy, 32*(18), 2007–2018. https://doi.org/10.1016/s0301-4215(03)00173-3

Sovacool, B. K., & Griffiths, S. (2020). The cultural barriers to a low-carbon future: A review of six mobility and energy transitions across 28 countries. *Renewable and Sustainable Energy Reviews, 119*, 109569. https://doi.org/10.1016/j.rser.2019.109569

United Nations Economic and Social Commission for Western Asia. (2020). *Arab Sustainable Development Report 2020*. https://asdr.unescwa.org/sdgs/pdf/en/ASDR2020-Final-Online.pdf. Accessed 18 May 2022.

United Nations Development Programme. (2018). *The Arab cities resilience report*. The United Nations Development Program (UNDP), Regional Bureau for Arab States (RBAS). NY: New York, United States.

United Nations Human Settlements Programme UN Habitat. (2020). *The state of Arab cities 2020: Financing sustainable urbanization in the Arab region*. https://unhabitat.org/sites/default/files/2020/06/sacr_2020_executive_summary_en.pdf. Accessed 18 May 2022.

United Nations Human Settlements Programme UN Habitat. (2021). *Smart sustainable cities and smart digital solutions for urban resilience in the Arab region: Lessons from the pandemic*. https://www.unescwa.org/sites/default/files/pubs/pdf/smart-sustainable-cities-digital-solutions-urban-resilience-arab-region-english.pdf. Accessed 16 May 2022.

Van der Werff, E., Taufik, D., & Venhoeven, L. (2019). Pull the plug: How private commitment strategies can strengthen personal norms and promote energy-saving in the Netherlands. *Energy Research & Social Science, 54*, 26–33. https://doi.org/10.1016/j.erss.2019.03.002

Wang, Z., Zhang, B., Yin, J., & Zhang, Y. (2011). Determinants and policy implications for household electricity-saving behaviour: Evidence from Beijing, China. *Energy Policy, 39*(6), 3550–3557. https://doi.org/10.1016/j.enpol.2011.03.055

Yigitcanlar, T. (2018). Smart city policies revisited: Considerations for a truly smart and sustainable urbanism practice. *World Technopolis Association, 7*(2), 97–112. https://doi.org/10.7165/wtr18a1121.19

Zaidan, E. (2019). Cultural-based challenges of the westernised approach to development in newly developed societies. *Development in Practice, 29*(5), 570–581. https://doi.org/10.1080/09614524.2019.1598935

Zaidan, E., & Abulibdeh, A. (2021). Master planning and the evolving urban model in the gulf cities: Principles, policies, and practices for the transition to sustainable urbanism. *Planning Practice & Research, 36*(2), 193–215. https://doi.org/10.1080/02697459.2020.1829278

Zaidan, E., Al-Saidi, M., & Hammad, S. H. (2019). Sustainable development in the Arab world—Is the Gulf Cooperation Council (GCC) region fit for the challenge? *Development in Practice, 29*(5), 670–681. https://doi.org/10.1080/09614524.2019.1628922

Zaidan, E., Ghofrani, A., & Dokaj, E. (2021). Analysis of human-building interactions in office environments: To what extent energy saving boundaries can be displaced? *Frontiers in Energy Research, 9*. https://doi.org/10.3389/fenrg.2021.715478

Zaidan, E., Ghofrani, A., Abulibdeh, A., & Jafari, M. (2022). Accelerating the change to smart societies—A strategic knowledge-based framework for smart energy transition of urban communities. *Frontiers in Energy Research, 10*. https://doi.org/10.3389/fenrg.2022.852092

Zelezny, L. C., Chua, P., & Aldrich, C. (2000). New ways of thinking about environmentalism: Elaborating on gender differences in environmentalism. *Journal of Social Issues, 56*(3), 443–457. https://doi.org/10.1111/0022-4537.00177

Zhang, X., Platten, A., & Shen, L. (2011). Green property development practice in China: Costs and barriers. *Building and Environment, 46*(11), 2153–2160. https://doi.org/10.1016/j.buildenv.2011.04.031

Esmat Zaidan works as Associate Professor in Policy, Planning, and Development in the Department of International Affairs, College of Arts and Sciences. She holds a Ph.D. in Geography and Environmental Management from University of Waterloo in Canada, a master degree in Urban Planning and Cities Design, College of Architectural Engineering, Birzeit University in Palestine, and a second master degree in Economic Development, University of Waterloo, Canada. She has twenty years progressive experience in academic institutions and international organizations such as the World Bank, the Norwegian Agency for Development Cooperation (NORAD), and the European Committee (EUC), while 11 years of her experience are in academic institutions: Qatar University and United Arab Emirates University. She has solid interdisciplinary scholarly work and activities; her research focuses on development, sustainable cities and societies, sustainable development policy and planning, energy policy, smart policies for urban sustainability, smart transformations, and the actions of communities and demand side in the transition to sustainability: the technological and human factors (society and advanced technology). In her publications, she integrates Artificial Intelligence (AI), Geographic Information Systems (GIS) techniques, spatial statistics, forecasting models, and future predictions into policy analysis and development particularly in the State of Qatar.

Chapter 16
Toward the Circular Qatari Zero-Waste Management Sector

Husameldin M. Talballa and Jonathan Gichuru

Abstract This chapter is an attempt to track the evolution of the solid waste management (SWM) system in Qatar including its recent technological leap of establishing Mesaieed's integrated waste management facility. A variety of factors that have contributed in shaping the sector are determined covering external and internal, hard and soft, ones in the political, economic, and social contexts. Despite the strides, the Qatari SWM sector has witnessed under the aspiration of the sustainability-conscious Qatar National Vision 2030 and its emanated two National Development Strategies (2011–2022) underperformance and lack of vertical and horizontal synchronization among stockholders are evident. A zero-waste framework is suggested as an ideal vehicle for upgrading the current waste management system into a circular one, backed with legislative, financial, social, and knowledge-based instruments. Successful local campus-scale zero-waste management models, like Qatar University, provide blueprints that can be amplified through a nexus systems thinking approach applied on an inclusive Energy-Water-Food-Waste urban setup.

Keywords Solid waste management · Zero waste · Circular economy · Policy instruments · Living lab · NEXUS/system thinking

16.1 Introduction

One of the very first objects that arrivals to Hamad International Airport in Doha encounter are the multi-compartment recycling bins. This would hint at the advanced level of the solid waste management system of the country. However, a quick glance inside any of these compartments and its content of haphazardly mixed waste fractions, along with the following unsegregated collection process would build a bit

H. M. Talballa (✉)
Qatar University, Doha, Qatar
e-mail: h.talballa@qu.edu.qa

J. Gichuru
Alliance of Biodiversity and International Center for Tropical Agriculture (CIAT), Nairobi, Kenya

L. Cochrane and R. Al-Hababi (eds.), *Sustainable Qatar*, Gulf Studies 9,
https://doi.org/10.1007/978-981-19-7398-7_16

different narrative of the actual recycling system functioning. This chapter describes the status of the municipal solid waste (MSW) management system in Qatar, its strengths and weaknesses elements in their political, social contexts, and finally considers challenges and opportunities for potential scenarios of development to a more circular, fully functioning one.

16.2 Solid Waste Management Outlook

16.2.1 Global Waste Generation

Solid waste generation is a worldwide problem that is crucial and one that requires prioritizing and application of effective management practices that foster environmental sustainability. The household solid waste generation rate is estimated at 1.7–2.7 kg per capita/day in Gulf Cooperation Council (GCC) States (Zafar, 2016), 1.2 in the United States, and 1.4 in European Union countries (MDPS, 2018). According to the Environmental Performance Index (EPI) 2016 data set, municipal solid waste generation around the world collectively exceeded two billion tons, less than half of which (42%) is handled within a formal waste management system. The population and economic status are key factors in determining the quantity of waste generated in a country. In low-income countries, the rate of waste generation ranges between 0.3 and 0.9 kg/per capita/day, whereas in high-income countries, the waste generation rate ranges between 1.4 and 2.0 kg/per capita/day (Chandrappa & Brown, 2012).

16.2.2 Waste Collection and Disposal

In recent decades, much progress has been made by some developing countries for waste management, but aspects such as collection coverage and controlled disposal rates remain low in the poorest countries. Around two billion people lack access to regular waste collection worldwide, while three billion people lack access to controlled disposal services for Municipal Solid Waste (UNEP and ISWA, 2015). The lack of waste collection and waste disposal services for this many people in the world creates a global public health and environmental imperative that demands a globally coordinated approach. A number of risks emerge when solid waste is disposed of improperly, including polluting water and the air, contaminating soils, creating human risks due to exposure to hazardous materials, among others. These systems also contribute to greenhouse gas (GHG) emissions via the decomposition of organic matter, particularly methane (Wendling et al., 2020). Concerted international action to provide solutions for the increasing waste generation associated with rapidly growing urban populations is much needed to avoid the current situation getting worse (Wilson & Velis, 2015).

16.2.3 From Business-Oriented to Multidimensional Waste Recovery and Recycling

The process of waste harvesting and recovery to generate secondary resources has increasingly become popular, and global businesses involved in this process are generating multimillion-dollar revenues from scouring the world for waste to recycling (Gregson & Crang, 2015). The global recycling industry has an estimated total revenue of US$500 million annually, the industry that employs the most number of people following agriculture (Minter, 2013). The main reason that businesses in the sector are reaping such huge profits is realized by the cheap costs of shipping containers carrying materials considered as waste from West to East or North to South within globalized shipping networks. The aspect of containerization in shipping has enabled global logistics for production and many small-scale entrepreneurs are entering the market for global trade of waste goods through the ability to ship relatively small consignments of discarded goods back in containers, rather than hiring bulk carrier ships (Minter, 2013). Another reason that attracts entrepreneurs to enter the trade of waste goods is the cheap labor costs once imported in combination with weak environmental regulations. Environmental regulations play a key role in enabling the whole process of resource recovery and recycling; for example, the degree of contamination is key to the acceptance of material before undergoing processing by manufacturers.

Economic production and consumption around the world have followed a common linear industrial model ever since the industrial revolution. The linear industrial model focuses on introducing manufactured goods from raw materials into the market, then enters the consumption phase where consumers utilize the benefits of the products and a later stage of disposal or incineration as waste when the goods reach end-of-life (Geissdoerfer et al., 2017). The "business as usual" model threatens a sustainable future, especially with the global demographic growth and other associated effects impacting finite resources. A shift from the current linear model to a circular model has been gaining traction significantly owing to the attention from governmental, academic, and organizational stakeholders in recent years (Ranta et al., 2018). A popular definition of circular economy (CE) is presented as "an economic model wherein planning, resourcing, procurement, production, and reprocessing are designed and managed, as both process and output, to maximize ecosystem functioning and human well-being" (Murray et al., 2017). In circular economy modalities, commodities are reused, recycled, or upcycled enabling the whole or parts to be reintegrated into the supply chain with minimal energy input and reducing the need to create new materials (Velenturf & Purnell, 2017).

16.2.4 Global and Regional Partnerships

Knowledge sharing and capacity building are key to ensure that zero-waste targets and circular economy approaches are utilized and expanded. For this to occur, partnerships and networks on regional and global scales are key for governments and their various private sector and non-profit sector partners. One of the examples of this is international organizations and municipalities sharing knowledge about standards for measurements within the waste management sector. Sharing experiences and lessons learned from successful policies and initiatives implemented on sustainable waste management and food systems is one way that support is provided to guide in developing standards that fit a country's context. Global partnerships and networks often play a huge role in highlighting the challenges encountered within the waste management sector through developing and diffusing knowledge across the world. An example is the United Nations Environment Programme (UNEP) Global Opportunities for Sustainable Development Goals (GO4SDGs) initiative. A number of stakeholders have come together to launch Regional Working Groups within this initiative, particularly focusing on food waste (see Chap. 12). This is a notable example of capacity building across scales, wherein peer-to-peer networks enable knowledge sharing between countries (UNEP, 2021).

16.2.5 Qatar Context

On June 05, 2017, the neighboring nations of Qatar (Saudi Arabia, United Arab Emirates, and Bahrain) as well as Egypt severed political ties with the country and imposed a blockade of the land, air, and sea, a blockade that would last for four years (Al Jazeera, 2020). The blockade kick-started Qatar's economic diversification efforts, and the country developed short-term strategies to become self-sufficient and cushion itself against the economic shock. Among the strategies implemented include developing expanding domestic manufacturing and food production as well as strengthening its infrastructure to enhance resilience to this kind of shock (Kabbani, 2021). Qatar's economic diversification and a rapid increase in infrastructure development, also fueled by the upcoming FIFA World Cup 2022, have led to an increase in waste generation, which is a challenge that has to be addressed as well.

The commitment by Qatar to increase its self-sufficiency by increasing its domestic production is the right step taken toward stability owing to the uncertainty that the region faces. However, the process of rapid expansion is quite intensive and can demand the use of massive quantities of resources which could ultimately impact food security. Qatar can incorporate the use of circular technologies and practice in the quest to exploit resources for production to ensure that it is done sustainably (Wesley, 2019).

The political will to engage further in waste management actions and strategies will be strengthened significantly with the idea that waste management can be an entry point to tackle difficult sustainable development issues, for example, environment and climate change, social domain aspects such as good governance of waste in cities and creating sustainable livelihoods, and enterprises covering the economic aspect.

16.3 Current Status of the Qatari SWM Sector

16.3.1 Waste Generation

Given that Qatar is geographically and demographically small on global terms, its total waste generation is low compared to most nations, however when analyzed on a per capita basis, it has one of the highest waste generation rates in the world (Al-Thani & Park, 2019). The most significant solid waste categories generated in Qatar include municipal solid waste (MSW), construction and demolition waste (CDW), bulky waste, end of life tires (ELT), and end of life vehicles (ELV). The first two categories; CDW and MSW are the major contributors to waste generation in the country.

CDW is the largest category of waste generated in Qatar, accounting for approximately 80% of all solid waste generated (Al Tamimi & Company, 2021). This form of waste includes the wastes that emerge in relation to construction projects, and include a wide range of demolition waste to excess, unused materials (Lu et al., 2015). CDW is connected with a construction boom that has been occurring for the better part of the last two decades.

MSW is the second largest waste category by source (Al-Maaded et al., 2012). Qatar produces more than 2.5 million tons of MSW annually (Al Tamimi & Company, 2021). The daily waste generation per capita as of 2015 in Qatar was at 1.8 kg (Zafar, 2016). MSW is generated from households or commercial establishments. The majority of substances comprising MSW include paper, vegetable matter, plastics, metals, textiles, rubber, and glass (USEPA, 2013). This does not include chemical or hazardous waste. Organic waste contributes the largest share of municipal waste in Qatar, which amounts to 57% of the municipal waste generated, while glass, papers, metals, and plastics have a smaller but valuable share of the MSW generated (QDB, 2017).

Total solid waste generation fluctuated according to the implementation of major development projects in the country; the total increased from a rate of 8 million tons in 2008 to 12 million in 2013, then decreased to below 10 million in 2014 (Planning & Statistics Authority, 2019). During this time period, major infrastructure projects were implemented, which explains these trends.

16.3.2 Collection

Waste collection in Qatar is carried out directly by municipalities through planned logistics by use of a large fleet of trucks that access MSW from the various collection points, or indirectly through private sector contracts (Zafar, 2016). There are four waste transfer stations that make the destination of all the urban waste generated from the seven municipalities in Qatar. The transfer stations are situated in South Doha, West Doha, Mesaimeer, and Dukhan. Q-Kleen is a private company responsible for the transportation of waste from the collection points to the transfer stations. The wastes are then compacted from those transfer stations and transferred to the domestic solid waste management center or landfill situated in Mesaieed through transfer trailers (Ahmad, 2016).

16.3.3 Management Facilities

Until recently, the waste management infrastructure in Qatar was not sufficient to handle waste generated per day. A private sector company was granted the contract to increase waste management capacity, at the Domestic Solid Waste Management Company (DSWMC), which includes recycling facilities as well as an integrated system for waste processing. The initial capacity, as of 2012 (Al-Maaded et al., 2012), was to handle 1,550 tons per day. The improvement of waste management is connected with the objectives set forth in the National Development Strategy (see Chaps. 2 and 3). Other projects that were launched include a tire recycling facility in 2012. This occurred alongside the expansion of regulation of waste, including in the construction sector as well as specific forms of waste handling (e.g., medical and radioactive wastes; MDPS, 2018). More recently, in 2018, the government began working with the Qatar Primary Materials Company to strengthen the handling and recycling of construction waste, with a specific landfill being established at Rawdat Rashed (with a capacity of 40 million tons; Al Tamimi & Company, 2021).

16.3.4 Waste Disposal

Solid wastes are largely disposed of in landfills within Qatar. Depending upon the waste type (as noted above), solid wastes are transferred to specific landfills, namely to Umm Al-Afai for bulky and domestic waste, Rawdat Rashed for construction and demolition waste, and Al-Krana for sewage wastes (Bello, 2018). The need to improve waste management is exemplified by the percentage of waste sent to landfills ten years ago, in 2012, which was 95%, while the DSWMC handled only 5%. The landfill disposal method is not sustainable because of its associated health and environmental problems (Ahmad, 2016). Waste spotted to be disposed of in open dumps sometimes does not only lead to severe environmental degradation but also results in loss of natural resources and arising of social issues (Report of the ARAB, 2008). Furthermore, landfill space is an issue in a geographically small country, as Qatar is, which has a size less than 12,000 km^2.

16.3.5 Challenges and Driving Forces

Qatar has a number of opportunities and challenges that have contributed to the current position of its SWM sector. These have been dispensed as a result of various major political, economic, and social circumstances affecting the country, such as the launching of the Qatar National Vision 2030 (QNV 2030) in 2008, winning the 2009/2010 competition to host the FIFA World Cup 2022, and the blockade posed by several neighboring countries which took place between 2017 and 2021.

In line with the QNV 2030 that focuses on economic diversification and aims to place the country on a sustainable development track, there has been rapid industrial development and mega construction projects in Qatar. This has been associated with large employment opportunities evidenced by increase in immigration rates and economic growth.

Despite the challenges it presented, the blockade has also had several strategic positive effects on Qatar in its quest to be self-sufficient through the development of several key industrial and food security projects. Among the new possibilities that the blockade unlocked was the fact that Qatar was able to unleash its local industrial and commercial capacities, particularly small and medium enterprises (SME), by suspending regional and economic commitments and partnerships in regard to trade and commerce relations. Qatar also implemented immediate measures and actions to deal with the situation by enabling businesses to skip bureaucratic procedures normally required for establishing industrial and commercial entities. Opportunities for governmental finance were also made widely accessible. For example, during the first year of the blockade, Qatar Development Bank (QDB) introduced the single window services for new industrial facilities. Proprietors could register a facility in the high potential sectors in Qatar and get financing within days.

The FIFA World Cup 2022 is another grand opportunity for Qatar that has brought about the development of multiple projects such as stadiums, training sites, and transport infrastructure in anticipation of the revered sporting event. The FIFA World Cup 2022 has a sustainability strategy that shall enable the organizers to deliver a tournament that sets new benchmarks for social, human, economic, and environmental development. This creates an opportunity for the mega-sports event to have a lasting legacy in the country, as it works to improve waste management and transition toward a more sustainable future. Delivering a carbon neutral FIFA World Cup target will be achieved through best construction practices and sustainable building standards that are already being implemented. The state has been transforming its building standards toward higher sustainability levels through the adoption of the Global Sustainability Assessment System (GSAS) standards, thereby facilitating the transition toward green buildings (Kader, 2016) (see Chap. 4). This certification system assesses and rates green buildings and infrastructure, which FIFA required to be "GSAS 4 Stars". A key initiative that the GSAS certification focuses on is waste minimization and recycling. The sustainability assessment requirements are extended to suppliers and host institutions that will be associated with the World Cup by implementing the FIFA Sustainable Sourcing Code. An example was the sustainability assessment recently conducted at Qatar University's student housing which will be used to host some of the participating teams. These practices have significantly improved waste management standards on organizational level.

There are several challenges that have arisen from the rapid industrial development and general growth associated with the QNV 2030 initiatives, the blockade strategies to be self-sufficient, and the construction of facilities to host the 2022 FIFA World Cup. The SWM sector in Qatar is among the key sectors that have been highly impacted. The economic growth and development have led to an increase in immigration rates from associated employment opportunities and an increase in construction sites leading to an unprecedented amount of MSW and CDW generated in the country.

MSW generated from households and/or commercial establishments is generated at the rate of 1.8 kg per capita daily, with organic waste taking the larger share of 57% while metals, paper, plastic, glass, and rubber having a smaller share in the MSW generated (QDB, 2017). One of the main challenges for MSW is that there is no source segregation (with few exceptions). As a result, the amount of waste that is recovered and recycled is limited. The lack of source segregation creates contamination, resulting in an ability to segregate many recyclable or reusable wastes at the end of the waste supply chain and thereby low recovery rates (QDB, 2017).

CDW management has become a critical challenge for construction companies as the building and construction industry is one of the key areas for diversification in the QNV 2030. The Qatar MME estimates 11,700 tons of waste generated out of the 21,000 tons generated daily is from the construction and demolition sector (MDPS, 2017). Majority of the waste generated was composed from excavated soil and sand, concrete, and other mixed non-hazardous waste.

Qatar has made significant strides toward attaining a circular economy (CE) in the construction sector by implementing recycling and reusing construction waste. Perhaps the achievements of CE in the construction sector may have not been intentional and well planned, and maybe they were realized out of special circumstances like the increased need for construction materials but focus on all the fractions of waste management is paramount. Qatar can easily take advantage of the accumulated local experiences and try to cover other fractions in a more systematic way, although they have originally evolved due to special circumstances. Some of the most sustainable strategies that can be adopted to complement the existing laws, regulations, and strategies to achieve CE include regeneration of natural systems, supply chains, and waste management cycles. Key stakeholders can use innovative solutions and best practices could be applied in development and implementation including procurement, construction, operations, and recycling.

16.3.6 Role of Stakeholders in Shaping SWM in Qatar

16.3.6.1 Government-Led Initiatives

The role of the government in ensuring the waste management system is efficient, effective, and enabling sustainable transitions is critical. While the private sector and individuals in society are producing waste, governments can set targets, create accountability mechanisms, and put in place (dis)incentives for action. Examples of these initiatives include instituting regulations (e.g., standards, bans, mandatory requirements), putting in place market-based mechanisms for change (e.g., taxes, fines, subsidies) and/or improving the infrastructure (e.g., collection, composting, recycling) and capacity support (knowledge, finance, information). An example of a government-led initiative in Qatar is the MME supported nationwide initiative et al. Khor Park to turn food waste and tree leaves into organic fertilizer through a recycling machine (MME, 2018).

The State of Qatar has implemented several strategies and instruments that provide solutions to the challenges posed by waste management. The use of new technologies in recycling, implementing specific legislation on waste management, and collaboration with relevant stakeholders to create new innovations in waste management are some of the bold steps that Qatar has taken (Al Tamimi & Company, 2021). Some of the major initiatives and strategies being implemented include the following:

Qatar National Vision 2030

Qatar adopted the QNV 2030 in 2008 to maintain development and sustainable growth (see Chaps. 2 and 3). The national vision was adopted to diversify the economy and build a skilled society capable of sustaining its own development and ensuring a high living standard. QNV 2030 has four key pillars including the environmental development pillar which forms the main focus for experts in the green sector. The environmental development pillar aims to establish the management of the environment such that there is harmony between economic growth, social development, and environmental protection. The environmental development pillar recognizes that preserving and protecting the environment will be realized through environmental awareness of the population, creating an agile and comprehensive legal system, and strengthening the capacity of environmental institutions involved. Emphasis is also set on the encouragement of regional cooperation to put in place preventive measures to mitigate the negative environmental effects of pollution arising from development activities. Qatar's National Vision provides the foundation for the formulation of a National Strategy which has incorporated solid waste management as one of the key components.

National Development Strategy, 2011–2016 (NDS-1)

The National Development Strategy 2011–2016 (NDS-1) adopted the QNV 2030 aspirations (see Chaps. 2 and 3). Within NDS-1, there are targets and initiatives specific to sustainability, within which waste management is a part. In alignment with QNV 2030, the country seeks to find a balance between economic and social development while seeking to protect and preserve the environment. The Environmental Sustainability Strategy 2017–2022 (ESS) does similarly, considering the environment as important within developmental processes to ensure it is protected for future generations. To do this, the ESS includes outcomes for six sectors and has twelve specific targets. Waste reduction is a key outcome, as is increased recycling rates and a more efficient waste management system (NDS-1 targets included: establishing a solid waste management plan, emphasizing the need to increasing solid waste recycling rates (from 8 to 38%), and maintaining domestic waste generation at 1.6 kg per capita per day). The recycling target was not met and was adjusted in NDS-, however the per capita waste generation target was achieved (and reduced to 1.3 kg per capita per day).

National Development Strategy, 2018–2022 (NDS-2)

The NDS-2 (2018–2022) built on the progress of NDS-1, seeking to expand infrastructure for waste handling, improving wastewater treatment and handling, managing electronic waste, among others. Under the NDA-2, the Environmental Sustainability Strategy ESS 2018–2022 was formulated (see Chaps 2 and 3). A key target is to achieve 15% recycling of solid waste over the course of the NDS-2. The ESS-2 complements the ESS-1 (2011–2016) that had some relatively slow progress, which prevented desired outcomes. The implementation of the ESS-2 is a responsibility of MME which requires close cooperation with supporting partners. Committing to the implementation process will undoubtedly lead to the fulfillment of the various commitments and activities addressed in the ESS.

The NDS-1 had great outcomes among them including the inauguration of a state-of-the-art Mesaieed facility. This institution has one of the world's largest composting plants. Another great outcome was the establishment of transfer stations for waste (five in total), through which wastes are collected and deposited into by government trucks and private contractors (Clarke et al., 2017). Part of Qatar's approach was to also close landfill sites to domestic waste. The operation of the DSWM plant began in December 2010. The plants full capacity processing potential is 2,300 tons of commingled MSW per day. The waste obtained can be separated into 627 tons for recycling, 857 tons for composting, and 922 tons for input to a waste to energy plant. Unfortunately, Qatar has had high volumes of domestic waste generated in the recent past leading to the new facility being overwhelmed. This resulted in the opening of a landfill close to the Mesaieed Plant in 2012 to cope with the overflow, pending either expansion of the plant or additional waste management facilities elsewhere.

Qatar does not lack the financial resources to be utilized in waste management and usually the main challenges stem from inadequate systematic approaches, planning, and management of the waste sector. The recent split of the MME might be an example of an improper systematic approach. There are now two separate ministries from the former MME including the Ministry of Municipality (MM) and the Ministry of Environment and Climate Change (MOECC). The technical challenge is manifested where all the waste-related issues are placed under the MM. The main reason for this move might have been due to the fact that the local municipalities are directly responsible for collecting solid waste in Qatar but technically, there is a direct correlation between waste management, the environment, and climate change. Waste management contributes a small but significant quantity to global GHG emissions with gases such as methane being emitted impacting the environment negatively. Furthermore, overreliance of disposal methods such as landfills create an environmental challenge especially for a small-sized country such as Qatar. Therefore, this creates a strong basis for handling all waste management related issues under the MOECC.

16.3.6.2 Society

According to the Qatar 2020 census, the major cities with the highest population include Doha with 46% of the total population, Al Rayyan with 29% of the total population, and Al Wakra with 9.3% of the total population (Planning and Statistics Authority, 2020). Therefore, the three cities are the major waste producing municipalities in Qatar due to the direct link of population with waste generation as mentioned earlier.

The behavioral change of Qatar's society should be prioritized in relation to waste management in Qatar. As mentioned earlier, Qatar has plenty of resources to employ technologies and infrastructure to deal with waste management, but this does not guarantee a functioning system, if the society is not incorporated and this is the main challenge that Qatar is facing. On reaching Qatar, a keen individual would notice the disconnect between Qatar's capability and the situation on the ground.

The initiative of source segregation using the multiple-compartment recycling bins widely placed in locations (e.g., shopping malls, educational institutions, healthcare centers) is usually not achieved because the bins contain mixed waste creating contamination and recovery becomes a challenge.

Grassroots Initiatives

The government cannot lead all changes, nor should it be expected to do so. Community-based initiatives are critical in ensuring that system transformation takes place—from reducing consumption to changing disposal habits (Grabs et al., 2016; Mariam et al., 2020). Within Qatar, there are a number of such initiatives, such as efforts seeking to reduce plastic consumption as well as efforts to support companies to transition away from plastic usage. Examples of this include efforts by the non-state actor Qatar Foundation (see Chap. 5) as well as a wide range of community organizations, such as Greener Future, the Doha Environmental Actions Project, and Activists in Action (a student group at Qatar Academy Doha on a mission to ban single-use plastic bags in Qatar. The private sector is also leading change, with the establishment of organic farms and new recycling initiatives (e.g., Elite and Seashore Recycling). Relatively less has been written on these efforts, which is an area for future research to contribute toward.

16.3.6.3 Private Sector

As stated earlier, waste management is majorly a function of the different municipalities in Qatar. However, a common feature in the sector is public private partnerships (PPP) which are formed to coordinate activities within the sector. PPP are collaborative arrangements between a public authority and private-sector companies used as a mechanism by the government to implement services or infrastructure using the resources and expertise of the private sector. Slightly more than 30% of waste management services, from primary collection to treatment and disposal, are provided through PPPs, even though such partnerships can be complex to structure and implement (Kaza et al., 2018). The PPPs can also include other stakeholders such as nongovernmental organizations and research institutions (Swaffield et al., 2018). A great example is the agreement between Qatar University and Al Meera supermarkets chain. Al Meera shall provide Qatar University with reverse vending machines to be installed at strategic spots in the campus. Qatar University students will recycle their bottles to get points at the supermarket and this is part of the CSR initiative by Al Meera to serve the community starting with Qatar University.

Qatar's recent privatization impetus comes as part of its Second National Development Strategy 2018–2022, which falls in line with its overarching economic development plan, QNV 2030 (MDPS). Private sector investors focusing on the waste management systems are keen to participate in PPP opportunities in Qatar such as waste to energy projects, waste collection, and management systems. Some of the major private sector waste management players in Qatar include Seashore Group of companies, Dulsco Qatar, Power Waste Management and Transport Co. WLL, Green Waste Management, Al Haya Enviro. The government's focus to reduce more than 90% of the waste from landfills by 2022, has seen a race by international companies to gain a share of the market. More than half of the companies that were granted a license to carry out waste management operations in the country are foreign companies and partner with Qatari companies in managing factories.

16.4 Zero-Waste-Based Roadmap to the Circular MSW Sector

A fully circular MSW system encompasses complicated and multidimensional interactions among a large number of technological choices, economic mechanisms, and control frameworks. To deal with this issue in Qatar, the current state of waste management has been addressed in the previous sections, here future prospects and possibilities for improvements in the country are presented.

The issue of limited resources is a reality in many developing countries which places them in a disadvantaged position where little investments in solid waste management systems are established. This is not the case in Qatar. This is proven in a number of ways for example through the many investments made so far in terms of waste management, for example, the establishment of the Mesaieed facility with a total cost of 3.9 billion Qatari Riyal for design, construction, and operation for 20 years among many others (Ng, 2013). However, the allocation of resources needs to be done in a more effective, efficient, and systematic way. This can be done by utilizing the experiences and lessons learnt from the special circumstances that the country has undergone. The measures that Qatar took for example during the blockade, can be integrated in the practices for developing the waste management sector and not only be used for emergency situations. The example of establishing the state-of-the-art facility which gets overloaded in very few years shows that there is need for proper short and long terms planning and implementation.

There are several initiatives and strategies being implemented in existing institutions in Qatar that have the potential to be scaled up to fit the country's waste management context despite the lack of a nationwide agreed upon assessment metrics that cover all their components. The state can still borrow from and extend initiatives, strategies, and mechanisms on waste management from organizations that offer testing beds for developing sets of solutions proven to fit same local technological and social circumstances. This can be utilized to enhance existing policy and economic instruments to create a more circular approach to solid waste management.

16.4.1 Qatar University Case Study

Qatar University (QU) has a progressive perspective of waste as a resource, rather than the normal disposing it of as waste. This is done in accordance with the waste management hierarchy. Waste prevention is the most favored option in the waste management hierarchy, followed by reuse and recycling, and the least preferred option is disposal. In other words, QU is moving from a make-use-dispose linear economy to a make-use-reuse/recycle CE. This is made possible through the integration of Qatar University Zero Waste (QUZW) into all realms of QU operations.

The Zero-Waste Initiative in QU is an important effort that aims to serve as a living lab for various interneally or externally developed sustainability practices, which can be replicated at a wider level in the country as part of the Ministry of Municipality and Environment's Zero-Waste programs.

QU has formulated the Qatar University's Zero-Waste (QUZW) Vision 2025 Action Plan, in response to the unsustainable levels of waste production. The action plan stipulates an ambitious plan for creating a circular campus serving as a model for sustainable consumption in line with the SDG 12. The Action plan unites and extends existing QU waste prevention efforts into systematic ambitious and exciting opportunities. The QUZW Vision 2025 Action Plan operates under four central pillars with the aim of achieving an ambitious waste target by 2025. The four central pillars of the Action Plan include: (1) creating governance and monitoring structure with strong partnerships; (2) developing a waste measurement and prevention system; (3) using a communications strategy to engage campus employees and students; and (4) fostering a culture of innovation and social change within QU.

There are various strategies and tools to support the four pillars of the action plan. One of the key tools is assigning deliverables and indicators to priority areas. The deliverables and indicators support the governance and monitoring pillar including developing partnerships with key waste management stakeholders, as well as new financial resources to fund new waste initiatives. An example is the partnership between QU and paper recycling contractors to implement a paper and cardboard waste recycling program that has achieved the milestone of recycling 100 tons of paper waste since 2017. Another tool is the annual campus waste audit along with frequent data collection initiatives which will be necessary to measure more precisely the waste production at QU, as well as the success of waste prevention policies. The data collected is important for evaluating current and future strategies. Additionally crucial is that QU developed a platform to disseminate important information to students, employees, and external stakeholders through the QUZW Communications Program. Important information to be disseminated through a good communication strategy include the Zero-Waste Progress Reports, waste audits, and assessments of environmental and financial benefits conducted regularly. The Sustainability Vibes series communicated through email braodcast, phone displyes and QU social media is an examples (Fig. 16.1).

Key projects around waste practices aimed at engaging students and employees to be agents of change have been started in the campus in support of fostering a culture of innovation and social change. This is achieved by using both the Communications Program and the academic resources on campus. An example of an upcycling innovation that is changing the mindset of individuals is the use of palm waste fiber generated in the QU campus to produce smart eco-friendly recycling dumpster. Such applications enlighten the QU community to view waste as a valuable resource and gradually shift toward a more sustainable lifestyle (Fig. 16.2).

QU represents a small potential Circular Doha with its population of 25,000 people and hosting several activities. Through a waste audit, QU identified the four most important fractions in terms of volume and addressed each of them in accordance with the circumstances. Focus areas of QU waste management fractions include paper, plastic, food, and green waste. For paper, food, and plastics, QU has focused on the generation hotspots and plans on placing reverse vending machines in the

Fig. 16.1 Example of the SUSTAINABILITY VIBES Series posters digitally communicated with QU community

The Smart Recycling Dumpster facilitates recycling while being itself a unique model of QUZW sustainable products with the following features:

- 40% by weigh of the dumpster body is made of recycled date palm waste collected from QU farm.
- 3W solar powered system is used to automatically watering the plants on the sides.
- 70% fill level is monitored by sensors to optimize collection.
- 1.9 Kg CO2 equivalent per kg of palm residues are avoided by diverting palm waste from landfill, using local materials and avoiding unnecessary transportation.
- 50% cut of environmental impact and cost by fabricated locally inside QU campus.

Fig. 16.2 SMART Recycling Dumpster Factsheet

food court, students housing premises, and student activities building which are the highest generation points for each of the waste.

QU is implementing the zero-waste concept which means following the waste management hierarchy. An example is shifting from paper-based systems to digital-based systems in many applications at the campus and recycling whatever paper waste is unavoidable. There are desk-side recycling boxes installed in offices and bigger recycling bins placed next to photocopier stations to collect wastepaper. The use of multiple compartments bins placed in different locations in the campus had been in use for long and it did not function as intended. The pilot of the small desk-side recycling boxes is proving efficient. For food waste, two units of dehydrators are placed at the male and female housing generating clean fertilizer. This is important because it is directly related to the concept of a circular campus. The organic loop is closed with the use of the dehydrators when it comes to food waste. When segregation of waste is made possible in the easiest way, people tend to comply. Each waste fraction is dealt with at the closest point of the source to make it convenient for the end-user.

QU has developed an integrated system to address all components together. However, and despite the solid achievements so far, it is logistically challenging to deal with waste management from a single standpoint in the institution, as the various waste streams in the campus are generated and handled by different departments. The QUZW Vision 2025 Action plan gives a clear roadmap theoretically, but the main challenge is full implementation. This is generally what is happening in Qatar in regard to policies and its implementation. Qatar has a well-articulated QNV 2030 and many other strategies that are set by the government, but implementation is still a major issue.

In the strive for creating a test bed for sustainability in QU, a study by the College of Engineering has been presented to both the Facilities and General Services as well as Capital Projects departments. It has recommended a Food-Water-Energy Nexus for urban planning at the campus. For developing a more sustainable campus of the future, a more holistic set is to include the waste resource sector in a Food-Water-Energy-Waste NEXUS approach. This shall transform the campus into an inclusive living lab for proven practices, especially when the organic materials cycle (i.e., food waste) is already a closed loop on campus.

16.4.2 Qatar Foundation Case Study

- Qatar Foundation for Education, Science and Community Development (QF) has established several research centers to actively contribute to the transition to a CE and knowledge-based economy. Qatar Environment and Energy Research Institute (QEERI) is one of the research centers established to provide academic research programs in the field of sustainable energy and sustainable environment (see Chap. 8). QF also has the Qatar Green Building Council (QGBC) which conducts environmentally sustainable practices for green building design and development in Qatar through leadership and collaboration. QF does not only represent a model to showcase the what's and how's CE elements (i.e., energy and materials) effectively interconnect on the organizational level, but actively catalyze change on the national level under a committed political will (see Chap. 5).

16.4.3 Synthesis and Key Messages

The concept of waste has evolved and is no longer viewed as mass that lacks value. Waste generation results in materials that can be recycled into valuable materials. This brings in the concept of a circular economy (CE) where materials from manufactured goods that have reached end of life can be recovered and recycled with minimal energy to produce valuable goods.

The zero-waste concept can be considered as a useful instrument to achieve a CE. It is a people-centered approach to the waste problem that encourages resource recovery, efficiency, and conservation of natural resources. Thus, allowing materials recovery to be injected back to the production process in a CE setting. Understanding the opportunities that are presented with embracing a CE and zero-waste concepts such as mitigation of climate change and minimizing the cost associated with cleaning and recovering can be beneficial especially for a country like Qatar that is grappling with waste associated with a surge in population and increased economic diversification. Sustainable future trends can be developed to guide in achieving the SDG 12 that calls for responsible production and consumption.

Qatar will have to transform the way it deals with its waste in terms of source segregation, collection, sorting, and regulations to be able to improve the recovery of the waste generated. This change would significantly reduce the need to extract raw materials, reduce carbon emissions, and bring in several other environmental benefits.

16.4.3.1 Driving Factors of Waste Generation in Qatar

The main driving factors of waste generation in Qatar is the rapid industrial development and economic diversification associated with the QNV 2030, the upcoming FIFA World Cup 2022, and the blockade. These factors have led to an increase in demographic growth and an increase in construction and development projects which in turn means a rise in waste generation rates. However, the lessons learned from these special circumstances can be of great value when integrated to complement existing laws and regulations in waste management. Poor systemic planning by government authorities on waste management, little awareness and information sharing affecting source segregation, and little incentives for recovery facilities are some of the major challenges experienced in the waste management sector in Qatar.

16.4.3.2 Transforming Waste Management Using Actors, Partnerships, and Instruments

An integrated approach to waste management, combining government effort with diverse stakeholders, can transform systems toward greater sustainability. The above case studies have also proven that waste management in Qatar can be tackled systematically, and solutions achieved. Top-down and bottom-up approaches can be effective approaches in unlocking the potential of technology to achieve the expected results. Therefore, this means that municipalities have a great opportunity to transform themselves from key areas of waste generation to ones that foster innovation for testing and advancing effective approaches for waste management.

The government, private entities, and the citizens all have an important part to play in waste management in Qatar. It is really important that every stakeholder is fully "on board" through awareness and knowledge sharing, and an unending commitment to ensuring equal access to affordable services for all. Through stakeholder engagement mechanisms, roles and responsibilities for every player will be assigned clearly. Participation can also be integrated into the policymaking process for effective outcomes. Partnerships are effective for proper service delivery to all citizens and involving the private sector for example can be of an advantage where there is better access to greater experience in operating modern technologies.

16.4.3.3 Achieving Behavioral Change

The importance of involving the community in the MSW management process is because they have a key role in waste generation and achieving behavior change would be largely beneficial in tackling the problem at the source through ways such as effective segregation. A functioning waste management system not only depends on technological advances but also relies on strategies and mechanisms deployed to influence behavior and attitude change toward the environment. Community engagement in MSW management processes can be achieved through awareness raising and information dissemination on the various waste generation streams and how to prevent and reduce waste. While a 2017 survey confirmed that the people of Qatar have an apparent willingness to change their waste handling practices to be more responsible, the same study indicated that 24% of the respondents did not know what happens to their collected waste. The top hurdle identified by the respondents themselves are deficiencies in the training of people (Clarke et al., 2017).

A possible explanation why compartment recycling bins located in different spots across the country are not achieving the intended purpose besides the insufficient level of awareness is the distance that people need to travel to reach refuse and recycling collection bins. This factor also influences their willingness to separate waste; as the distance to the recycling bins decreases, the number of fractions that citizens separate and collect at home increases (Gonzalez-Torre et al., 2005). This was clearly evident in Qatar through the outcomes of adopting the desk-side box alternative at Qatar University as explained in the case study above. The government and civil society groups among others can achieve community engagement through engaging volunteers and supporting innovative initiatives on waste management. An example of how information-based approaches could be used to influence the behavior of the individuals in society with regard to reducing waste generation is the Sustainable Ramadan Campaign, led by UNEP. This initiative focuses upon the holy month to encourage people to live more sustainable and reduce waste. These nudges can help the adoption of expanding infrastructure.

It is vital to note that behavioral changes within communities is gradual and takes time while applying different sets of legal, economic, and educational incentives to reach desired results. This is true even in more environmentally conscious societies. When Germany started its recycling system in 1991 with 3% recycling rate, it took them more than 25 years to achieve a recycling rate of 66% in 2017. 20 years was needed for the British country household recycling rate to increase from 5 to 64% (PFC, 2020). The Alpine region took the same period to reap the result of increasing citizen awareness through training and national policies to get volumes of non-recyclable waste generated to decline by about 90 kilos per person per year. Outstanding recycling rates was achieved eventually, like 93% of glass bottles and 91% of aluminum cans.

16.4.3.4 Instruments

Several instruments such as direct legislation, public policy, and regulations are effective in waste management and especially in influencing the society's behavior related to achieving desired targets in waste management. The State of Qatar through its relevant ministries takes lead in formulating legislation, regulations, strategies, and initiatives for reducing waste followed by implementation. However, there is a need to combine long-term strategies and roadmaps with short-term targets and concrete measures locally to ensure effective implementation. Also, incentives that have the negative effects of preventing or discouraging actions to reduce waste need to be reviewed and upgraded to working ones.

Formulation and implementation of waste management interventions need to be tailored to local circumstances addressing the social and cultural factors such as values and norms, gender, food security, health, and equality that play a vital role in ensuring their success and impacts. Initiatives that address individual or system-level factors of waste and also include the social practices related to the waste generated are more likely to be effective, and technologies can be incorporated to accelerate the intervention.

16.4.4 Conclusion and Recommendations

Inspired by the QNV and its emanated NDSs Qatar has gained ground on the path of realizing the goal of sustainable development. However, in regard to waste management, a thorough review is required to improve and harmonize inputs of different stockholders. Zero waste is suggested as an ideal vehicle for renovating the current waste management system into a circular one, key recommended means are as follows:

- Public participation should be at the center of waste management processes to encourage collaboration with members of the community and businesses while establishing clear strategic goals.
- The relevant stakeholders involved must be able to secure political commitment to the set goals and targets and ensure that there will be continued support beyond political terms of office.
- Continued awareness raising and information sharing is paramount to address behavior change by communicating clearly to waste generators on what is required of them and the roles that they can take up to allow for an effective waste management system.
- Establishing additional partnerships between municipalities and private sector players to deliver efficient and sustainable services is crucial as waste generation is rapidly increasing in Qatar. The nongovernmental sectors and research institutions shall also be involved to provide technical assistance and capacity building.
- Encourage sound institutions with clear roles and responsibilities to be able to capture the existing gaps in the waste management sector and to avoid overlapping of duties and activities. An example would be a well-resourced waste department that has been assigned an appropriate level of authority and autonomy.
- Developing solutions to existing gaps within the sector by integrating sets of instruments including policy instruments such as direct legislation, economic such as subsidization, and exemptions, and social instruments like voluntary agreements and information nudging.
- There is a need to allocate more authority to the existing environmental regulator in Qatar to allow for full enforcement of regulations and ensure compliance in an effective manner.
- Qatar should develop and enforce a zero-waste framework law to guide waste management processes. There should also be an exclusive long-term resources management strategy to provide a long-term stable framework to secure embedding circular waste-related elements in future investments in infrastructure upgrading. A holistic Energy-Water-Food-Waste NEXUS for development planning would be an ideal comprehensive approach.

References

Ahmad, F. I. (2016). *Sustainable solutions for domestic solid waste management in Qatar* (Thesis, Qatar University).

Al Jazeera. (2020). Five things to know about the blockade against Qatar. *Al Jazeera*. https://www.aljazeera.com/news/2020/6/5/qatar-blockade-five-things-to-know-about-the-gulf-crisis

Al Tamimi and Company. (2021). *Waste management—A Qatari perspective*. https://www.tamimi.com/law-update-articles/waste-management-a-qatari-perspective/

Al-Maaded, M., Madi, N. K., Kahraman, R., Hodzic, A., & Ozerkan, N. G. (2012). An overview of solid waste management and plastic recycling in Qatar. *Journal of Polymers and the Environment, 20*(1), 186–194.

Al-Thani, S., & Park, S. (2019). The case for sustainable concrete waste management in Qatar. *Sustainability in Environment, 5*, 1. https://doi.org/10.22158/se.v5n1p1.

Bello, H. (2018). Impact of changing lifestyle on municipal solid waste generation in residential areas: Case study of Qatar. *International Journal of Waste Resources, 8*, 335. https://doi.org/10.4172/2252-5211.1000335

Chandrappa, R., & Brown, J. (2012). *Solid waste management: Principles and practice*. Springer Science & Business Media.

Clarke, S. F., Nawaz, W., Skelhorn, C., & Amato, A. (2017). Towards a more sustainable waste management in Qatar: Retrofitting mindsets and changing behaviors. *QScience Connect* (1). https://doi.org/10.5339/connect.2017.qgbc.4

Geissdoerfer, M., Savaget, P., Bocken, N. M., & Hultink, E. J. (2017). The circular economy—A new sustainability paradigm? *Journal of Cleaner Production, 143*, 757–768.

Gonzalez-Torre, P. L., & Adenso-Díaz, B. (2005). Influence of distance on the motivation and frequency of household recycling. *Journal of Waste Management; 25*(1), 15–23.

Grabs, J., Langen, N., Maschkowski, G., & Schäpke, N. (2016). Understanding role models for change: A multilevel analysis of success factors of grassroots initiatives for sustainable consumption. *Journal of Cleaner Production, 134*, 98–111.

Gregson, N., & Crang, M. (2015). From waste to resource: The trade in wastes and global recycling economies. *Annual Review of Environment and Resources, 40*, 151–176.

Kabbani, N. (2021). *The blockade on Qatar helped strengthen its economy, paving the way to stronger regional integration*. Brookings. https://www.brookings.edu/blog/order-from-chaos/2021/01/19/the-blockade-on-qatar-helped-strengthen-its-economy-paving-the-way-to-stronger-regional-integration/

Kader, S. A. (2016). *Trash to treasure: Recycling construction waste in Qatar*. www.qatarconstructionnews.com/trash-treasure-recycling-construction-waste-qatar.

Kaza, S., Yao, L., Bhada-Tata, P., & Van Woerden, F. (2018). *What a waste 2.0: A global snapshot of solid waste management to 2050*. World Bank Publications.

Lu, W., Chen, X., Peng, Y., & Shen, L. (2015). Benchmarking construction waste management performance using big data. *Resources, Conservation and Recycling, 105*, 49–58. https://doi.org/10.1016/j.resconrec.2015.10.013

Mariam, N., Valerie, K., Karin, D., Angelika, W. R., & Nina, L. (2020). Limiting food waste via grassroots initiatives as a potential for climate change mitigation: A systematic review. *Environmental Research Letters, 15*(12), 123008.

Ministry of Development Planning and Statistics. (2017). *Environment statistics in the state of Qatar, 2015*. https://www.mdps.gov.qa/en/statistics/Statistical%20Releases/Environmental/EnvironmentalStatistics/Environment_QSA_EN_2015.pdf

Ministry of Development Planning and Statistics. (2018). Qatar Second National Development Strategy 2018–2022.

Ministry of Development Planning and Statistics. Qatar Environment Day. (2017). https://www.mdps.gov.qa/en/statistics1/pages/topicslisting.aspx?parent=Environmental&child=EnvironmentalStatistics

Ministry of Municipality and Environment of Qatar. (2018). *MME launches initiative to turn organic waste into fertilizer*. http://www.mme.gov.qa/cui/view.dox?id=702&contentID=5485&siteID=2.

Minter, A. (2013). *Junkyard planet: Travels in the billion dollar trash trade*. Bloomsbury.

Murray, A., Skene, K. N., & Haynes, K. (2017). The circular economy: An interdisciplinary exploration of the concept and application in a global context. *Journal of Business Ethics, 140*(3), 369–380.

Ng, S. (2013). Managing and Monitoring National Development Strategy 2011–2016 Outcomes & Qatar National Vision 2030 Goals, Reduced waste, More Recycling and More Efficient Use. In Waste Statistics Training Workshop, Doha.

Planning and Statistics Authority. (2019). *Environment statistics report in the state of Qatar 2017 Doha—Qatar*. https://www.psa.gov.qa/en/knowledge/Documents/NDS2Final.pdf

Planning and Statistics Authority. Main results of the general census of population, housing and establishments 2020 Doha—Qatar.
Plastic for Change. (2020). *Doing it right—Which countries recycle the most?* https://www.plasticsforchange.org/blog/doing-it-right
Qatar Development Bank. (2017). *Materials recovery.* https://www.qdb.qa/en/Documents/Materials_Recovery.pdf
Qatar Electricity and Water Company (KAHRAMAA). About. https://www.qewc.com/qewc/en/about/
Ranta, V., Aarikka-Stenroos, L., Ritala, P., & Mäkinen, S. J. (2018). Exploring institutional drivers and barriers of the circular economy: A cross-regional comparison of China, the US, and Europe. *Resources, Conservation and Recycling, 135*, 70–82.
Report of the ARAB. (2008). Forum for environment and development, Arab environment future challenges. Waste Management, chapter 8, NEFISA ABOU-ELSEOUD.
Swaffield, J., Evans, D., & Welch, D. (2018). Profit, reputation and 'Doing the Right Thing': Convention theory and the problem of food waste in the UK retail sector. *Geoforum, 89*, 43–51.
United Nations Environment Programme (UNEP) and International Solid Waste Association (ISWA). (2015). Global waste management outlook. In D. C. Wilson, L. Rodic, P. Modak, R. Soos, A. Carpintero, C. A. Velis, M. Iyer & O. Simonett (Eds.), UNEP International Environment Technology Centre: Osaka, September 2015. http://www.unep.org/ietc/InformationResources/Events/GlobalWasteManagementOutlookGWMO/tabid/106373/Default.aspx
UNEP DTU Partnership and United Nations Environment Programme. (2021). *Reducing consumer food waste using green and digital technologies.* Copenhagen and Nairobi.
United States Environmental Protection Agency. (2013). *Environmental report.* United States Environmental Protection Agency.
Velenturf, A., & Purnell, P. (2017). Resource recovery from waste: Restoring the balance between resource scarcity and waste overload. *Sustainability, 9*, 1603.
Wendling, Z. A., Emerson, J. W., de Sherbinin, A., Esty, D. C., Hoving, K., Ospina, C. D., Murray, J. M., Gunn, L., Ferrato, M., & Schreck, M. (2020). *Environmental performance index.* Yale Center for Environmental Law & Policy. epi.yale.edu.
Wesley. (2019). *How Qatar's food system has adapted to the blockade.* Chatham House—International Affairs Think Tank. https://www.chathamhouse.org/2019/11/how-qatars-food-system-has-adapted-blockade
Wilson, D. C., & Velis, C. A. (2015). Waste management-still a global challenge in the 21st century: An evidence-based call for action. *Waste Management & Research, 33*(12), 1049–1051.
Zafar, S. (2016). *Solid waste management in Qatar*. https://www.ecomena.org/solid-waste-management-in-qatar/

Husameldin M. Talballa has been a prominent leader in the sustainability space for over 17 years, while developing his career across the distinctive span of Europe, Africa, and Asia. Notably, he is recognized for his current work of designing and spearheading a campus-wide zero waste program as a Senior Sustainability Engineer at Qatar University. Additionally, he is also credited with significant contributions to various multinational companies and national governments in establishing projects and advancing initiatives in waste management, climate actions planning, environmental consultancy, water resources research, business development, and impacts mitigation of oil and gas operation sites. Furthermore, through his early work as a consultant for SWECO, Sweden's leading consulting firm, he has also contributed to the World Bank funded project for building capacity of the Iraqi hazardous waste management sector. Foundationally, he is a chemical engineer and holds a scholarship-merited MSc in Sustainable Technology from the Royal Institute of Technology (KTH) in Stockholm.

Jonathan Gichuru is a resourceful Environmental Specialist with over 3 years of extensive experience in climate change adaptation, environmental policy, environmental impact assessment, and waste management. His major strengths include climate risk assessment, climate-smart agriculture, climate adaptation planning, and waste water management. Throughout his career, he has gained a number of skills and has proven background in field experimentation, trade-offs analysis, gender-responsive research, participatory approaches, qualitative research methods, scientific writing, data dissemination, and facilitating workshops. In the recent past, he has been recognized for developing highly effective partnerships with stakeholders to establish and implement climate risk profiles for two counties in Kenya. This was a career highlight since he managed to successfully navigate existing challenges at the time including data collection at the height of the COVID-19 pandemic. He is a team player dedicated to motivating and mentoring colleagues within multicultural settings.

Chapter 17
Education for Sustainable Development in Qatar

S. Duygu Sever and M. Evren Tok

Abstract This chapter examines Education for Sustainable Development (ESD) in Qatar by scrutinizing the actors, initiatives, and policies at global and national scales. The study presents an overview of Qatar's national policy on integrating sustainable development into the education system and highlights nation-specific characteristics. The chapter further analyzes major formal, informal, and non-formal educational learning spaces in Qatar which have been harnessing national policies, global trends, and local specificities to effectively serve Qatar National Vision 2030 as well as Qatar's commitment to the UN Sustainable Development Goals.

Keywords Education for Sustainable Development (ESD) · Education policy · Qatar National Vision · Sustainable development goals · Localization

17.1 Introduction

Sustainable development is a complex term referring to a multitude of dimensions pertaining to sustainability. Environmental, human, social, and economic tenets of sustainability, albeit differential in many ways, pinpoint a convergence around the centrality of education for sustainability. This transformation can be driven by equipping the citizens with knowledge, skills, and competencies. Formal and informal education both have a role in shaping twenty-first-century skills, competencies, and mindsets needed in todays and the future's global citizens.

In November 2021, the Economic and Financial Committee (Second Committee), of the 76th session of the UN General Assembly adopted a resolution on Education for Sustainable Development in the framework of the 2030 Agenda for Sustainable Development. The resolution encouraged governments and other stakeholders to

---

S. D. Sever (✉) · M. E. Tok
Hamad Bin Khalifa University, Doha, Qatar
e-mail: smehmetoglu@hbku.edu.qa

M. E. Tok
e-mail: etok@hbku.edu.qa

L. Cochrane and R. Al-Hababi (eds.), *Sustainable Qatar*, Gulf Studies 9,
https://doi.org/10.1007/978-981-19-7398-7_17

"increase efforts to systematically integrate and institutionalize education for sustainable development in the education sector and other relevant sectors" by increasing the capacities of policymakers, institutional leaders, and educators, strengthening research and innovation, and monitoring Education for Sustainable Development (ESD) in order to support the scaling-up of good practices (United Nations, 2021). What was striking about the resolution is the role played by Qatar as the coordinator of negotiations on the resolution, together with Tajikistan.

ESD relates to the Qatar National Vision 2030 (QNV 2030) in many ways. For instance, environmental protection is one of the core values for the future envisioned for Qatar, and QNV 2030 acknowledges that education will also work as a catalyst for social progress. Similarly, 2011–2016, as well as 2017–2022 and 2023–2028 National Development Plans (NDS) exhibit direct and indirect references to the role of education in fostering sustainability (see Chaps. 2 and 3). Qatar embraces the importance of education as a tool for empowerment, civic participation, human development, and sustainable societies laden by ethical and moral frameworks in education.

This chapter examines ESD in Qatar by scrutinizing the actors, initiatives, and policies at global and national scales. The study presents an overview of Qatar's national policy on integrating sustainable development into the education system and highlights nation-specific characteristics. The chapter further analyzes major formal, informal, and non-formal educational learning spaces in Qatar, which have been harnessing national policies, global trends, and local specificities to effectively serve QNV 2030 as well as Qatar's commitment to the UN Sustainable Development Goals (SDGs).

17.2 The Importance of ESD for Qatar

Sustainable development necessitates simultaneous and balanced progress in the spheres of environment, economy, and society. This brings the ESD to the fore, which holds great potential for making a comprehensive transformation in the society and economies. ESD targets to empower learners with knowledge, skills, values, and attitudes necessary for making informed decisions and undertaking responsible actions for environmental integrity, economic viability, and a just society. ESD is expected to transform society and thus to enable SDGs, by empowering people of all genders, ages, and present and future generations, while respecting cultural diversity. In this regard, ESD is a lifelong learning process and an integral part of quality education (UNESCO, 2022a).

The 2030 Agenda for Sustainable Development includes the United Nations fourth Sustainable Development Goal (SDG4), dealing with quality education. SDG4 is critical for meeting the SDGs and implementing the 2030 global agenda. Education is a shared responsibility among individuals, education and training institutions, and governments (Boeren, 2019). National education policies and strategies are, in fact, the necessary starting point to achieve the SDG4 and the SDGs more generally.

Although, the state regulates education and controls its delivery, in some cases, state-led educational models may coexist (Pak & Lee, 2018) with initiatives from private and civil society actors (Marshall, 2018).

ESD is concerned with global responsibilities to address the century's challenges for human survival and security at the intersection of ecology/environment, equity/social development, and finance/economy. As the threats are not conventional anymore, it is not possible to expect conventional teaching and training methods to meet the necessities of the century. Accordingly, progress can only be achieved by responsible, problem-solving global citizens who are ready and capable to act at the local, national, and international levels with unconventional and innovative mindsets informed by values, ethical standpoints, and morally informed youth. ESD also includes the component of "global citizenship" in order to attain major analytical, cognitive, and social skills, such as empathy, conflict resolution, communication, critical-creative-innovative thinking, and information literacy, as also highlighted by Education 2030 of UNESCO. This approach recognizes that the answers to non-traditional challenges our world faces today in the environmental, economic, and political spheres, global or national, rest at the dignity, and compassion not only for each other but also for our planet. Therefore, the alignment of global citizenship and ESD with lifelong quality education for all emerges as a key strategy and priority for national and global prosperity.

UNESCO (2016) outlined specific learning objectives for sustainable development based on cognitive, socio-emotional, and behavioral domains. While the cognitive domain refers to knowledge and thinking skills, the socio-emotional aspect refers to attitudinal and value-driven motivational aspects. Finally, the behavioral dimension is related to practical, actionable forms. While this tripartite framework is useful in principle, it does not recognize the local context, the specific relationship between education and a given society, and the need for policymakers, curriculum developers, and teachers to work with and critically think about ESD at several levels (Welply et al., 2019, 85). While the UNESCO framework is a useful starting point, it does not provide sufficient insights on practical implementations and, consequently on potential trajectories of localization.

Being holistic and transformational, ESD encompasses learning content and outcomes, pedagogy, and the learning environment itself, and therefore, must be integrated to every educational and training level, with specificities applying to different ages, types of training, and localized situations, in order to increase students' awareness and capacities in relation to the SDGs (Dür & Keller, 2018). Important preconditions of ESD include educated staff, aware and capable to teach what is relevant to learn about sustainability (Pak & Lee, 2018), but also relevant pedagogical resources, adapted to the specific context in which they are used.

Hopkins and Mckeown's (2005, 13) revealed that there is a need for a more holistic approach combining both curriculum and institutional change to engage in effective education on sustainable development. The curricula should address current and potential environmental and sustainable development issues and equip the students with "lifelong learning skills applicable not only for examination purposes but also for their entire life, especially at home, in their communities and at workplaces"

(Mawonde & Togo, 2020, 3). Accordingly, Kurz (2020, 285) highlights that the programs and curricula should be modified so as to depict, first, what kind of new qualifications and attitudes are needed in today's and future's global citizens. This requires redesigned teaching methods which blend social responsibility activities and real cases in the curriculum (Elobeid et al., 2016; Gomez et al., 2018, 227) and combine them with experiential learning as well as interdisciplinary projects (Kurz, 2020, 285). In these efforts, principles such as "educating and empowering students to be leaders who impact the world sustainably, translating research and teaching into practice, using the campus as a living lab, institutionalizing best practices in sustainable operations, and amplifying impact by sharing successes with others" are already taking their place in the sustainability plans of leading universities such as Harvard (Harvard University, 2019). This transformation, with no doubt, also requires a strategy for faculty development since new content and methods necessitate retraining of faculty members as well as incentives to support research on sustainable development (Kurz, 2020).

Zooming into the specific case of Qatar reveals that education is key in order to transform the country into an advanced society capable of achieving sustainable development and the SDGs. Qatar is a small coastal country facing some unique challenges with regard to the impact of climate change. Diminishing water and hydrocarbon resources, pollution, and environmental degradation, increasing salinity in groundwater sources for agriculture, declining air quality, and threats to biodiversity and marine life are only few among many challenges which risk human well-being at the food, water, and environment nexus (Qatar General Secretariat for Development Planning, 2008, 34; 2011, 229). Chapters in this edited volume address some of these key issues, such as sea-level rise (see Chap. 9), water (see Chap. 11), food security (see Chap. 12), and biodiversity (see Chap. 13) in detail. Acknowledging these challenges, the QNV 2030 has the primary goal to attain economic diversification and achieve sustainable development in Qatar. The QNV 2030 aims to transform Qatar "into an advanced country by 2030, capable of sustaining its own development and providing for a high standard of living for all of its people for generations to come" (Qatar General Secretariat for Development Planning, 2008, 2). To this extent, it underlines the crucial roles of education, training, and capacity building. Qatar's active role in fostering sustainable development implementation is evident: the fact that Her Highness Sheikha Moza is an SDG's Advocate is a clear evidence of this commitment.

The Ministry of Education and Higher Education (MOEHE), which is the government entity charged with supporting and regulating education in Qatar (including the full array of K-12 public and private schools and institutions of higher education), sets itself the mission to: "regulate and support high-quality learning opportunities across all educational levels to provide individuals in Qatar's society with the needed knowledge, skills, and attitudes according to their abilities and in line with the national values needs". MOEHE's strong emphasis on "twenty-first-century skills, values, and attitudes" as well as on the need to keep up with world trends in teaching and learning significantly resonates with the global discourse on the necessity and importance of ESD.

17.3 Political Economy Framework for ESD in Qatar

Economic diversification away from hydrocarbons has increasingly become a critical priority for Qatar (D'Alessandro et al., 2019). Enhancing the role of the private sector and creating a knowledge economy based on innovation, entrepreneurship, efficient infrastructure, and good governance is at the heart of diversification efforts. This need for transformation is not only in response to the changing global economic dynamics but also arises due to the challenges Qatar faces at the intersection of environment, development, and society. For a knowledge-based, environmentally sustainable economy, Qatar needs to invest in knowledge products, creativity, and innovation, meaning that the cultivation of skills and know-how among its citizens and the youth is critical (Du & Chaaban, 2020, 2; Weber, 2014).

On the sustainable development front, although Qatar showcases a strong political commitment to global efforts setting ambitious targets for sustainable development, it is facing significant challenges. To illustrate increasing electricity and water consumption, rising dependence on automobiles, household air cooling consumption, industrial and urban activities which result in large-scale waste, air, and water pollution, and causing threat to the marine ecosystems are among some issue areas that Qatar needs to address (Planning and Statistics Authority, 2018, 318).

The progress of Qatar for economic diversification and sustainable development is very much dependent on the political economy context of the country, availability of resources, and citizens' commitment to such a transformation, replacing unsustainable consumption patterns with responsible solutions. With this background, this section briefly overviews Qatar's key strategy documents to detect the potential role of the ESD in Qatar's development path.

17.3.1 Qatar National Vision 2030

QNV 2030, is the fundamental strategy document that lays the ground for Qatar's target of a diversified and innovation-driven knowledge-based economic development, with four main pillars, namely, human development, social development, economic development, and environmental development (Qatar General Secretariat for Development Planning, 2008; see Chaps. 2 and 3) (Table 17.1).

As part of its national vision, Qatar aims to build a modern world-class educational system which will provide its citizens necessary skills and knowledge for addressing changing complex dynamics of the world, by also encouraging analytical and critical thinking, strong sense of citizenship and belonging, as well as creativity and innovation, and promoting respect for Qatari society's values and heritage (Qatar General Secretariat for Development Planning, 2008, 17).

For sustainable development, QNV 2030 acknowledges that promoting moral and ethical grounds for preserving and protecting the environment, including air, land, water, and biological diversity, and having an environmentally aware population are

Table 17.1 Pillars of Qatar National Vision 2030

Four pillars of Qatar National Vision			
Human development	Social development	Economic development	Environmental development
Development of all its people to enable them to sustain a prosperous society	Development of a just and caring society based on high moral standards, promoting Arab-Islamic identity and capable of playing a significant role in the global partnership for development	Development of a competitive and diversified economy capable of meeting the needs of, and securing a high standard of living for, all its people for the present and for the future	Seeking harmony between economic growth, social development, and environmental protection

pre-requisites. Therefore, educational curricula and training programs responding to current and future needs of Qatar are identified as crucial (Qatar General Secretariat for Development Planning, 2008, 20).

17.3.2 Qatar National Development Strategy 2011–2016

Building on QNV 2030, the first Qatar National Development Strategy 2011–2016 (NDS-1) identifies key themes, strategies, and targets for diversifying Qatar's economy and ensuring the country's social and economic progress. To this end, reforms, extensive stakeholder consultation, and benchmarking of international best practices are critical (Qatar General Secretariat for Development Planning, 2011, 139). An important theme that repeats through the strategy document is to prepare the Qatari citizens to become part of the country's economic progress and to make sure that Qatar has a capable and motivated workforce, equipped with skills and knowledge for managing sustainable development. Therefore, education is identified both as a goal and an enabler to achieve the aspirations of QNV 2030.

Under the theme of education, the main focus of NDS-1 revolves around the need to prepare the Qatari citizens for an economic and social vision beyond the country's wealth and economic activities based on hydrocarbons, which directly corresponds to the efforts for economic diversification. In this regard, developing and improving quality education is a key element to ensure that the education system in Qatar addresses the needs of current and future labor markets (Qatar General Secretariat for Development Planning, 2011, 139). Another key element for the theme of education is the emphasis on providing the students a solid ground for Qatari religious, moral, and ethical values, national identity, traditions, and cultural heritage. Designing education programs which will strengthen Qatari national values so as to instill a sense of belonging, responsibility, and citizenship among the students is identified as a priority.

Just like education, environmental sustainability is a repeated key component. When it comes to the theme of environmental sustainability, and how it intersects with education, the main concept that crystallizes in the document is "environmental awareness". Education is an enabler for environmental sustainability and improved environmental management. In this context, the development strategy involves key policies including: to increase environmental awareness and encourage sustainable production and consumption practices, to improve behavior for environmental sustainability (e.g., use of water and electricity), to support and increase environmental studies programs in schools, to establish school curriculum containing principles of sustainability, and to link local and international education initiatives (Qatar General Secretariat for Development Planning, 2011, 245–252). Although these targets are not coined around the exact wording of "education for sustainable development", the strategy corresponds to the essence of ESD.

17.3.3 Qatar National Development Strategy 2018–2022

Qatar's second National Development Strategy 2018–2022 (NDS-2) follows the path of QNV 2030 and NDS-1, with the targets of:

> Sustaining economic prosperity through economic infrastructure development, economic diversification and private sector development, and management of natural resources; promoting human development through a comprehensive and integrated healthcare system, quality education and training, efficient and committed workforce; and a sound social development through social protection, public security and safety, cultural enrichment and sports excellence. It also seeks to achieve a sustainable development that preserves the environment. (Planning and Statistics Authority, 2018, 5)

The political leadership in Qatar showcases a determination to use oil and gas investment revenues to finance the long-term sustainable development vision of the country and to develop institutional, financial, and human capacities to serve this purpose (Planning and Statistics Authority, 2018, 11). Accordingly, while the emphasis on economic diversification and private sector development continues to be integral to Qatar's development strategy, the necessity to develop and implement sustainable environmental practices in all pillars of QNV 2030 is acknowledged for enabling the country to develop and sustainably prosper (Planning and Statistics Authority, 2018, 302). To this end, environmental sustainability requires active and effective participation by all groups and members of society, and environmental awareness should form the basis and drive of human relation to the environment, as stated by NDS-2.

The theme of education in the document mainly addresses the target of developing tools to increase the youth's understanding of Qatari values, culture, and heritage as well as global citizenship. On the other hand, strong emphasis on the need to promote environmental awareness and to build an environmentally aware and supportive society for environmental sustainability (Planning and Statistics Authority, 2018, 318) implies an important role for education. Nevertheless, the reference to this dimension in the document does not only target the formal education system per se, but heavily underlines initiatives, efforts, or projects in the informal or non-formal forms of education, outside the school system such as the promotion of community-based environmental conservation initiatives or "environmental awareness campaigns to familiarize individuals with environmental considerations, requirements, sensitivity and conservation methods; change the behaviors and practices harmful to the local environment; and promote proper management, especially for water and electricity resources" (Planning and Statistics Authority, 2018, 318). Kahramaa Awareness Park and the Tarsheed campaign, launched by the Qatar General Electricity and Water Corporation, are two specific references in the document given as good examples of sustainable awareness efforts which target to promote a culture of responsible consumption (Planning and Statistics Authority, 2018, 305). The strategy document interestingly states that "it is impossible to build such awareness within a few years" (Planning and Statistics Authority, 2018, 305) signaling the need for a long-term and comprehensive "environmental education" supported through "curricula, broadcast and printed media, and modern social media" (Planning and Statistics Authority, 2018, 318).

17.4 Historical Context

Over the last decades, Qatar's education system has been going through several reforms. Although the initial attempts of reform target broader strategies such as achieving critical thinking, it is possible to detect an increasing reference to sustainable development and ESD in the overall education policy and tools.

Especially between the 1960s and 1980s, significant reforms have been in place for the education system in Qatar.[1] In 1965, the country replaced imported course material from Egypt, Saudi, and other Arab countries with its first national curriculum. Additionally, Qatar has become a member of key education-related regional and international networks including ALECSO (Arab League Educational, Cultural, and Scientific Organization) in 1962 and UNESCO in 1972 (Alkhater, 2016).

The establishment of Qatar Foundation is another important benchmark in the history of the education system in Qatar. Qatar Foundation for Education, Science and Community Development (QF) was established in 1995 as an initiative for investing in education, research, and capacity development in Qatar (see Chap. 5). The efforts of QF for contributing to a knowledge-based economy for the nation have been further consolidated with the inauguration of the Education City on October 13, 2003. Importing best regional and global practices, accelerating local education reform and policies, internationalization of education and strengthening Qatar's capacity in innovative education have been among the major motivations of Education City (Khodr, 2011, 1; Qatar General Secretariat for Development Planning, 2011).

[1] For a review of the history of Qatar's modern K-12 education, please see Alkhater 2016.

With regard to the education reforms, in the early 2000s, Education for a New Era (EFNE) has been one of the most significant ones. In 2001, RAND, a renowned American non-profit research organization, was invited by the Qatari government to conduct a comprehensive examination of Qatar's K-12 education system and to provide recommendations and strategy options for creating "a world-class system that would meet the country's changing needs" (Alkhater, 2016; Romanowski & Nasser, 2012, 2). EFNE was initiated in 2002 covering educational services, the curriculum, the quality of teachers, and the availability of educational resources (Khodr, 2011, 4). EFNE was highly motivated to make sure that the education system renders the students "active and self-responsible learners and teachers to act as guides and facilitators for knowledge seekers" (Al-Thani et al., 2021; Brewer et al., 2007) as opposed to the legacy of a memorization-based system (Weber, 2014, 3). EFNE targeted to cultivate "questioning and critical thinking skills" among the students, to ensure that "the teachers are promoting critical thinking skills by providing opportunities for students to learn them" (Romanowski & Nasser, 2012, 2) and eventually to provide qualified employees to the Qatari workforce equipped with critical thinking, communication, and teamwork skills (Romanowski & Nasser, 2012, 3; Stasz et al., 2007). In the curriculum standards introduced with EFNE, in the K-12 education system, it was detected that only the science curriculum had references to the issue areas concerning environment, climate change, and the importance of practices such as recycling (Zguir et al., 2021). Throughout the years, the reforms of EFNE have been subject to significant changes and its proposed system has been replaced by a more centralized model with the establishment of MOEHE in 2016 (Al-Thani et al., 2021; Zguir et al., 2021).

Currently, the General Framework for National Education of MOEHE (MOEHE, 2022b) sets the curriculum standards in Qatar. Building on the premises of QNV 2030 and subsequent NDS documents, the framework repeats the need to give the students the skills and competencies required for the twenty-first century, in order to prepare them for the changing dynamics of our complex world. In this regard, commitment to sustainable development emerges in the document as one of the key values along with peace, tolerance, mutual respect, and cultural understanding. In the curriculum framework, two major themes, which are also at the core of ESD, stand out: the reference to effective and responsible citizenship and commitment to sustainable development. Being a global citizen in an intertwined world, achieving SDGs, balancing economic growth with responsibilities toward future generations and combining global practices with local cultural values emerge as main premises communicated by MOEHE. Accordingly, with this recent curriculum framework, the historical evolution of ESD in Qatar comes to a point where a significant policy discourse finds its place in a fundamental official document for the education system (MOEHE, 2022b).

17.5 Current Situation of ESD in Qatar

As briefly reviewed in the previous section, Qatar has been undertaking reforms and initiatives to modernize the educational system toward a world-class quality education and has been mobilizing its resources in this effort: in 2020, the government allocated QAR 22.1 billion for the education sector, representing 10.5% of its total budget (Qatar Development Bank, 2021, 5). As part of these efforts, increasing reference to the importance of aligning the curriculum with SDGs, integrating ESD into teaching and learning activities, and designing education programs which can serve to the purpose of raising awareness on sustainability challenges in especially locally critical issues such as water, food, energy, are in place at the discourse level (Al-Kuwari et al., 2021; Zguir et al., 2021). At the practice level, the implementation of ESD emerges as a complex sphere with a combination of official, top-down initiated efforts, diverging global frameworks and extracurricular school activities in the formal and non-formal forms. This section overviews current efforts for ESD in Qatar.

In the context of formal school structure, UNESCO Associated Schools Network is one of the key anchors for ESD in Qatar. The UNESCO Associated Schools Network (ASPnet) aims to link educational institutions across the world around the ultimate goal of reaching SDG target 4.7 on Global Citizenship Education (GCED) and ESD. The network operates with the principle that ESD is necessary to enable the students to "address present and future global challenges constructively and creatively and to create more sustainable and resilient societies" (UNESCO, 2022b). ASPnet aims to transform learning and training environments in a way to integrate local and global sustainability issues into the curriculum, to reinforce interactive, participatory teaching and learning, to foster critical thinking, and to bring about changes in attitudes. ASPnet provides its associated schools with innovative educational materials, new teaching and learning approaches, capacity building activities as well as opportunities to connect and exchange experiences, knowledge, and good practices with schools, individuals, communities, policymakers, and society as a whole. Furthermore, in this system, the school campus itself is considered as a learning environment: all practices are targeted to be linked with environmentally friendly activities in line with the local community's needs, values, and sustainability requirements, with the mission to educate responsible agents of change in their schools, communities, and societies, who are capable of taking informed decisions for sustainable development. Over 11,500 ASPnet member schools in 182 countries work in collaboration for international understanding, peace, intercultural dialogue, sustainable development, and quality education in practice (UNESCO, 2022b).

Qatar has been a member of UNESCO since January 27, 1972 and the country joined the UNESCO Associated Schools Network in 1983, with initially one school, which is Doha Secondary School for Boys School (Qatar National Commission for Education, Culture and Science, 2022). Currently, according to the official website of UNESCO Associated Schools Network, 82 schools in Qatar are listed as members (UNESCO, 2022b).

The National Commission in Qatar identifies the tasks of UNESCO schools in Qatar as including, but not limited to, the environmental protection and conservation, the promotion and encouragement of mutual learning of cultures, the promotion of the principle of resolving conflicts without resorting to violence, and the promotion of the spirit of solidarity, integration, and helping others. The National Commission's Plan also includes Green School Program for the students of Primary, Preparatory, and Secondary UNESCO Associated Schools with the following objectives: (1) promoting positive behavior among students about the environment and conservation, (2) transforming the school environment into green oases, (3) strengthening students' sense of responsibility toward the environment and belonging to society, and (4) activating the role of the school toward sustainable development and environmental awareness (Qatar National Commission for Education, Culture and Science, 2022). The schools are provided with material and activities, they can also become part of regional and international projects and competitions.

Another major program contributing to ESD in Qatar is the Eco-Schools Program. Eco-Schools Program is the largest international accreditation and education for sustainable development program in the world. Created in 1994 in Denmark, the program represents the vision of a sustainable world in which ESD creates positive change and empowers people. The program operates in 94 countries with more than 55,000 registered schools, 19.8 million students, and 1.4 million teachers. Eco-Schools align their themes and activities with SDGs and encourage the students to take action for tangible sustainability results.[2]

The main objectives of Eco-Schools Program include, but not limited to, developing the knowledge, skills, and motivation the students need to tackle contemporary challenges and increasing understanding as well as the commitment to SDGs. This is a student-led program, encouraging the students to address the problems of our world, producing an action plan, and implementing action to drive change toward sustainable development. In this regard, the program offers an innovative format allowing the students to use school campuses as educational laboratories for sustainability practices.

The Eco-Schools Program was launched in Qatar in 2018 by Qatar Green Building Council, as the first comprehensive program in Qatar implementing a structured framework for long-term ESD, as opposed to hosting individual sustainability events or initiatives. Currently, there are 30 registered schools and 7 green flag awarded schools in Qatar, with more than 28,000 students and 4,000 teachers. In addition to its global network, the program is conducted in partnership with some key local stakeholders including Tarsheed-Kahramaa, Quranic Botanic Gardens, Supreme Committee for Delivery and Legacy, QF, and The Center for Green Schools-USGBC.

In addition to UNESCO Associated Schools Network and Eco-Schools Program which mainly target the formal education system, there is a variety of local initiatives, organizations, informal or non-institutional actors in Qatar directly and indirectly contributing to the ESD with their initiatives, activities, and partnerships. Arab Climate Youth Movement, Qatar Youth Power, Akhlaqunna, Greener Future, Quranic

[2] For further information, please see https://www.ecoschools.global/.

Botanic Garden, SDGeneration Network, and Qatar National Library are only a few actors within a long list of ESD teaching and learning ecosystem in Qatar.

These stakeholders of ESD, along with the formal education system, play a critical role in sustainability awareness and in promoting a culture of responsible consumption and protection of nature. A significant example of this concerns Kahramaa Awareness Park (KAP), which is also mentioned in Qatar's Voluntary National Review for Agenda 2030. The park is opened in the context of the National Campaign for Global Citizenship and Sustainable Development Education and includes multiple facilities such as screen documentaries on energy rationalization, awareness-raising processes, and energy conservation simulation processes, as well as specialized tools to raise awareness of energy efficiency and rationalization (Planning and Statistics Authority, 2021, 108).

It is worth underlining that QF operates as a critical entity at the intersection of formal and informal efforts for ESD. QF has been a key actor in launching initiatives that instill the concept of sustainability in the education programs, as well as in community activities through awareness programs, research centers, and educational institutions. Along with influential research centers working on environment and sustainable energy, such as Qatar Environment and Energy Research Institute (QEERI), QF has launched and implemented several initiatives, platforms, and institutions, including WISE platform to foster innovation in international education, Silatech initiative contributing to youth empowerment, Education Above All Foundation providing educational opportunities for children living in low-income families or facing humanitarian crises. The recently launched Earthna, a non-profit policy research and advocacy center established under QF to address sustainability policy in the critical areas of arid climates, sustainable cities, and communities, has also an ambitious agenda to inform and influence national and global sustainability policy and to encourage behavioral change in the community. The overall strategy behind QF's efforts offers a great impetus to ESD in Qatar, combining progressive education, sustainability, and social progress. Creating knowledge in ESD practices, cultivating sustainable thinking and practices based on religion, local traditions, and values, examining global know-how and best practices in sustainable development education, investigating innovative solutions for value-based education targeting the protection of environmental ecosystem, protecting, and restoring the environment while balancing economic development and well-being are among the core action items of these efforts.

17.6 Key Issues for Implementing ESD in Qatar

17.6.1 Policy

Qatar's national education system has gone through rounds of reform and modifications due to various reasons outlined earlier in the chapter. Education policy has always been at the epicenter of major policy changes leading to structural reforms and reshufflings. The most immediate manifestation today is the lack of alignment between policy and implementation "between the assessment practices, educational goals, and the SDGs" (Al-Kuwari et al., 2021).

Scholars argue that "Qatar has introduced several reforms to its education system in the past two decades, but due to the lack of detailed implementations of these reforms, the whole system has been weakened, resulting in unexpected outcomes" (Al-Thani et al., 2021). On the other hand, it should also be acknowledged that education system in Qatar has been exposed to deeper and urgent challenges especially vis a vis the hyper modernizing, globalizing, and transforming taking place. Education, as well as some other sectors such as health and economy, became the immediate focus of attention in policymaking given that increasing numbers and types of schools necessitated a multifaceted approach.

It is striking to observe that Qatar's national policymaking in education, with special reference to QNV 2030, became deeply associated with Qatar's external image and role in promoting sustainable development. Therefore, education policy, reform, and redesign of curriculum had to respond to multiple, and often conflicting objectives. For instance, QNV 2030 and the development planning it brought in train made strong references to identity, Arab/Islamic values, and language. However, increasing number of nationals and school structures made it even more difficult to configure a holistic and strategic approach to integrating ESD and SDGs.

17.6.2 The School System

The school system in Qatar is based on a fragmented structure. Currently, in the education system in Qatar (from pre-primary to secondary schools), there are 318 public schools including more than 126,000 students: and more than 686 private schools serving approximately 200,000 students (MOEHE, 2022a). These schools include a variety of nationalities, and their education is based on differing curricula. Figure 17.1 demonstrates different school types operating in Qatar's education system.

Different types of schools lead to a variety of practices and levels of ESD. The commitment, resources, and engagement in ESD changing across different types of schools make the integration of ESD into the official education system challenging. Furthermore, the schools demonstrate a highly multicultural teaching and learning environment, including both Qatari and Non-Qatari students. During the 2019–2020 term, in public schools, 53% of the students were Qatari while in private schools, 81% of the students were Non-Qatari (Planning and Statistics Authority, 2020). While this can be a socially enriching experience for the students, from the perspective of ESD, it brings the necessity of offering a common ground of understanding, values, and ethics for sustainability, as well as a carefully designed curriculum for balancing the students' national identities and their commitments as global citizens.

Another key issue area in the school system concerns the teachers, who play a crucial role "in the conjunction of ideal, designed, and actual teaching and learning experience delivery" (Al-Thani et al., 2021). Globally, 69 million additional teachers

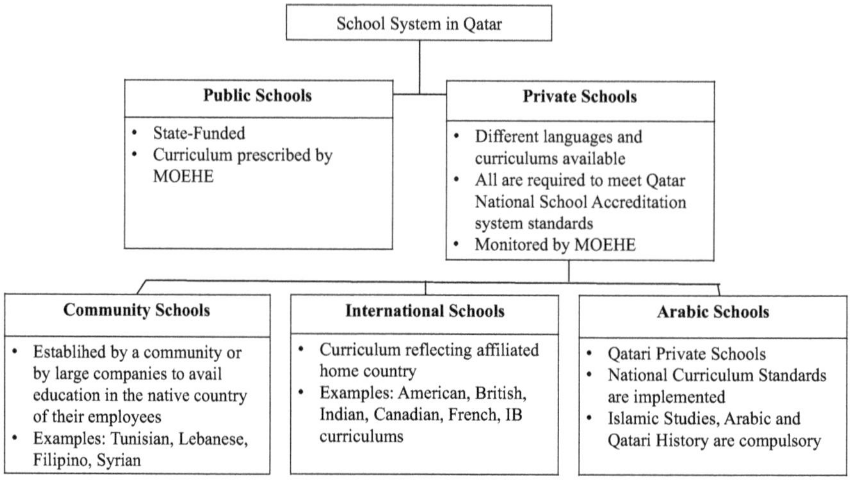

Fig. 17.1 Schools system in Qatar (Qatar Development Bank, 2021)

are needed in order to achieve the educational goals of the 2030 Agenda (Bamber, 2020). There is also the need to educate existing teachers to transform their teaching activities from extracurricular series of sustainability events, into ESD effectively integrated within the course curricula. Several studies conducted in Qatar revealed that the teachers need to be better prepared with training and professional development programs for delivering ESD, in order to increase their levels of awareness regarding sustainable development and how it can be connected with education, and in order to equip them with the skills and teaching materials to fulfill an efficient ESD (Al-Thani et al., 2021; Du & Chaaban, 2020, 2; Zguir et al., 2021).

The need for comprehensive assessment tools and data is another key component of the education system to assess ESD strategies, tools, and their effectiveness both for student learning and for the system's efficiency. Currently, even for ESD related SDGs, namely SDG4 Quality Education Target 4.7 stating, "by 2030, ensure that all learners acquire the knowledge and skills needed to promote sustainable development, including, among others, through education for sustainable development and sustainable lifestyles" and SDG 13 Climate Action, target 13.3 targeting to "improve education, awareness-raising and human and institutional capacity on climate change mitigation, adaptation, impact reduction and early warning" Qatar has no available data (Qatar Planning and Statistics Authority, 2021). This is an area open for progress in order to detect deficiencies in the system and to trace the improvement.

17.6.3 Localization of ESD

Although ESD targets to create a shared vision for sustainable development and global citizenship, there is no one model of ESD that can fit into all settings (Bui& O'Brien, 2020, 88). Various case studies have already proved that not paying attention to the dynamics of local community, indigenous knowledge, and culture hamper progress in ESD (Atkinson & Wade, 2015, 231). To ensure ownership, involvement of relevant stakeholders, and achievement of sustainable development, ESD must be localized and adapted to local realities, challenges, needs, and capacities of societies. In this regard, understanding and localizing the moral bases of ESD will better help to address the challenges identified in the QNV 2030 and create a generation of informed, morally cognizant, independent, innovative, and creative youth. Romanowski and Nasser (2012, 15) underline that "Qatar has to determine how its traditions, values and beliefs can coexist with the shifting winds of cultural and educational change…Without a sound understanding of how change and culture interact, educational reform might lose direction suffering unwanted economic, educational and cultural consequences". This means that ESD must be aligned with Qatari traditions and values. Localization of teaching content, creation of materials in the local language, and inclusion of local sustainability issues which can directly relate to students' environment is crucial.

17.7 Concluding Remarks

The leadership of Qatar acknowledges that the success of the country in its efforts for a sustainable, diversified, knowledge-based economy will be significantly determined by the development of its human capital. A world-class education system providing the citizens with the necessary knowledge, skills, and behaviors to develop their capabilities is a priority for Qatar in order to achieve a sustainable development path. Accordingly, at the policy level, the country is strongly committed to the aim of building an environmentally aware and responsible society, while promoting sustainable environmental practices and environmental awareness. As indicated earlier, this commitment is a result of both domestic policymaking and external anchors such as hyper-globalization, Qatar's soft power tools which focus extensively on SDGs and various other national branding strategies which position Qatar as an exemplar of a small state with impactful and large-scale ambitions.

The transformation of policy into actual practice is a rather complicated task. For a successful ESD plan, the SDG Accord Annual Report 2019 highlights that the education system needs to have an explicit action plan for the support of sustainable development, by integrating the SDGs into all levels of education, incorporating the SDGs into the whole curriculum by stressing their relevance to every course and their applicability to everyday scenarios (The SDG Accord, 2019). Moreover, there is a need to improve the availability of ESD-specific methods and innovations in teaching and learning available to teachers and students (Preston, 2014, 44). This means that although Qatar has a vibrant ESD ecosystem with political willingness and capacity, the comprehensive integration of ESD into the education system is still an area open for progress. Qatar still needs to design its own version of ESD and balance domestic and global forces that affect the policy world in Qatar.

Acknowledgements This research was made possible in part by NPRP grant #12C-0804-190009, entitled SDG Education and Global Citizenship in Qatar: Enhancing Qatar's Nested Power in the Global Arena, from the Qatar National Research Fund (a member of the Qatar Foundation).

References

Al-Kuwari, M. M., Du, X., & Koç, M. (2021). Performance assessment in education for sustainable development: A case study of the Qatar education system. *Prospects*, 1–15.

Al-Thani, W. A., Ari, I., & Koç, M. (2021). Education as a critical factor of sustainability: Case study in Qatar from the teachers' development perspective. *Sustainability, 13,* 11525. https://doi.org/10.3390/su132011525.

Alkhater, L. R. (2016). Qatar's borrowed K-12 education reform in context. In *Policy-making in a transformative state* (pp. 97–130). Palgrave Macmillan.

Atkinson, H., & Wade, R. (Eds.). (2015). *The challenge of sustainability: Linking politics, education and learning*. Policy Press.

Bamber, P. (Ed.). (2020). *Teacher education for sustainable development and global citizenship.* Routledge.

Boeren, E. (2019). Understanding Sustainable Development Goal (SDG) 4 on "quality education" from micro, meso and macro perspectives. *International Review of Education, 65*(2), 277–294. https://doi.org/10.1007/s11159-019-09772-7

Brewer, D. J., Augustine, C. H., Zellman, G. L., Ryan, G. W., Goldman, C. A., & Ryan, G. (2007). *Education for a new era: Design and implementation of K-12 education reform in Qatar*. Rand Corporation.

Bui, K., & O'Brien, W. (2020). Sustainability in higher education implementation in Bulgaria. In U. Azeiteiro & P. Davim (Eds.), *Higher education and sustainability opportunities and challenges for achieving sustainable development goals* (pp. 72–88). Taylor & Francis.

D'Alessandro, C., Sever, S. D., & Tok, M. E. (2019). Global environmental governance and the GCC: Setting the agenda for climate change and energy security. In L. Pal & M. E. Tok (Eds.), *Muslim perspectives to global governance* (pp. 197–227). Palgrave Macmillan.

Du, X., & Chaaban, Y. (2020). Teachers' readiness for a statewide change to PjBL in primary education in Qatar. *Interdisciplinary Journal of Problem-Based Learning, 14*(1), n1.

Dür, M., & Keller, L. (2018). Education for sustainable development through international collaboration. A case study on concepts and conceptual change of school-students from India and Austria on gender equality and sustainable growth. *Education Sciences, 8*(4), 187. https://doi.org/10.3390/educsci8040187

Elobeid, D. E., Kaifi, B. A., & Lele, U. (2016). Corporate social responsibility in higher education institutions: The experience of the college of business and economics at Qassim University. *International Leadership Journal, 8*(1), 120–143.

General Secretariat for Development Planning. (2008). Qatar National Vision 2030.

Gómez, L., Pujols, A., Alvarado, Y., & Vargas, L. (2018). Social responsibility in higher educational institutions: An exploratory study. In D. Crowther et al. (Eds.), *The goals of sustainable development responsibility and governance* (pp. 215–231). Springer.

Harvard University. (2019). *Sustainability report*. http://report.green.harvard.edu/. Accessed 28 March 2022.

Hopkins, C., & McKeown, R. (2005). *Guidelines and recommendations for reorienting teacher education (Technical Paper N° 2),* UNESCO. http://unesdoc.unesco.org/images/0014/001433/143370e.pdf. Accessed 28 March 2022.

Khodr, H. (2011). The dynamics of international education in Qatar: Exploring the policy drivers behind the development of education city. *Journal of Emerging Trends in Educational Research and Policy Studies, 2*(6), 514–525.

Kurz, R. (2020). UN SDGs: Disruptive for companies and for universities? In S. Idowu, O. Samuel, R. Schmidpeter, & Z. Liangrong (Eds.), *The future of the UN sustainable development goals* (pp. 279–291). Springer.

Marshall, K. (2018). Global education challenges: Exploring religious dimensions. *International Journal of Educational Development, 62*, 184–191. https://doi.org/10.1016/j.ijedudev.2018.04.005

Mawonde, A., & Togo, M. (2020). The role of SDGs in advancing implementation of sustainable development: The case of University of South Africa, South Africa. In U. Azeiteiro & P. Davim (Eds.), *Higher education and sustainability opportunities and challenges for achieving sustainable development goals* (pp. 1–18). Taylor & Francis.

Ministry of Education and Higher Education. (2022a). *Annual Statistics of Education in the State of Qatar*.

Ministry of Education and Higher Education. (2022b). *General framework for National Education of MoEHE*. https://www.edu.gov.qa/Documents/Publications/CurriculumStandards/GeneralFrameworkNationalEducation.pdf?csf=1&e=PfpExd. Accessed 6 April 2022.

Pak, S., & Lee, M. (2018). Hit the ground running: Delineating the problems and potentials in state-led Global Citizenship Education (GCE) through teacher practices in South Korea. *British Journal of Educational Studies, 66*(4), 515–535. https://doi.org/10.1080/00071005.2018.1533101

Planning and Statistics Authority. (2018). Qatar National Development Strategy 2018–2022.

Planning and Statistics Authority. (2020). *Education statistics*. https://www.psa.gov.qa/en/statistics1/StatisticsSite/Pages/default.aspx. Accessed 6 April 2022.

Planning and Statistics Authority. (2021). Qatar Voluntary National Review 2021: Report on the implementation of the 2030 Agenda for Sustainable Development.

Preston, R. (2014). Education for sustainable development, peace and global citizenship: Towards a framework. In *Education for peace and sustainable development* (pp. 41–59). MGIEP/UNESCO.

Qatar Development Bank. (2021). *Education sector in Qatar*. https://www.qdb.qa/en/Documents/Education_Sector_in_Qatar_EN.pdf. Accessed 28 March 2022.

Qatar General Secretariat for Development Planning. (2008). Qatar National Vision 2030.

Qatar General Secretariat for Development Planning. (2011). Qatar National Development Strategy 2011–2016: Towards Qatar National Vision 2030.

Qatar National Commission for Education, Culture and Science. (2022). *Associated schools*. https://qnc.edu.gov.qa/En/AssociatedSchools/Pages/default.aspx. Accessed 28 March 2022.

Qatar Planning and Statistics Authority. (2021). *Progress of SDG indicators in Qatar*. https://sdg-en-psaqatar.opendata.arcgis.com. Accessed 28 March 2022.

Romanowski, M. H., & Nasser, R. (2012). Critical thinking and Qatar's education for a new era: Negotiating possibilities. *The International Journal of Critical Pedagogy, 4*(1).

Stasz, C., Eide, E., & Martorell, F. (2007). *Postsecondary education in Qatar: Employer demand, student choice, and options for policy*. Rand Corporation.

The SDG Accord. (2019). *Annual SDG Accord Report 2019: Progress towards the Global Goals in the University and College Sector*.

UNESCO. (2022a). *What is education for sustainable development?* https://en.unesco.org/themes/education-sustainable-development/what-is-esd. Accessed 25 March 2022.

UNESCO. (2022b). *UNESCO associated schools network*. https://aspnet.unesco.org/en-us/Pages/Education%20for%20Sustainable%20Development.aspx. Accessed 25 March 2022.

UNESCO. (2016). *Education for people and the planet: Creating sustainable futures for all.* Global Education Monitoring Report: UNESCO Publishing. Retrieved from http://unesdoc.unesco.org/. Accessed 25 March 2022.

United Nations. (2021). Sustainable development: education for sustainable development. Resolution no: A/C.2/76/L.40.

Weber, A. S. (2014). Education, development and sustainability in Qatar: A case study of economic and knowledge transformation in the Arabian Gulf. In *Education for a knowledge society in Arabian Gulf countries*. Emerald Group Publishing Limited.

Welply, O., Taamouti, A., & Font, G. B. (2019). *Evaluating the impact of global citizenship Education young people's attitudes towards equality, diversity, and tolerance* (WISE Initiative Research Report). Available at https://www.edu.gov.qa/Documents/HigherEdTracks/Annual%20Statistics%2020202021_%20معتمد[2083].pdf. Accessed 25 March 2022.

Zguir, M. F., Dubis, S., & Koc, M. (2021). Embedding Education for Sustainable Development (ESD) and SDGs values in curriculum: A comparative review on Qatar, Singapore and New Zealand. *Journal of Cleaner Production, 319*, 128534.

S. Duygu Sever is a Post-Doctoral Researcher at Hamad Bin Khalifa University, Doha. She holds a PhD in Political Science and International Relations from Koç University, İstanbul. She has formerly worked for IFRI (Institut Français des Relations Internationales), Center for Energy in Paris as a Visiting Researcher and for EU Horizon 2020 Project: FEUTURE (The Future of EU-Turkey Relations: Mapping Dynamics and Testing Scenarios) for Energy and Climate Drivers, as a Project Assistant. Her research agenda addresses energy transition, sustainable development, and policymaking, by combining global-national dynamics with climate change, geopolitics, and international relations. Her dissertation titled as "The Domestic and Global Context of Energy Preferences: Explaining the Choice for the Nuclear," seeks to explain the complex dynamics which constitute pushing and constraining factors for a country's decision to engage in nuclear energy generation. Along with her academic activities and publications, she has also been a delegate in major platforms including Next Generation Delegates by Chicago Council on Global Affairs, Women in Energy, Young Leaders in Energy Initiative, Likeminds: German-Turkish Junior Expert Initiative by Robert Bosch Stiftung, and Young Academics Program by Global Relations Forum. She has received numerous scholarships and awards including Young Visiting Researcher Fellowship by the Embassy of France, Koç University Fellowship for Outstanding Success, TÜBİTAK (Technological and Research Council of Turkey) Scholarship, KUTEM 2020 Best Energy Paper of the Year Award and Koç University 2021 Academic Excellence Award.

M. Evren Tok is Associate Professor and Assistant Dean for Innovation and Community Advancement, Program Coordinator for Islam and Global Affairs Program at College of Islamic Studies, Hamad Bin Khalifa University. He also acts as the regional coordinator for the Executive Master's Program (Development Policies and Practices) jointly offered by Graduate Institute of Geneva and Hamad Bin Khalifa University). He is the co-editor of Policy Making in a Transformative State: The Case of Qatar, published by Palgrave MacMillan, 2016. His recent publications "Global Governance and Muslim Organizations" published by Palgrave MacMillan and Innovating South-South Cooperation by University of Ottawa Press in 2020. He is the director of a QNRF-funded cluster (2021–2026) entitled "SDG Education and Global Citizenship in Qatar." He is the Lead Project Investigator for Qatar National Research Fund Priority Program for 3 years on "Localizing Entrepreneurship Education in Qatar." He is the founder of the maker space entitled Maker Majlis: Islam in a Global World, which received Qatar Sustainability Award under Best University Initiative in 2020.

Chapter 18
Developing a Vibrant Entrepreneurship Ecosystem in Qatar: A Sustainable Pathway Toward the Knowledge-Based Economy?

Tarek Ben Hassen

Abstract For countries that depend on nonrenewable resources such as oil and gas, economic diversification is vital to ensure sustainable and resilient economic development. In July 2008, the government of Qatar launched the long-term national strategy Qatar 2030 (QNV 2030), intending to transform Qatar into a knowledge-based economy capable of attaining sustainable development by 2030. Since then, the Qatari government has invested continually and implemented many policies to encourage innovation, entrepreneurship, the private sector, and the advancement of human capital competencies to turn the economy into a knowledge-based one. Indeed, entrepreneurship is a crucial engine for economic development, one of the essential engines of economic diversification and building a knowledge-based economy. Qatar has, therefore, made a significant determination to develop a flourishing entrepreneurial ecosystem, including establishing key institutions and organizations to support entrepreneurs, such as incubators and financial framework, as evidenced by the increasingly strong performances in key international indices published by multiple global organizations. Nevertheless, despite Qatar's desire to diversify its economic base, entrepreneurs still have to contend with some challenges. Some of these difficulties are inextricably linked to Qatar's features as a rentier state. This chapter aims to analyze the present state of the knowledge-based economy in Qatar with a focus on the entrepreneurial ecosystem. It has two objectives. Firstly, the chapter intends to investigate the features and recent development of Qatar's entrepreneurial ecosystem, especially after the COVID-19 pandemic. Second, the article intends to examine the dynamics and many challenges that shape this ecosystem.

Keywords Knowledge-based economy · Entrepreneurial ecosystem · Economic diversification · Policy · Governance · Qatar

T. Ben Hassen (✉)
Program of Policy, Planning, and Development, Department of International Affairs, College of Arts and Sciences, Qatar University, Doha, Qatar
e-mail: thassen@qu.edu.qa

L. Cochrane and R. Al-Hababi (eds.), *Sustainable Qatar*, Gulf Studies 9,
https://doi.org/10.1007/978-981-19-7398-7_18

18.1 Introduction

The concept of a "knowledge-based economy" refers to industrialized countries' increasing dependence on knowledge, skilled labor, and innovation (OECD, 1996). Many perspectives hold that information and communication technology (ICT), innovation, research, and development (R&D), education, entrepreneurship, and the economic and institutional framework are critical pillars and drivers of the knowledge-based economy (Arundel et al., 2008). The pillars are intertwined, and establishing a solid knowledge economy requires high performance on each of them (Hvidt, 2015, 30). Entrepreneurship is rapidly acknowledged as a critical component of economic development, job creation, and competitiveness (Kirkwood, 2009).

Entrepreneurship may be described as the process of seeking, analyzing, and utilizing opportunities to develop new products and services (Shane & Venkataraman, 2000). Establishing a company is influenced by two major elements: personal factors connected to the entrepreneur's personal characteristics and environmental ones relating to the location (Martin & Osberg, 2007). Accordingly, the local context is crucial in understanding entrepreneurship since it is a socially rooted phenomenon resulting from a collaborative organizational effort (Thornton et al., 2011). Entrepreneurial activity is not an isolated occurrence but a network of people with unique functions (Drakopoulou & Anderson, 2007). Entrepreneurs are socially dependent on their operating settings (Granovetter, 1985). Most recently, in policymakers circles, the concept of the entrepreneurial ecosystem has gained considerable traction (Alvedalen & Boschma, 2017; Isenberg, 2010; Malecki, 2011). According to Mason and Brown (2014), an entrepreneurship ecosystem is a "network of interconnected entrepreneurial actors, entrepreneurial organizations (e.g., firms, venture capitalists, business angels, banks), institutions (universities, public sector agencies, financial bodies), and entrepreneurial processes that work together to connect, mediate, and govern the performance of the local entrepreneurial environment" (p. 5). An entrepreneurial ecosystem includes collaborative and fruitful interactions between various firms. Thus, the entrepreneurial ecosystem method starts with the enterprising person rather than the organization, highlighting the importance of the entrepreneurship setting (Stam, 2015; Stam & van de Ven, 2021).

Developing a knowledge-based economy and a robust entrepreneurial ecosystem are presently top priorities for economic diversification policies, especially in emerging and developing countries such as Arab and Gulf Cooperation Council (GCC) countries (Ben Hassen, 2021). Economic diversification is vital to guarantee sustainable economic development, particularly for countries that rely on nonrenewable natural resources like oil and gas, such as the GCC countries (Al Naimi, 2022; Al-Qahtani et al., 2022). Indeed, economic diversification is a vital component of sustainable development since it promotes structural and long-term change in the economy as well as other development pillars, such as social institutions (Jolo et al., 2022). Even though hydrocarbon industries still drive the GCC's economy, questions have been raised regarding the sustainability of resource-reliant models due to systemic shocks, resource depletion, and changing demographics and consumer preferences (Mohamed et al., 2022). It is no longer certain that oil revenues will be sufficient to sustain oil economies in the short- to medium-term due to fluctuating oil prices and a fast rate of energy transition (Luciani & Moerenhout, 2021). For the GCC economies, sustainability involves adapting to changing conditions, keeping prior advances in income per capita for their citizens, and perhaps reducing the gap with the wealthiest countries (Luciani, 2021). Over the long term, the hydrocarbon sector's prospects are gloomy, and economic diversification is necessary (IMF, 2021). It is widely accepted that diversified economies are more sustainable than oil-exporting nations in the Gulf (Ben Hassen, 2022a; Luciani, 2021).

Moreover, in the context of the 2030 Agenda for Sustainable Development and the SDGs, economic diversification is also essential to reach Goal 8: "Promote inclusive and sustainable economic growth, employment and decent work for all." Economic diversification is especially tied to target 8.2, which intends to "Achieve higher levels of productivity of economies through diversification, technological upgrading and innovation, including through a focus on high value added and labour-intensive sectors." Simultaneously, entrepreneurship is linked to SDGs 4 and 8. Indeed, SDG target 4.4 aspires to significantly expand the number of young people and adults with appropriate skills, including technical and vocational skills, for employment, decent jobs, and entrepreneurship. Further, SDG target 8.3 aims to "Promote development-oriented policies that support productive activities, decent job creation, entrepreneurship, creativity and innovation, and encourage the formalization and growth of micro-, small- and medium-sized enterprises, including through access to financial services" (United Nations, 2015).

In fact, several GCC countries have made structural changes in recent decades to diversify their economies and transform them into knowledge-based ones (Ben Hassen, 2022b). In July 2008, the long-term national plan Qatar National Vision 2030 (QNV 2030) was established by the Qatari government to transform Qatar into a developed country ready for sustainable development by 2030 (GSDP, 2008). Since then, the Qatari government has invested continually and implemented many policies to encourage innovation, entrepreneurship, the private sector, and the advancement of human capital competencies to turn the economy into a knowledge-based one. Indeed, entrepreneurship is a crucial engine for economic development, one of the essential engines of economic diversification and building a knowledge-based economy. Qatar has, therefore, made a significant determination to develop a flourishing entrepreneurial ecosystem, including establishing key institutions and organizations to support entrepreneurs, such as incubators and financial framework, as evidenced by the increasingly strong performances in key international indices published by multiple global organizations.

This chapter aims to analyze the present state of the knowledge-based economy in Qatar with a focus on the entrepreneurial ecosystem. It has two objectives. Firstly, the chapter intends to investigate the features and recent development of Qatar's entrepreneurial ecosystem, especially after the COVID-19 pandemic. Second, the chapter intends to examine the dynamics and many challenges that shape this ecosystem.

18.2 The Current Qatari Economic Context and the State of the Economic Diversification Process

Since the early 1990s, Qatar has invested extensively in hydrocarbon production, particularly in liquefied natural gas (LNG) (Qatar National Bank (QNB), 2018). Qatar's per capita income has risen as the country has become a significant player in the LNG global industry (IMF, 2019). Furthermore, most industrial sectors, such as petrochemicals, are connected to oil and gas. Accordingly, oil and gas are crucial to most businesses and other economic activities. Even though non-hydrocarbon activity has expanded as a proportion of real GDP over the last decade, hydrocarbon-funded government projects continue to play a significant role (IMF, 2019). Consequently, Qatar's economy relies on oil and gas revenues to finance government expenditures and drive economic development since there are no taxes (Tok et al., 2020).

However, fluctuations in oil prices may cause a rise or fall in capital surplus and capital accumulation. Oil prices have fluctuated over the last 50 years (Baffes & Kshirsagar, 2015). For example, Brent's (oil) price plummeted below $50 in 2016 after doubling from 2009 to 2011 (IMF, 2019). In 2020, due to the COVID-19 pandemic, prices dropped 33%, from 61,41 dollars a barrel in 2019 to 41,26 dollars a barrel (World Bank, 2021). As a result, the existing economic situation is unsustainable (Ben Hassen, 2021). The recent drop in oil prices during the COVID-19 pandemic posed a significant challenge to the Qatari economy. Indeed, following a 12% increase between 2008 and 2012, growth slowed to 4.4% in 2013 and was expected to fall to 2.6% in 2019 (World Bank, 2020). Qatar transitioned from a budget surplus of 10.3% in 2015 to a deficit of 4.07% in 2016, its first deficit in 17 years. Meanwhile, government debt climbed from 35.8 to 47.6% of GDP (Schwab, 2017). Qatar faced its first recession in a generation in 2019, although by just −0.2%, because of the COVID-19 pandemic. Consequently, after a −3.6% recession in 2020, Qatar's economy started to recover in 2021, with a 1.5% growth rate and an estimated growth rate of 3.4% in 2022. Meanwhile, Qatar's budget deficit in 2021 is expected to be 0.9% of GDP, down from 3.6% the previous year, due to a rebound in hydrocarbon prices. Furthermore, the recent normalization of relations with its neighbors will benefit Qatar by increasing profits from its worldwide state-owned airline when transport connections resume. The resolution of the rift may potentially resurrect the hope of deeper GCC integration and regional crisis burden-sharing (IMF, 2021).

Globally, national lockdowns and social distancing measures have resulted in a considerable drop in natural gas use in 2020 (International Energy Agency, 2020). Qatar's export income has suffered due to its dependence on oil-linked gas contracts. Overall Qatari exports fell by 42.8% in June 2020 compared to the previous year (World Bank, 2020). There is also uncertainty in the LNG market, with fears of oversupply with new reserves coming from Australia and the USA, which could challenge Qatar's position (Focus Economics, 2020). Prior to the COVID-19 pandemic, Qatar was already in a period of economic transition.

Given the limitations of the traditional oil-based economy, Qatari authorities believe that the country needs a long-term strategy to lessen its reliance on hydrocarbon revenues and establish a more stable and diverse economy (Baabood, 2017). Indeed, diversification is vital for a significant commodity exporter like Qatar because it helps handle short-term shocks and prepares for long-term changes in the economic environment. Qatar's economy would be more resilient and better prepared for the future decades if it had a diversified range of economic activities, exports, and revenues (IMF, 2019).

In July 2008, the Qatari government established the long-term national plan Qatar 2030 (QNV 2030) to transform Qatar into a developed country capable of achieving sustainable development by 2030. It aims "at transforming Qatar into an advanced country by 2030, capable of sustaining its own development and providing for a high standard of living for all of its people for generations to come" (GSDP, 2008, 2). QNV 2030 identifies the country's long-term priorities and provides a context for establishing national strategies and action plans. The strategy rests on four interrelated and mutually reinforcing pillars of development: environmental, economic, human, and social (Table 18.1).

QNV 2030 was followed by two National Development Strategy (NDS), five-year plans, Qatar's first National Development Strategy 2011–2016 (NDS-1) and the second National Development Strategy (2018–2022) (NDS-2) (GSDP, 2011, 2018). The strategies were prepared to define concrete steps and results to resolve obstacles and advance QNV 2030 objectives (Ben Hassen, 2019).

Table 18.1 QNV 2030 pillars (GSDP, 2008)

Economic development	**Social development**
This pillar seeks to build a sustainable and diversified economy capable of maintaining a high quality of life for its people now and in the future The national vision will govern the Qatari economy, strike a balance between a knowledge-based and an oil-based economy, attract new investors, boost competitiveness, and encourage growth	This pillar aims to create a fair and supportive society founded on strong moral principles and to be willing to play an essential role in the partnership for sustainable development It requires a social welfare and protection system and women's empowerment by offering equal opportunities for all people in different fields such as education and employment
Human development	**Environmental development**
QNV 2030 aims to educate all Qatari citizens so that they can sustain a prosperous society This involves a high standard educational system and a modern healthcare system	To ensure a compromise between developmental needs and environmental preservation for future generations It involves managing the environment based on harmony between social development, economic growth, and environmental protection

18.3 Characteristics of the Entrepreneurship Ecosystem in Qatar: Strengths and Shortcomings

Since 1997, Qatar has established numerous institutions, incubators, organizations, and funding structures to aid entrepreneurs and create an effervescent entrepreneurship ecosystem (Ben Hassen, 2019). However, despite this success, entrepreneurs in Qatar still face some barriers and difficulties (Ben Hassen, 2021).

18.3.1 Strengths of the Entrepreneurship Ecosystem in Qatar

Firstly, the main strength of the entrepreneurship ecosystem in Qatar is the strong governmental implication. Indeed, government actions have influenced a variety of facets of the knowledge-based economy. As Rubin (2012) mentioned, "Qatar has set the bar high with its goal of becoming a knowledge-producing economy at record speed. However, the country holds some strong cards: a clear vision, highly committed leadership, and abundant resources to devote to the cause" (4). Qatar has intensely focused on the relevance of small and medium-sized businesses (SMEs) and entrepreneurship in the state's overall development goals. Encouraging SMEs and entrepreneurs was made a major priority by the highest levels of the state institutions (Tok et al., 2020).

Consequently, the Qatari entrepreneurial ecosystem is primarily the product of substantial government involvement, with the government serving as the primary driver. This leadership is motivated by a desire to diversify the economy, as stated in QNV 2030 and NDS-1. QNV 2030 recognizes the significance of entrepreneurship in achieving economic diversification and reducing Qatar's reliance on hydrocarbon sectors (GSDP, 2018). According to NDS-1, diversification entails strengthening the business environment and private sector development, fostering entrepreneurship, and restructuring the labor market (GSDP, 2011). As Qatar attempts to meet the objectives outlined in QNV 2030 and transition from a carbon-export-based economy to a knowledge-based one, the private sector is expected to play a key role (Mehrez, 2019). Indeed, since 1997, Qatar has established numerous institutions, incubators, organizations, and funding structures to aid entrepreneurs and create an effervescent entrepreneurship ecosystem, such as the Qatar Development Bank, Qatar Business Incubation Center (QBIC), Social Development Center, Silatech, Center for Entrepreneurship (Qatar University), Qatar Science and Technology Park (QSTP), Enterprise Qatar, and Qatar Foundation (QF) (Ben Hassen, 2019, 2020) (Table 18.2).

In fact, the vast majority of organizations in Qatar dedicated to fostering entrepreneurship were founded or are funded by the government as part of a complex web of initiatives and policies (Ennis, 2015). Further, the Qatari government is making significant efforts to improve the legal and economic environment to decrease business risk and encourage start-ups (Sahli, 2021). Consequently, start-ups and

Table 18.2 The leading entrepreneurship support organizations in Qatar (Ben Hassen, 2019)

Organization	Est. Date	Mission and goals
Social development Center (Nama Center)	1996	Nama Center helps broaden the opportunities available to young people, develop their capacity, and empower them
Qatar Development Bank (QDB)	1997	Supporting Qatari entrepreneurs to help contribute to the diversification of the local economy by successful small and medium-sized businesses able to participate in international markets
Qatar Science and Technology Park (QSTP)	2005	Enabling an open innovation, research, and entrepreneurship ecosystem
Injaz Qatar	2007	Creating an SME environment conducive to a creative entrepreneurial spirit and business development helps SME achieve business excellence
Enterprise Qatar	2008	Being the focus of SMEs and the locomotive of Qatar's economic diversification process
Silatech	2008	Silatech is a global development organization that links young people to economic opportunities and jobs through creative business development and employment programs
Digital Incubation Center (DIC)	2011	Promoting ICT innovation in Qatar, especially among millennials at the early stages
Bedaya Center	2011	The Bedaya Center gives Qatari young people access to several programs, including career advice, developing employment skills, and entrepreneurship
Qatar Business Incubation Centre (QBIC)	2013	The QBIC is a special incubation center offering support resources for entrepreneurs that have an idea of starting a new company or are willing to expand an established business
Center for Entrepreneurship (QU)	2013	Help students and the QU community to grow and turn business ideas into successful start-ups

SMEs in Qatar already have access to a robust ecosystem of public assistance from various government departments and organizations. As a result of careful planning, Qatar has identified high-potential industries such as smart manufacturing, fintech, sports-tech, and fashion and design as areas where the private sector can use its capabilities to build global competitiveness and produce new revenue streams (Oxford Business Group, 2021).

More recently, the COVID-19 pandemic prompted a significant reaction from the Qatari government, which has made efforts to prevent the loss of new and developing businesses while also aiding entrepreneurs in adapting to the new economic reality. Government action to safeguard employees and consumers of new and expanding businesses has also been taken, and a significant increase in the digital and online delivery of regulations for entrepreneurs (QDB, 2021a).

Since 1995, these policies and measures have contributed to developing entrepreneurial activity in Qatar and facilitating the process. From 2016 to 2020, the Global Entrepreneurship Monitor (GEM) data collection provides a comprehensive yet brief assessment of Qatar's entrepreneurial development. There was a noticeable rise in the number of people in Qatar launching or operating new firms (Hawi et al., 2022). Further, the number of persons in Qatar engaged in early-stage entrepreneurship has consistently climbed since 2017, and the percentage of established firm ownership has hit an all-time high since the benchmark survey was done in 2016. Furthermore, the rate of Entrepreneurial Employee Activity has almost doubled since 2019. As the rate of company discontinuation in Qatar decreases, so does the rate of entrepreneurial activity (QDB, 2021b).

Consequently, in 2020, Qatar was ranked third in the GCC member states and the Middle East and North Africa (MENA) region, and eighth globally, according to the National Entrepreneurship Context Index (NECI).[1] It scored slightly lower than the United Arab Emirates (UAE) and Saudi Arabia, which ranked fourth and seventh globally, respectively (QDB, 2021a). Moreover, in 2021, start-ups in Qatar received record-level funding, and venture investment in Qatar reached a new high, with QAR 69 million invested in start-ups, a 92% increase over the previous year's amount (KPMG, 2021; QDB, 2021b). Also, 58% more money was invested in technology-based initiatives in 2021 than in the previous year. Further, there has been an increase in interest and investment in Qatar's venture capital (VC) ecosystem from private individuals, corporations, and international corporations. In 2021, more than 90% of all start-up investments in Qatar came from private and foreign investors. While private and international corporations made up about half of all VC investments in 2016, there was an encouraging rise in interest and engagement from non-governmental entities (QDB, 2021b). Investors in Qatar decided to fund businesses such as e-commerce, delivery services, and fintech, which witnessed increased demand throughout the pandemic and kept the top three rankings in terms of transaction volume and value. Additionally, entrepreneurs in the e-commerce industry received almost 60% of the total capital obtained in Qatar (KPMG, 2021).

Second, entrepreneurship in Qatar is fueled by opportunities. Qatar was placed eighth out of 54 participating countries for opportunity-driven early-stage entrepreneurs in 2017. Although 82.4% of early-stage enterprises in Qatar are driven by opportunity, this is more than any other MENA area, including the UAE (QDB, 2017). Indeed, in 2020, even though many businesses have been negatively affected

---

[1] GEM launched the National Entrepreneurial Context Index (NECI) in 2018, which measures an economy's entrepreneurship environment. The NECI informs policymakers, practitioners, and other important stakeholders on the overall strength of the entrepreneurial ecosystem (QDB, 2021a).

by the pandemic, COVID-19 has sparked an uptick in new company ventures in Qatar. According to the GEM Qatar Report 2020, the Total Early-stage Entrepreneurial Activity (TEA) rate rose from 14.7% to 17.2% throughout 2020. About 45.6% of the respondents who are not already participating in entrepreneurial activity want to become so within the next three years. Half of these people cited COVID-19 as an influence. Four out of ten early-stage entrepreneurs (41.9%) and one-third of experienced company owners feel that COVID-19 has presented new business prospects they want to explore. One out of every four persons polled in the GEMS' Adult Population Survey knows at least one person in Qatar who has launched a company due to the coronavirus outbreak (QDB, 2021a). Two variables might account for this. First, Qatar has one of the lowest unemployment rates globally among Qataris and expats. During the first quarter of 2021, the number of job seekers in Qatar reached 3,138. In the first quarter of 2021, the unemployment rate was 0.2%, compared to 0.1% in the first quarter of 2020 (Planning & Statistics Authority, 2021). Second, few jobless expats remain in the country since residency permits are tied to work (QDB, 2017).

18.3.2 Obstacles of the Entrepreneurship Ecosystem in Qatar

Firstly, human capital is the most significant impediment to entrepreneurship in Qatar. The lack of human resources and the mismatch between the capabilities needed by the sector and those offered by the educational system are the two most significant constraints faced by start-ups in this respect.

Indeed, start-ups face a shortage of technology-related human resources, particularly engineers. Most new businesses voiced concerns about lacking skilled human resources in their respective industries (Ben Hassen, 2020). As a matter of fact, this is a recurring problem in Qatar's educational system. Since 1995, Qatar has made significant efforts to strengthen its educational system, investing heavily and implementing several reforms (Koc & Mohamed, 2017). Nonetheless, Qatar continues to lag in educational achievement, particularly in science, technology, engineering, and mathematics (STEM) sectors (Said, 2016), which are considered the main components of achieving a knowledge-based economy (Durazzi, 2019). Although Qatari 4th and 8th-grade students showed growth in scientific performance in the Trends in International Mathematics and Science Study (TIMMS) and the Program for Worldwide Student Assessment (PISA), they remain far behind the international average (Said, 2016). Furthermore, an increasing number of high school graduates cannot enter university since their academic credentials fall short of most universities' requirements. Many secondary institutions use Arabic as a teaching medium (Berrebi et al., 2009). As a result, many students have lost opportunities to pursue their education due to language barriers, which significantly setback the country's aim of growing human capability (Mohamed et al., 2022).

In addition, though just a tiny fraction of Qatari students are interested in a career in the STEM fields, most of them are driven to work in the government (Sellami et al., 2017). There are several advantages to working in the public sector, including better pay, less stress, and less responsibility (Forstenlechner & Rutledge, 2010). Consequently, Qataris have a low level of involvement in the private sector. Hence, students are no longer enrolling in STEM professions at a pace that will contribute to Qatar's economic advancement in the future (Said, 2016). A decline in science and math enrollment at the postsecondary level must be reversed to meet better the demands of knowledge-based economy companies (GSDP, 2011, 52).

Second, according to the GEM study of 2016 (QDB, 2017), 77.3% of the Qatari adult population feels that entrepreneurs have a high degree of social standing and respect, and 65.9% believe that establishing a new firm is a suitable career option. However, in 2020, over four out of ten persons who perceive the strong potential for entrepreneurship said they would be discouraged from establishing a firm due to a fear of failure. Qatari citizens are more likely than expats to say they would be hesitant to establish a company because they fear failure (44.8% and 39.6%, respectively) (QDB, 2021a). The fear of failure may deter some would-be entrepreneurs from beginning a firm. In fact, company ownership is seen highly by Qatari society, but only as a source of additional revenue. In Qatar, the "passive entrepreneur" is the most popular kind of entrepreneur. The entrepreneur has a stable full-time job in the public sector and runs a side company to supplement his income. It is risky for him to quit his work and launch his own firm as a full-time entrepreneur.

Thirdly, access to finance is a major barrier for many Qatari entrepreneurs. Indeed, according to IMF (2019), SMEs get just 2% of total credit in Qatar, indicating that there is still room for improvement in inclusion and fostering entrepreneurial activity in the country. Indeed, in 2020, financing options for new and expanding businesses via private lenders such as crowdsourcing (3.6) and initial public offerings (IPOs) got the lowest scores from experts (QDB, 2021a). In fact, business owners are often unable to get appropriate financing. In Qatar, most new businesses are backed by informal financing channels (ex., family, relatives, friends, work colleagues, neighbors, strangers, etc.) (Kebaili et al., 2015). In 2020, 74.4% of early-stage entrepreneurs utilized personal funds to fund their businesses, 28.8% used personal loans, and 14.8% used commercial loans (QDB, 2021a). However, the fact that Qatar's GDP is the highest in the world could make it seem contradictory that the country has restricted access to funding (Mehrez, 2019). This problem might be explained by the culture of banks in Qatar and the GCC in general. Most banks in Qatar prefer to invest in more secure businesses, such as real estate. As explained by the senior regional economist at HSBC: "most banks in GCC states are traditionally unwilling to lend to small, little-known firms, preferring instead the security and predictability of lending to large firms, such as those with state connections" (Arabian Business, 2012). Simultaneously, new businesses in Qatar confront exorbitant costs. Rent, workforce, and the initial investment in materials and equipment are all costly (Kebaili et al., 2015). Further, to Komalasari (2016), several external factors impact entrepreneurship in Qatar, including the difficulties of obtaining commercial facilities or offices at a fair price, the availability of finance, especially "angel investors," and the complexity of business regulations. Accordingly, 49% of the entrepreneurial experts surveyed in Qatar believe that giving more significant financial support to start-ups and expanding enterprises may boost the country's entrepreneurial environment (QDB, 2021a).

18.4 Conclusion

This chapter aims to analyze the present state of the knowledge-based economy in Qatar with a focus on the entrepreneurial ecosystem. It has two objectives. Firstly, the chapter intends to investigate the features and recent development of Qatar's entrepreneurial ecosystem, especially after the COVID-19 pandemic. Secondly, the article intends to examine the dynamics and many challenges that shape this ecosystem.

Economic diversification is vital in achieving sustainable economic development, especially for countries relying on nonrenewable natural resources, such as oil and gas, in the case of the Gulf Cooperation Council (GCC) countries. Moreover, the global crisis caused by the COVID-19 pandemic underscored the significance of boosting resilience to adverse shocks (Ben Hassen, 2022b). Indeed, the pandemic highlighted the need to promote non-hydrocarbon sectors by strengthening the fundamental pillars of the knowledge-based economy: ICT, innovation, R&D, education, entrepreneurship, and the economic and institutional regime. Since 1995, the Qatari government has invested continually and implemented several policies to encourage innovation, entrepreneurship, the private sector, and the advancement of human capital competencies to turn the economy into a knowledge-based one. Meanwhile, Qatar has made a significant attempt to build a dynamic entrepreneurship ecosystem, including establishing important institutions and organizations to assist entrepreneurs, such as incubators and funding structures, as evidenced by rankings published by various international organizations. Consequently, Qatar is outperforming the other Middle Eastern countries regarding entrepreneurship development, offering more chances for both aspiring and full-time entrepreneurs. Furthermore, the COVID-19 pandemic increased the country's entrepreneurial drive, resulting in more start-ups and many projects. Therefore, the Qatari entrepreneurial ecosystem is primarily the product of substantial government involvement, with the government serving as the primary driver. Meanwhile, the majority of early-stage technical enterprises are motivated by opportunities. Nevertheless, despite Qatar's desire to diversify its economic base, entrepreneurs still have to contend with several challenges. Some of these difficulties are inextricably linked to Qatar's features as a rentier state.

Firstly, human capital is the primary impediment to entrepreneurship. A minority of Qatari students are considering a future career within one of the STEM disciplines, while many remain interested in public sector employment. Governments are the primary employer of local labor in the rentier state because they provide access to rentier capital for a portion of the population (Beblawi, 1987; Gray, 2011). Further, there is a gap between the skills demanded by technical start-ups and those offered by the educational system. Accordingly, the development of STEM knowledge, skills, and competencies must be prioritized if Qatar's educational system provides a workforce capable of fulfilling the demands of contemporary society and the country's fiercely competitive job market (Sellami et al., 2017).

Secondly, in terms of social values and culture, our findings show that company ownership is seen positively by Qatari society, but only as an additional source of income. The "passive entrepreneur" is the most popular kind of entrepreneur in Qatar. The entrepreneur has a stable full-time job in the public sector and a side company to supplement his income. Furthermore, there is a significant disparity between people who wish to start a business and those who actually do so. This may be explained by a fear of failure, preventing some prospective entrepreneurs from launching a firm.

Thirdly, access to finance is a significant barrier for many Qatari entrepreneurs. The culture of Qatari banks explains this challenge. Most Qatari banks prefer to invest in more secure ventures, such as real estate. This problem is also linked to the rentier model, which encourages high- and quick-return investments in real estate and financial speculation over productive investments in prospective value-added industrial sectors, which often take longer to reap the rewards (Ennis, 2015; Gray, 2011).

According to Ennis (2015), the development of Qatar's entrepreneurial activities is entangled between an international capitalist policy agenda and a national economy hampered by two interrelated dependencies challenging to correct: hydrocarbons and foreign labor addictions. The author also points out that entrepreneurship has perpetuated the rentier state structure, causing a contradiction between economic reform and the existing structural challenges. Further, Qatar's transition to a knowledge-based economy is facing a critical structural challenge: the rentier system, which is hegemonic within the socio-economic and political system in Qatar and the GCC region in general, exposes many limitations (Ennis, 2015). With little private-sector participation, the rentier model makes the government the key player in the knowledge-based economy, which aligns with the government's position in the rentier state (Beblawi, 1987).

The shift to a knowledge-based economy is a complex process that requires the participation of several stakeholders from all sectors (innovation, R&D, education, entrepreneurship, companies, etc.). As a result, a successful transition requires a national plan that avoids the "silos mindset," which involves all stakeholders with defined aims, actions, and missions for each, encouraging collaboration and information sharing. In reality, inadequate sectoral coordination and integration is a critical institutional issue in Qatar owing to a lack of "planning culture" and "teamwork," as well as a silo mindset (GSDP, 2011). Further, as the IMF (2019) outlined, exports and activity diversification may benefit from structural reforms and sector-specific initiatives. In sectors with opportunities for exports and innovation, sector-specific policies should be implemented. Developing knowledge in specific clusters, such as food and water security, should be the focus since these are two significant challenges in Qatar and the GCC region (see Chaps. 11 and 12).

References

Al-Qahtani, M., Fekih Zguir, M., Al-Fagih, L., & Koç, M. (2022). Women entrepreneurship for sustainability: Investigations on status, challenges, drivers, and potentials in Qatar. *Sustainability, 14*(7), 4091. https://doi.org/10.3390/su14074091

Al Naimi, S. M. (2022). Economic diversification trends in the Gulf: The case of Saudi Arabia. *Circular Economy and Sustainability, 2*(1), 221–230. https://doi.org/10.1007/s43615-021-00106-0

Alvedalen, J., & Boschma, R. (2017). A critical review of entrepreneurial ecosystems research: Towards a future research agenda. *European Planning Studies, 25*(6), 887–903. https://doi.org/10.1080/09654313.2017.1299694

Arabian Business. (2012). *GCC private sector boom still "vulnerable"* . https://www.arabianbusiness.com/gcc/saudi-arabia/saudi-arabia-industries/gcc-private-sector-boom-still-vulnerable-480365

Arundel, A., Van Cruysen, A., Hansen, W., Kanerva, M., & Kemp, R. (2008). *Workpackage 1 defining the knowledge-based economy: Final synthesis report*. https://www.uni-trier.de/fileadmin/fb4/projekte/SurveyStatisticsNet/KEI-WP1-D1.5.pdf

Baabood, A. (2017). *Qatar's resilience strategy and implications for state—Society relations*. Istituto Affari Internazionali (IAI) and Foundation for European Progressive Studies (FEPS). https://www.iai.it/sites/default/files/iaiwp1736.pdf

Baffes, J., & Kshirsagar, V. (2015). Sources of volatility during four oil price crashes. *Applied Economics Letters*, 1–5. https://doi.org/10.1080/13504851.2015.1078435

Beblawi, H. (1987). The rentier state in the Arab world. *Arab Studies Quarterly, 9*(4), 383–398. https://www.jstor.org/stable/41857943?seq=1

Ben Hassen, T. (2019). Entrepreneurship, ICT and innovation: State of Qatar transformation to a knowledge-based economy. In H. Alkhateeb (Ed.), *Qatar political, economic and social issues* (pp. 193–209). Nova Publiser.

Ben Hassen, T. (2020). The entrepreneurship ecosystem in the ICT sector in Qatar: Local advantages and constraints. *Journal of Small Business and Enterprise Development, 27*(2), 177–195. https://doi.org/10.1108/JSBED-04-2019-0119

Ben Hassen, T. (2021). The state of the knowledge-based economy in the Arab world: Cases of Qatar and Lebanon. *EuroMed Journal of Business, 16*(2), 129–153. https://doi.org/10.1108/EMJB-03-2020-0026

Ben Hassen, T. (2022a). A transformative state in the wake of COVID-19: What is needed to enable innovation, entrepreneurship, and education in Qatar? *Sustainability, 14*(13), 7953. https://doi.org/10.3390/su14137953

Ben Hassen, T. (2022b). The GCC economies in the wake of COVID-19: Toward post-oil sustainable knowledge-based economies? *Sustainability, 14*(18), 11251. https://doi.org/10.3390/su141811251

Berrebi, C., Martorell, F., & Tanner, J. C. (2009). Qatar's labor markets at a crucial crossroad. *The Middle East Journal, 63*(3), 421–442. https://doi.org/10.3751/63.3.14

Drakopoulou Dodd, S., & Anderson, A. R. (2007). Mumpsimus and the mything of the individualistic entrepreneur. *International Small Business Journal: Researching Entrepreneurship, 25*(4), 341–360. https://doi.org/10.1177/0266242607078561

Durazzi, N. (2019). The political economy of high skills: Higher education in knowledge-based labour markets. *Journal of European Public Policy, 26*(12), 1799–1817. https://doi.org/10.1080/13501763.2018.1551415

Ennis, C. A. (2015). Between trend and necessity: Top-down entrepreneurship promotion in Oman and Qatar. *The Muslim World, 105*(1), 116–138. https://doi.org/10.1111/muwo.12083

Focus Economics. (2020). *Qatar economic outlook*. https://www.focus-economics.com/countries/qatar

Forstenlechner, I., & Rutledge, E. (2010). Unemployment in the Gulf: Time to update the"social contract." *Middle East Policy, 17*(2), 38–51. https://doi.org/10.1111/j.1475-4967.2010.00437.x

Granovetter, M. (1985). Economic action and social structure: The problem of embeddedness. *American Journal of Sociology, 91*(3), 481–510.
Gray, M. (2011). *A theory of "late rentierism" in the Arab States of the Gulf*. https://repository.library.georgetown.edu/bitstream/handle/10822/558291/CIRSOccasionalPaper7MatthewGray2011.pdf
GSDP. (2008). *Qatar national vision 2030*. https://www.psa.gov.qa/en/qnv1/Pages/default.aspx
GSDP. (2011). *Qatar national development strategy (2011–2016) towards Qatar national vision 2030*. https://www.psa.gov.qa/en/nds1/Documents/NDS_ENGLISH_SUMMARY.pdf
GSDP. (2018). *Qatar second national development strategy 2018–2022*. https://www.psa.gov.qa/en/knowledge/Documents/NDS2Final.pdf
Hawi, A., Al-Kuwari, F., & Garonne, C. (2022). Entrepreneurship development in Qatar. In *Entrepreneurial rise in the Middle East and North Africa: The influence of Quadruple Helix on technological innovation* (pp. 29–51). Emerald Publishing Limited. https://doi.org/10.1108/978-1-80071-517-220221003
Hvidt, M. (2015). The state and the knowledge economy in the Gulf: Structural and motivational challenges. *The Muslim World, 105*(1), 24–45. https://doi.org/10.1111/muwo.12078
IMF. (2019). Policies to drive diversification for Qatar. *International Monetary Fund*. https://doi.org/10.5089/9781498317863.002.A001
IMF. (2021). *Economic prospects and policy challenges for the GCC countries*. https://www.imf.org/en/Countries/QAT#countrydata
International Energy Agency. (2020). *Gas 2020 analysing the impact of the COVID-19 pandemic on global natural gas markets*. https://www.iea.org/reports/gas-2020/2020-meltdown#abstract
Isenberg, D. (2010, June). The big idea: How to start an entrepreneurial revolution. *Harvard Business Review*.
Jolo, A. M., Ari, I., & Koç, M. (2022). Driving factors of economic diversification in resource-rich countries via panel data evidence. *Sustainability, 14*(5), 2797. https://doi.org/10.3390/su14052797
Kebaili, B., Al-Subyae, S. S., Al-Qahtani, F., & Belkhamza, Z. (2015). An exploratory study of entrepreneurship barriers: The case of Qatar. *World Journal of Entrepreneurship, Management and Sustainable Development, 11*(3), 210–219. https://doi.org/10.1108/WJEMSD-03-2015-0014
Kirkwood, J. (2009). Motivational factors in a push-pull theory of entrepreneurship. *Gender in Management: An International Journal, 24*(5), 346–364. https://doi.org/10.1108/17542410910968805
Koc, M., & Mohamed, B. H. (2017). Challenges and drivers for Qatars transformation into a knowledge-based economy ans society work in progress in education system reforms. *Pressacademia, 4*(1), 281–288. https://doi.org/10.17261/pressacademia.2017.545
Komalasari, S. D. (2016). *The role of regulations in promoting growth of SMEs and start-ups in Qatar*. Qatar University.
KPMG. (2021). *Tech—Start up investments round up*. https://home.kpmg/qa/en/home/media/press-releases/2021/08/vc-investments-in-qatar-startups-surge-by-182.html
Luciani, G. (2021). Framing the economic sustainability of oil economies. In L. Giacomo & T. Moerenhout (Eds.), *When can oil economies be deemed sustainable?* (pp. 9–30). https://doi.org/10.1007/978-981-15-5728-6_2
Luciani, G., & Moerenhout, T. (2021). Introduction: Economic sustainability of oil economies. In G. Luciani & T. Moerenhout (Eds.), *When can oil economies be deemed sustainable?* (pp. 1–8). https://doi.org/10.1007/978-981-15-5728-6_1
Malecki, E. J. (2011). Connecting local entrepreneurial ecosystems to global innovation networks: Open innovation, double networks and knowledge integration. *International Journal of Entrepreneurship and Innovation Management, 14*(1), 36. https://doi.org/10.1504/IJEIM.2011.040821
Martin, L. R., & Osberg, S. (2007). Social entrepreneurship: The case for definition. *Stanford Social Innovation Review, 5*(2), 29–39.

Mason, C., & Brown, R. (2014). *Entrepreneurial ecosystems and growth-oriented enterprises.* https://www.oecd.org/cfe/leed/entrepreneurial-ecosystems.pdf

Mehrez, A. (2019). Investigating critical obstacles to entrepreneurship in emerging economies: A comparative study between males and females in Qatar. *Academy of Entrepreneurship Journal, 25*(1), 1–15. https://search.proquest.com/docview/2239574856/abstract/FE482344F0A948ABPQ/1?accountid=13370

Mohamed, B. H., Disli, M., Al-Sada, M., Bin, S., & Koç, M. (2022). Investigation on human development needs, challenges, and drivers for transition to sustainable development: The case of Qatar. *Sustainability, 14*(6), 3705. https://doi.org/10.3390/su14063705

OECD. (1996). *The knowledge-based economy*. OECD.

Oxford Business Group. (2021). *What are the post-pandemic prospects for Qatar's start-ups and SMEs?* https://oxfordbusinessgroup.com/news/report-what-are-post-pandemic-prospects-qatar-start-ups-and-smes

Planning and Statistics Authority. (2021). *Quarterly Bulletin—Labor force survey, First quarter (Q1) 2021.* https://www.psa.gov.qa/en/statistics1/pages/topicslisting.aspx?parent=Social&child=LaborForce

Qatar National Bank (QNB). (2018). *Qatar economic insight—September 2018.* https://www.qnb.com/sites/qnb/qnbqatar/document/en/enQatarInsight092018

QDB. (2017). *Global Entrepreneurship Monitor—Qatar report 2016.* https://www.qdb.qa/en/publications

QDB. (2021a). *Global Entrepreneurship Monitor GEM report 2020.* https://www.qdb.qa/en/publications

QDB. (2021b). *Qatar venture investment report 2021.* https://www.qdb.qa/en/publications

Rubin, A. (2012). *Higher education reform in the Arab World: The model of Qatar*. Middle East Institute. https://www.mei.edu/publications/higher-education-reform-arab-world-model-qatar

Sahli, S. (2021). Entrepreneurship in Qatar. In *Entrepreneurship in the Gulf cooperation council region* (pp. 79–109). World Scientific (Europe). https://doi.org/10.1142/9781786348081_0005

Said, Z. (2016). Science education reform in Qatar: Progress and challenges. *EURASIA Journal of Mathematics, Science and Technology Education, 12*(8). https://doi.org/10.12973/eurasia.2016.1301a

Schwab, K. (2017). *The global competitiveness report 2017–2018.* https://www.weforum.org/reports/the-global-competitiveness-report-2017-2018

Sellami, A., El-Kassem, R. C., Al-Qassass, H. B., & Al-Rakeb, N. A. (2017). A path analysis of student interest in STEM, with specific reference to Qatari students. *EURASIA Journal of Mathematics, Science and Technology Education, 13*(9). https://doi.org/10.12973/eurasia.2017.00999a

Shane, S., & Venkataraman, S. (2000). The promise of enterpreneurship as a field of research. *The Academy of Management Review, 25*(1), 217. https://doi.org/10.2307/259271

Stam, E. (2015). Entrepreneurial ecosystems and regional policy: A sympathetic critique. *European Planning Studies, 23*(9), 1759–1769. https://doi.org/10.1080/09654313.2015.1061484

Stam, E., & van de Ven, A. (2021). Entrepreneurial ecosystem elements. *Small Business Economics, 56*(2), 809–832. https://doi.org/10.1007/s11187-019-00270-6

Thornton, P. H., Ribeiro-Soriano, D., & Urbano, D. (2011). Socio-cultural factors and entrepreneurial activity. *International Small Business Journal: Researching Entrepreneurship, 29*(2), 105–118. https://doi.org/10.1177/0266242610391930

Tok, E., Koç, M., & D'Alessandro, C. (2020). Entrepreneurship in a transformative and resource-rich state: The case of Qatar. *The Extractive Industries and Society*. https://doi.org/10.1016/j.exis.2020.04.002

United Nations. (2015). *Transforming our world: The 2030 agenda for sustainable development.* https://sdgs.un.org/2030agenda

World Bank. (2020). *Qatar's economic update—October 2020*. https://www.worldbank.org/en/country/gcc/publication/economic-update-october-2020-qatar#:~:text=Theannualgrowthratefor,deterioratingnon-energymarketsentiment.&text=Thefiscaldeficitisexpected,stimulustomitigate COVID-19

World Bank. (2021). *COVID-19 pandemic and the road to diversification*. https://documents1.worldbank.org/curated/en/748461627924058675/pdf/Gulf-Economic-Update-COVID-19-Pandemic-and-the-Road-to-Diversification.pdf

Tarek Ben Hassen is an Assistant Professor of Policy, Planning, and Development at the Department of International Affairs, College of Arts and Sciences, at Qatar University. He holds a MSc. in management of innovation (University of Quebec in Chicoutimi, Canada; 2006) and a Ph.D. in Urban Studies, specialized in economic geography, from the University of Quebec in Montreal, Canada (2012). Previously, he was a lecturer in geography at the University of Burgundy in France and the University of Quebec in Montreal. His research over the past 16 years is clustered around the following three interdisciplinary themes/topics. The first research theme falls into the subfield of economic geography, emphasizing the study of the territorial components of innovation and entrepreneurship, the impacts of global industrial and technological changes, and the emerging of the knowledge-based economy and the creative economy. The second research theme focuses on the Gulf Region's current development dynamics and economic diversification initiatives. The third research theme focuses on studying the dynamics of sustainability and innovation within food systems. He has several publications in reputed international journals, including Sustainability, the Canadian Journal of Urban Research, the Regional and Urban Economics Journal, Environmental Science and Pollution Research, Journal of Small Business and Enterprise Development, and Frontiers in Nutrition.

Chapter 19
Pathways for a Sustainable Future

Leslie A. Pal, Reem Al-Hababi, and Logan Cochrane

Abstract The strategic vision of the State of Qatar seeks to pursue development while aiming for a balanced approach to the social, human, economic, and environmental pillars of the vision. The legal and governance mechanisms supporting this transition are only effective when implemented, and can be hindered by limited access to data for decision-making. In the last decade, significant changes have taken place, catalyzed by commitments made to host the FIFA World Cup 2022 as well as innovations led by the Qatar Foundation. Yet, as a country with heavy reliance on hydrocarbon resources, the transition toward a more sustainable future involves trade-offs, the options for which present different pathways (from pragmatic to transformative). Climate change will present significant challenges for Qatar, including sea level rise and increasing temperatures, but also impacts on terrestrial and marine biodiversity. Domestically, key areas of the economy (energy, water, food, urban development, waste management) require integrated, systems approaches for moving toward greater sustainability. This future needs to be enabled by new ways of teaching and learning as well as new ways of thinking about and doing business. Not all issues could be covered in this collection (most notably, transportation, heating and cooling systems, desalination, health, and air quality, among others). However, this book has provided a wealth of evidence on diverse subjects, and this concluding chapter brings these diverse options and recommendations together.

Keywords Qatar · Sustainability · Transition · Policy · Pathways

L. A. Pal (✉)
College of Public Policy, Hamad Bin Khalifa University, Doha, Qatar
e-mail: lepal@hbku.edu.qa

R. Al-Hababi
Gulf Studies Program, Qatar University, Doha, Qatar

L. Cochrane
Hamad Bin Khalifa University, Doha, Qatar
e-mail: LCochrane@hbku.edu.qa

L. Cochrane and R. Al-Hababi (eds.), *Sustainable Qatar*, Gulf Studies 9,
https://doi.org/10.1007/978-981-19-7398-7_19

19.1 Introduction

Sustainability is a comforting word, pliable enough to embrace the environment, society, and economy—all should be "sustainable"—but sufficiently ambiguous to seem uncomplicated. It was coined as a policy concept in 1987 in the Brundtland Commission's *Our Common Future*, where it was joined to "development" as a guide for economic policy, then further extended to embrace the environment with the Rio Declaration in 1992. As the environment broadened to include climate, and then the various dimensions that both contribute to and are affected by climate change, the concept retained its earlier roots in economic policy, but now includes almost everything—societies and economies are to be sustainable; water, fish, and wildlife are to be managed sustainably; cities have to be smart and sustainable; lifestyles and even the most minute personal behaviors should be calibrated for sustainability, and moreover for their contributions to sustainable futures (which are plural, since they vary by locale). More tellingly, this is not simply a conceptual or philosophical framing—it has institutional foundations in the United Nations system and global agreements, prominently the 2015 Paris Climate Accord as well as the 2030 Agenda and the Sustainable Development Goals. With international agreements and institutions come processes, and with those processes come national commitments to meet global standards and contribute to global outcomes. As this book shows, "sustainable Qatar" is a massive and complicated undertaking, as is a true "sustainability agenda" for any country.

Our chapters discuss almost every aspect of sustainability (though as the next section argues, there are still some areas that require more analysis), and three broad themes emerge from the analysis. The first is that while sustainability is a global challenge, the responses have to be tailored to regional, national, and local conditions. Localizing responses to these conditions are not only environmental or climatic, they include social and economic factors. A sustainability agenda for northern European countries will have to be different than one for equatorial nations. The agenda will also be affected by population size, level of economic development (and hence the financial resources to transition to clean energy), key industries and their energy demands, consumption patterns, and even culture. As noted in the introductory chapter, to understand sustainability in Qatar, we must first understand the unique conditions, opportunities, and challenges of Qatar as a nation.

A second theme embedded in many of the chapters is the current array of policy frameworks and legal instruments that give effect to Qatar's sustainability goals. The core of these frameworks focuses on environmental issues and on climate change, but since almost everything we do and all the ways in which we live affect the environment (not least through carbon emissions), a "sustainability agenda" in principle will cascade through almost every conceivable public policy—as an example, this book has chapters on smart cities and on education. That said, our chapters provide an impressive catalog of equally impressive policy and legislative efforts undertaken by the State of Qatar since the early 2000s, when the government began to address environmental issues, gradually evolving into a sustainability agenda.

Given the scope and breadth of that agenda, weaving together everything from bicycles to biodiversity, the coordination and capacity challenges can be immense. This is particularly true of larger countries, with multiple sub-national jurisdictions and different environmental and economic eco-systems (e.g., the United States, China, or Brazil), but our chapters show that it has been a challenge for even a small unitary state like Qatar. Exemplary of these challenges is that they require policy coherence across sectors and scales. Chapters on the domestic energy and water systems showcase the importance of multi-scale engagement, while chapters on the marine and terrestrial ecosystems highlight the importance of multi-sectoral responses.

The third theme that emerges from this collection is related to the implementation challenges, and in particular, the transition from a carbon-based society and economy to a post-carbon future. Even imagining what that might mean is daunting. Most projections of the global energy mix, say to 2050 or 2060, provide scenarios where hydrocarbons retain a key share of energy supply. If there is a mix, in what proportions? For which sectors? With what energy sources, and in what combinations? As noted earlier, these transitions will be tailored to local conditions, and thereby varied. Post-carbon implies post-industrial, and while "knowledge-based economy" trips off the tongues of policy wonks, specifying what that might mean both in general terms and specifically for any given country, workforce, or educational system, is something else. The particular challenge of this transition for Qatar is given by one single fact—its current economy is highly dependent on fossil fuel extraction and export (liquefied natural gas "LNG"). For Qatar, the transition is not simply from one set of energy sources to another, but a transformation of its entire economy.

We treat each of these themes below, then briefly consider some remaining gaps in our coverage of topics, before turning to a summary of recommendations for a sustainable Qatar.

19.2 Qatar: Unique Conditions, Challenges, and Opportunities

Sustainability as a policy destination depends on where you begin, and almost every chapter in this collection considers the unique conditions in Qatar. The overarching context is that the country is a relatively small, peninsular country in the Arab Gulf region, molded by four key factors—gas reserves, heat, sand, and Islam. Our chapters went well beyond this, and in the aggregate, they paint a detailed portrait of the country and the implications for sustainability policy. The following summarizes some of these key contextual factors, which mark the initial starting point for sustainability discussions:

1. *Size*: Small, with a land mass of only 11,521 km^2. Qatar is only 85 km wide, or about an hour's drive. While there are some islands, Qatar is primarily a

peninsula, surrounded by seawater. It is also flat, with only a few hills that are about 100 m above sea level. Up to 18% of the country is liable to inland flooding, with 96% of the population living in coastal areas (Chaps. 9 and 11). These factors highlight a key risk related to sea level rise (Chap. 9).

2. *Land surface*: Groundwater has high salinity, and there are few areas suitable for agriculture. Nearly all the country is sandy desert covered with scrub plants and loose gravel. A key challenge emerging from this is limited opportunities for agriculture (Chap. 11), at least in the forms we have commonly known it (future research may explore opportunities for vertical farming and aquaculture—both in their infant stages within the country).
3. *Population*: The total population is 2.8 million, of which about 88% are foreigners. In 1939, Qatar had a population of only 28,000. Almost the entire population resides in Doha and one or two adjacent communities (Al Wakra, Al Khor). As a result, urban planning is critical for reducing reliance upon personal vehicles, and creating community spaces that embody the sustainability values outlined in the national vision (Chaps. 11, 14 and 15).
4. *Biodiversity and eco-systems*: There are only 19 species of terrestrial wild animals in Qatar, with a possible six mammal species, but records are limited. Plant species (adapted to a hot, dry climate) are estimated at 400 (see Chap. 13). While these species are comparatively few compared to other biological "hotspots", they are of particular importance to this environmental setting. Maintaining this biodiversity requires integrated approaches, such as strengthening the tracking and record keeping of existing species.
5. *Water*: Scarce precipitation, limited groundwater, and high levels of salinity result in few sources of water, combined with no natural lakes or rivers as sources of freshwater. At the same time, demand for water is rapidly increasing alongside population growth and economic development, leading to Qatar having one of the highest per capita water consumption rates in the world (430 L/person per day; Chap. 11). Other than groundwater used for the agricultural sector, desalination is the primary source of water for the country, which is energy intensive. Research and innovation are required to increase efficiency, and more efficient technologies that are already available need to be more widely employed. Given the energy intensity of obtaining fresh water via desalination, water leakage or loss in the water system requires attention; water leakage is a major problem (real loss is about 5.28%). Water storage is also an issue, with the country having only two days of strategic water stock (Chap. 11).
6. *Temperature*: The temperature averages around 17 °C in both spring and autumn months, but goes up as high as 50 °C in July and August (Chap. 11). Although not explored in this book, temperature rise will have implications for energy and water demand, on human health, and on systems functionality during heatwaves. Climate projections, therefore, need to be integrated across sectors to ensure that planning takes into account not only the average temperature rise but also the potential for more intense and prolonged heatwaves.
7. *Energy*: Qatar has 24.29 trillion cubic meters (Tcm) of proven gas reserves, about 12% of the world total, and has the third-largest single reserve in the

world (Chap. 11). Currently, about 60% of its LNG exports go to Asia, but production capacity is planned to increase from 77 million ton to 126 million ton per year by 2027 (Chap. 6), and given the Russian invasion of Ukraine, the global energy market is re-orienting with Qatar finding new demand for its LNG. While this presents an economic opportunity, it also has an environmental cost. The implication is that for the coming decades, it is likely that Qatar will continue to be a major LNG producer and exporter. This is both an opportunity (for revenues) and a challenge (as countries increasingly move to renewables). Explorations of how Qatar could be a leader in transitioning to hydrogen highlight how opportunities also exist to utilize resources in ways that are cleaner and offer long-term economic potential.

8. *Emissions*: Qatar's has one of the world's highest emissions rate per capita annually, coming primarily from industrial activities related to LNG (extraction, manufacturing, and heat production), and power and water production (Chap. 6). On a per capita basis, this makes Qatar the top CO_2 emitter in the world, though only 0.28% of the world total. As the world transitions away from fossil fuels, there will increasingly be pressure to reduce emissions, for example through carbon capture (Chap. 7), which Qatar might actually lead, with suggestions that by 2035 the country has the potential to reduce the carbon intensity of Qatar's LNG facilities by 35%. These reductions will require research and infrastructural investment, but will make progress toward climate commitments, and help position Qatar's resources in a more favorable light.
9. *Electricity consumption*: Qatar has the fourth-highest per capita electricity consumption per capita in the world (Chap. 11). Virtually all of this energy is generated through fossil fuels (over 80% from natural gas). Despite abundant, blazing sunshine, Qatar only opened its first large-scale solar power plant in 2021. Despite having enough natural gas to power its national electricity needs, there is both a need and an opportunity to shift to renewables (especially solar).
10. *Food production and security*: Before the blockade of 2017, Qatar imported about 90% of its food. Despite climate and soil conditions, the government decided to make food security a national priority (see below), and there has been an expansion of local agricultural produce (including dairy), with an increase of 20% in the number of farms, organic vegetable production, and the potential for exports. Food waste was a challenge before 2017, however with new domestic production food loss is also a key challenge—particularly given the enormous resources needed to strengthen domestic food production (Chap. 12). Given the newness of the industry, research and development can enhance production and reduce food loss, while new and innovative policy approaches are required to reduce food waste. Redistributing edible food and collecting organic waste for composting or biochar are important, but addressing the source of food waste requires much more attention (including in revising curricula in the education system; Chap. 17).
11. *Solid Waste*: Qatar is one of the highest per capita waste-generating countries in the world, producing about 10 million metric tons per year. The good news

is that production peaked in 2013, and has been declining (largely construction waste emerging out of a construction boom and associated demolition waste; Chap. 16). Municipal waste is largely not segregated, resulting in large amounts of waste that could be re-used, re-cycled, or transformed into other value-adding commodities are ending up disposed in landfills. Remaking the waste management system has not attracted as much policy attention as it should, given the costs and opportunities available therein (Chap. 16).

12. *FIFA and World Cup 2022:* This may seem an odd thing to list with the preceding items, but ever since the World Cup was awarded to Qatar in 2010, it has affected national development in everything from construction (e.g., eight soccer stadiums, a completely new metro system, hotels, and destination venues), to population growth (massive recruitment of migrant labor to work on construction projects and in the hospitality sector), and hence energy consumption and emissions. At the same time, the FIFA World Cup 2022 Sustainability Strategy, with its five pillars and over 70 initiatives, has created a parallel policy framework to achieve the first carbon–neutral World Cup. The pillars, for example, comprise sustainable buildings, mitigation of greenhouse gas (GHG) emissions, air quality offsets, waste minimization, and water conservation (Chap. 4). Preparations for the World Cup over the last decade have driven changes in the construction sector, and hence energy and water consumption, population growth and waste production. The FIFA Sustainability Strategy, combined with the Qatar—FIFA joint venture of Qatar 2022 LLC, has highlighted sustainability as one of the key features of the World Cup, which offers the opportunity to have a legacy that accelerates and supports transitions to higher levels of sustainability.

To paraphrase Tolstoy, every national sustainability strategy is sustainable in its own way. Considering the unique combination of features above—geographical, geological, ecological, demographic, and economic—it is clear that sustainability strategies in Qatar cannot be the same as Sweden's or Samoa's. Within that context, we can better appreciate the policy challenges. How can the country achieve some measure of food security in an arid climate, without much arable land and next to no fresh water supplies? How will it reduce electricity consumption for air-conditioning when summer temperatures are routinely in the 40s and sometimes the 50s °C? How will the population mitigate the threat of rising sea levels when 96% of the population lives in coastal areas, and almost all of that population is clustered in and around one city (Doha)? Those are challenges. This book has provided feasible options for ways forward, but which will require commitment and consistency. Qatar also has some advantages. It is small, and size matters, in this case in the sense that Qatar is essentially a city-state, similar to Singapore. In principle, this can encourage greater policy coherence and focus (though capacity and implementation are always problems, as they are everywhere). It is rich, has abundant supplies of energy through its natural gas reserves, and despite a volatile global economy, looks well-positioned for continued growth in LNG exports over the next 20 years. Through FIFA, Qatar lashed itself willingly to the mast of sustainability, and as we will see below, has genuinely committed itself to the required policy frameworks.

19.3 Sustainability: Policy Frameworks and Legal Instruments

The second theme that emerges from our chapters is the country's existing sustainability policy frameworks and legal instruments. As Chap. 3 noted, Qatar ratified the Kyoto Protocol in 2005 and was one of the first countries in the Arab world to ratify the Paris Agreement, also accepting the Doha Amendment in 2020. Like all other countries in the world, Qatar also supported the United Nations Sustainable Development Goals (SDGs) and Agenda 2030. Together, these agreements form the highest international scaffolding for national sustainability strategies. Within that scaffolding, each country has additional agreements, as well as its own strategies, policies, and laws. The following are the most notable instruments of Qatari sustainability policy and law.

1. *International Agreements*

   Qatar is a signatory to United Nations Framework Convention on Climate Change and the Paris Agreement. It has enhanced its original Intended Nationally Determined Contribution (INDC), and in 2021 proposed a 25% reduction in GHG emissions, relative to its baseline scenario. We will not list them all here (see Table 3.1 in Chap. 3), but it is also a signatory to at least nine sustainability declarations and instruments in the Arab region. These are more than rhetorical flourishes—Qatar has shown leadership in some niche areas. As Chap. 3 notes, it has contributed $100 million to support climate action among small island states, and the Qatar Fund for Development is also focusing its support there (with the Gates Foundation, in 2022 it committed $200 million to support smallholder farmers with climate-adaptive agriculture and developing resilient food systems). Qatar is also a signatory to the Convention on Biological Diversity (Chap. 13). These international agreements not only set forth agendas, but in some cases involve reporting, as Qatar has done with the Voluntary National Review in 2021. This advances accountability and transparency regarding these agreements.
2. *Constitution, National Vision, National Development Strategy, and National Environment and Climate Change Strategy*

   Very few countries have a quartet of overlapping and reinforcing policy frameworks of the sort that Qatar does. Chapter 2 discusses the first three (Constitution, QNV, and NDS) in detail. A unique feature of the Qatari Constitution (adopted in 2005—the country only became independent of Britain in 1971) is its Article 33: "The State shall preserve the environment and its natural balance in order to achieve comprehensive and sustainable development for all generations." Another notable feature of the constitution and the culture of the country is that Islam is the official religion, and Qatar's principal (though not sole) source of law is *Shari'a* law. As Chap. 3 points out, this leads to an emphasis on stewardship rather than ownership (*khalifa*), and on fiduciary obligations (*waqf*). Further extensions of Islamic principles in Qatari law can be seen in Islamic finance and in *zakat*, or charitable giving, and both of these have been modernized to support environmental stewardship.

Most nations in the Global North have dispensed with formal vision statements or national economic and development strategies, leaving these for annual pronouncements by their presidents or prime ministers, and annual national budgets. But in the Global South, national development strategies coupled to some overarching and long-term vision are quite common. Qatar has both, and they do inform and shape national policy discussions. The QNV 2030 was announced in 2008, virtually at the beginning of the country's soaring LNG revenues and the rapid acceleration of its economic and social transformation (and just two years before being awarded the World Cup). It remains a national touchstone. As Chap. 2 explains, it has four pillars (social, human, economic and environmental), and the environmental pillar highlighted challenges specific to Qatar's ecosystem: rising sea levels, water, pollution, and environmental degradation. The pillar also projected the need for an environmentally aware population, adequate legal mechanisms, effective institutions to build awareness and stimulate green technologies, urban planning, and international cooperation. The NDS (there have been two to date, 2011–2016, 2018–2022, and a third (2023–2027) is being crafted) are detailed extensions of the QNV 2030, extending each of the four pillars into specific policy initiatives, complete with implementation strategies and Key Performance Indicators. Chapter 2 outlines the targets in the two NDS (which grew increasingly ambitious).

The fourth interlocking framework for sustainability is the National Environment and Climate Change Strategy, adopted in 2021. It identified five priorities (emissions and air quality, biodiversity, water, circular economy and waste management, and land use), and further reinforces the QNV 2030 and the NDS.

3. *Sectoral Instruments*

Both the international agreements and the overarching frameworks mentioned above are just that—agreements (with commitments) and frameworks. A further level of policy and legal specification is usually required through sector-specific instruments—e.g., food or transportation. In principle, those sectoral instruments will reflect and amplify the higher-level commitments enunciated in the framework documents, but institutional and collaborative challenges begin to arise the closer one gets to the ground. We will turn to these complications in a moment, but first, a partial list of some of the sectoral instruments discussed in our chapters.

- Qatar's Law No. 30 of 2002 on Environmental Protection (Chap. 3)
- The *Executive By-Law for the Environment Protection Law* contains provisions on reducing air pollution, and GHG emissions (Chap. 3)

- Executive By-Law, *Qatar's Resolution of the Council of Ministers* establishes a Committee on Climate Change and Clean Development Mechanism (CDM) (Chap. 3)
- Global Drylands Alliance (established in 2017), is an international agency addressing food security and dryland ecosystems (Chap. 3)
- National Biodiversity Strategy and Action Plan (NBSAP) (Chap. 2)
- The Environmental Sustainability Strategy (ESS), launched by the Ministry of Municipality and Environment with the first and second NDS for the periods 2011–2016, and 2018–2022 (Chap. 2)
- Law No 19 (2004) is dedicated to conserving wildlife and natural habitats and led to the establishment of several nature reserves (Lusail Reserve in 2005; Al Mashabiya, Al Eraiq, and Al Thakhira Reserves in 2006; Khor Al Adaid Reserve in 2007; Southern Area Reserve in 2018; and Al Rafaa Reserve in 2020) (Chap. 2)
- Qatar National Food Security Strategy (2018–2023), developed after the 2017 blockade. Aimed at boosting local production through farm support programs, increasing strategic storage, as well as domestic production of vegetables, fruit and dairy products (to secure 30–70% self-sufficiency) (Chaps. 11 and 12)
- Qatar National Master Plan and Municipal Spatial Development Plans set targets for population densities across the country (Chap. 14)
- National Campaign for the Conservation and Efficient Use of Water and Electricity *Tarsheed*. Launched by Kahramaa in 2012 to educate energy and water conservation, sustainability, and environmental awareness (Chaps. 11 and 16)
- District Cooling Program, aiming to reach 100% usage of non-potable water in cooling plants (Chap. 11).

By any objective standard, this is an impressive list of international commitments, national frameworks, and sectoral initiatives. And we have probably not captured them all. This illustrates our opening point in this chapter—sustainability policy is hard. It begins with pollution and environment and biodiversity, adds water and air, and as it moves to carbon and climate, begins to embrace industry, transportation, education, food and just about everything else. It's a blend of three-ring circus and multi-dimensional chess, and immediately poses issues of coordination, collaboration, capacity, and implementation. And on top of that, sustainability is not sustainability within current energy and economic configurations—it is a transition to a post-carbon future. Not surprisingly, our chapters have noted these challenges.

19.4 Implementation and Transition

Damilola Olawuyi and Elena Athwal get to the heart of the matter in Chap. 3. They note the "rapid profusion of sustainability policies and laws in Qatar" and the "rapidly increasing volume of ongoing programs and initiatives, as well as the number of administrative and governance bodies with supervisory functions in economic, social, and environment spheres." However, for these to be effective there is a need "to promote greater coordination amongst the various agencies and ministries in order to enhance greater cooperation, interoperability, and resource sharing in the implementation of sustainability programs" (what they call a "nexus approach"—see Recommendations below). Marcello Contestabile (Chap. 8) makes a similar point: "It is also a complex problem of systemic nature, as moving towards a low carbon, efficient energy system means that the different parts of the energy system (buildings, transport, power and water, and industry) will become increasingly interlinked and that decisions made today will have major consequences on future systems cost and environmental performance."

In Chap. 2, Reem Al-Hababi noted some of the problems with the NDS and the QNV, not for their ambition, but for their capacity to deliver on that ambition. The first NDS, for example, had admirable mid-term environmental sustainability targets, but by 2016 only two of them had been fully achieved. They were reduction of gas flaring in energy production (down by about 30%) and reduction in per capita domestic waste generation (down to 1.3 kilos). The second NDS reported on the shortcomings of the initial targets and admitted that there were policy planning and implementation gaps, unclear prioritization, and poor coordination at sectoral and cross-sectoral levels. Al-Hababi also notes some overlap in the responsibilities and mandates of the Climate Change and Clean Development Committee and the Climate Change Department.

Other chapters tended to agree that the apparatus of laws and policies is about as good as it gets for a full-bore national sustainability policy, but that effective implementation will be key. Chapter 16 notes that the second NDS had a target of 15% recycling of solid waste, something achievable in its view as long as the Ministry works with supporting partners. Chapter 11 described the Qatar Water Security Policy and Qatar Water Strategy, and the importance of its accompanying Implementation Plan. Sever and Tok (Chap. 17) explored the implementation of ESD (Education for Sustainable Development) and argue that it is a "complex sphere with a combination of official, top-down initiated efforts, diverging global frameworks and extracurricular school activities in the formal and non-formal forms." That the outcomes of the Key Performance Indicators and targets have been mixed exemplifies the challenges of capacity and complementarity for coherence.

While recognizing areas in need of attention, our book also shows some encouraging successes and promising examples of collaboration and real efforts to implement for impact. Given that Qatar is effectively a city-state, with almost all of its population living in Doha, the Ministry of Municipal Affairs was a de facto ministry of environment through urban planning and transportation policy. In 2021, the ministry was reshuffled and now is named Ministry of Environment and Climate Change. In a possibly good sign, its building is literally next door to the Ministry of Foreign Affairs and may provide some measure of coordination of local, municipal, national, and international initiatives. Chapter 11 concludes that the achievements in the water sector have been "impressive" and include both growing public awareness and involvement by civil society and local communities in water conservation projects. An example mentioned in both Chaps. 11 and 16 is *Tarsheed*, run by

Kahramaa (the Qatar General Electricity and Water Corporation), which has been effective in reducing water and electricity consumption in the country. In the waste management sector, the effort to reduce landfill waste by 90% by 2022 has led the government to bring in major private sector waste management players such as the Seashore Group of companies, Dulsco Qatar, Power Waste Management and Transport Co. WLL, Green Waste Management, Al Haya Enviro. Chapter 12 provided evidence of both non-governmental organizations (NGOs) (Hifz AlNeama) and the private sector (Wahab) in supporting food security and reducing food waste. Chapters 5 and 16 explained how the Qatar Foundation both supports research around sustainability and provides a test bed for sustainable practices in Education City (indeed, Sheikha Moza, the founder and chairperson of Qatar Foundation, is a UN SDG Advocate—Chap. 17).

Sustainability transitions require a full and integrated policy framework, and their implementation is a colossal challenge (often fraught with difficult trade-offs). Qatar's efforts and commitments, as shown in this book, have been remarkable and serious. Effective implementation and coordination are works in progress, but there are bright lights and successes, and despite its environmental challenges, the country has substantial advantages it can marshal as well. More work is to be done, and we list the book's recommendations below. But we close this section with yet another somber observation. Climate change forces sustainability policy into a higher register—it cannot be sustainability for the status quo of energy sources and consumption. We cannot continue to rely on fossil fuels, and just husband them, reduce and recycle. Sustainability is not about statis, but about transition and transformation—we must shift from fossil fuels to renewables, and with that shift will come a transformation in our economic systems, a fourth or fifth industrial revolution that will be as profound as the shift from wood and coal to petroleum.

Every country will transition (or not) in its own way. Qatar's unique context will drive a unique transition. To take just three of the unique features we cited earlier: tiny population; substantial wealth based on hydrocarbons and one industry (natural gas); a desert climate. Add to that, the economic and social transformation of the country has all happened in one generation, since about 1996 with the first shipment of LNG (Chap. 6 and 7). Even as it rapidly evolved into a "late-stage petro-developmental state" Qatar has recognized the need to diversify out of natural gas into a knowledge-based economy. The gas fields are endless, but the days of gas consumption are numbered as the world eventually shifts to renewables. Hence the emphasis on education and research (e.g., the Qatar Foundation), and the government's efforts to encourage both entrepreneurship and industries such as smart manufacturing, fintech, sports-tech, and fashion and design (Chap. 18). Winning the World Cup, and then using it as a catalyst for sustainability (the FIFA World Cup 2022 Sustainability Strategy), is an example of this strategy, since Qatar wants to become a global hub for elite sporting events, and the ancillary and supporting economic sectors they can stimulate. In this way, current sustainable practices, from soccer matches to food security, trace an arc that begins to bend to an alternate future.

19.5 Areas for Future Research Attention

This book has examined sustainability in Qatar from various angles, but some topics could not be covered. The key ones that deserve further analysis are:

- Domestic transportation: Some chapters have mentioned the recently operated public transit system, Doha Metro, as an example of major projects that were established with environmental considerations. Transportation is one of the core components that support the socioeconomic systems' interactions and development (Rodrigue, 2020). At the same time, the transportation sector contributes to environmental challenges, including "climate change, human toxicity, terrestrial acidification, and water depletion" (Al-Thawadi & Al-Ghamdi, 2019, p. 3). Transitioning to a more sustainable transportation system in Qatar—as in any other place—involves more than developing infrastructure "to provide an integrated, eco-friendly, reliable, and safe multimodal transportation system" (Al-Buenain et al., 2021, p. 620). Decarbonizing the transportation sector involves a change in society's perception of public transportation and electric vehicles, especially since private automobiles (almost all conventional gasoline) are the primary mode of transport in Qatar.
- Aviation sector: Aviation has undergone significant growth and massive infrastructural development over the past two decades. Although the carbon footprint of airline operations has been halved in comparison with those of the late 1990s, GHG emission values for airline activities are projected to increase drastically due to the expansion of the sector (Elhmoud & Kutty, 2020). On the one hand, the development of the sector in Qatar—and in the region generally—has been encouraged by several factors, beginning with geographical location, low operating costs, and the large expatriate population's travel requirements. On the other, the aviation sector—namely its key actor, Qatar Airways—has been contributing to the national economy by the improved connectivity that it has been providing, connecting the local market with foreign markets, and developing tourism. Moreover, in terms of contributing to the GDP, air transport represents half of the value added to the GDP of the entire transportation and communication sector (Petcu, 2021). Given the role aviation plays in Qatar, this is a notable omission.
- International shipping and supply chains: International logistics are significant for global industries since the "mobilization of commodities across the supply chain is propelled by air, land, or sea transportation" (Al-Enazi et al., 2022, p. 2). Shipping activities (e.g. maritime shipping) are responsible for 2.1% of global GHG emissions, and this percentage is projected to rise in the next several years, considering the new trade routes and links expected to be established and the increase in trading of various commodities. The natural gas trade experienced a 1.6% of growth rate on an annual basis between 2017 and 2022, with global demand reaching 300 billion cubic meters. Around 30% of the global seaborne trade is hydrocarbons shipping activities. This is significant to Qatar's gas industry and its corresponding supply chains (Al-Enazi et al., 2022), and it is also important given Qatar's reliance on imports for key commodities, including food.

- Cooling systems: Buildings (whether residential or commercial) are responsible for the majority of energy and electricity uses and, accordingly, energy-related GHG emissions. The demand for electricity is expected to increase, especially with the population growth in developing economies that are located in hot-climate regions, such as the Arab Gulf region. In such regions, the cooling energy consumption in typical buildings is three times higher than that of similar buildings in moderate climate regions. Moreover, in hot-climate regions, space cooling loads vary based on season and time of the day, which results in straining electricity grids. These challenges and cooling requirements will be exacerbated by climate change (Eveloy & Ayou, 2019). Much work has been done in Qatar, particularly by the Qatar Green Building Council (now within Earthna), but more work needs to be done on heating and cooling demands.
- Social aspects of sustainability: The book also lacks a chapter on the implications of sustainability efforts on social aspects in Qatar, such as gender equality and topics that relate to women's participation in the private sector workforce or science, technology, engineering and math (STEM) attainment and whether such transitions would yield an impact in this respect. Relatedly, a chapter on the implications of sustainability efforts (e.g., transitioning to smart cities, knowledge-based economy) on migration in Qatar is not included in this collection as well. The link between developing smart cities and migration trends is considered a novel research area; existing literature highlights "the relevance of planning for smart cities, city development and infrastructure projects, and the migration of educated workers" (Mouazen & Hernández-Lara, 2021, p. 123).
- Public health: The implications of environmental degradation and climate change on public health and the state's efforts toward them are not covered in this volume. Air pollution and rising temperatures are adding to existing human health concerns and the increased frequency of certain diseases and health issues (Longlong et al., 2018; Skelhorn, 2019).
- Emergency planning: Climate change may not be gradual, but might spike, with climatic tipping points, and so there is the need for climate-related emergency planning. The tipping points might be in the form of sudden sea level rise or intense heat stress, but preparatory work needs to be done on what these could mean for Qatar.

19.6 Recommendations

The chapters in this book converge on five broad recommendations for a sustainable Qatar.

1. From a policy governance and monitoring perspective, an oversight body should be created to evaluate the development of sustainability initiatives, and provided with the assessment tools and methodologies to serve this purpose. This includes ensuring that required capacity is available across sectors to enable nexus approaches for policy coherence.

2. On implementation, areas for progress include establishing a governance approach to enhance the coordination and coherence between various stakeholders in carrying out and delivering sustainability initiatives and projects, and improving the technical and institutional capacities of relevant entities. Enhancing transparency of data will allow for the implementation gaps to be more quickly identified and addressed. This can leverage the strong research environment that has been established in Qatar.
3. Sustainability policy planning should include developing a systems perspective on sustainability, adopting an integrated approach with a mix of policy instruments, and enhancing information sharing and data availability for policymakers and relevant stakeholders. There should be active engagement of various stakeholders during policy design, and regional and international collaboration to make sure that local or national policies and actions are aligned with those at higher levels.
4. Businesses and private actors should be encouraged to integrate sustainability into their operations and value chains and to engage with civil society and NGOs in sustainability efforts. This is also the case for public sector actors, including the Ministry of Education and Higher Education, within its leadership role for the education sector.
5. Sustainability of the gas sector and the overall economy requires investments in technologies and methods to reduce emissions, increasing the share of renewables in the energy mix, and investing in new energy sectors while utilizing gas and oil revenues in diversification strategies. There are opportunities to take a global leadership role in the transition to new energies, such as hydrogen, an option for which the State of Qatar is well positioned.

References

Al-Buenain, A., Al-Muhannadi, S., Falamarzi, M., Kutty, A., Kucukvar, M., & Onat, N. (2021). The adoption of electric vehicles in Qatar can contribute to net carbon emission reduction but requires strong government incentives. *Vehicles, 3*(3), 618–635.

Al-Enazi, A., Bicer, Y., Okonkwo, E., & Al-Ansari, T. (2022). Evaluating the utilisation of clean fuels in maritime applications: A techno-economic supply chain optimization. *Fuel, 322*, 124195.

Al-Thawadi, F., & Al-Ghamdi, S. (2019). Evaluation of sustainable urban mobility using comparative environmental life cycle assessment: A case study of Qatar. *Transportation Research Interdisciplinary Perspectives, 1*, 100003.

Elhmoud, E., & Kutty, A. (2020). Sustainability assessment in aviation industry: A mini- review on the tools, models and methods of assessment. *Proceedings of the 2nd African International Conference on Industrial Engineering and Operations Management*, Zimbabwe (pp. 1228–1238).

Eveloy, V., & Ayou, D. (2019). Sustainable district cooling systems: Status, challenges, and future opportunities, with emphasis on cooling-dominated regions. *Energies, 12*(2), 235.

Longlong, T., Ryouta, I., Koji, T., & Norihiro, I. (2018). Development of human health damage factors related to emissions by considering future socioeconomic scenarios. *The International Journal of Life Cycle Assessment, 23*(12), 2288–2299.

Mouazen, A., & Hernández-Lara, A. (2021). A bibliometric review of smart cities and migration. In A. Visvizi, M. D. Lytras, & N. R. Aljohani (Eds.), *Research and innovation forum 2020: Disruptive technologies in times of change*. Springer.

Petcu, C. (2021). The role of Qatar Airways in the economic development of Qatar: Before and during the Gulf crisis. In M. Zweiri, M. M. Rahman, & A. Kamal (Eds.), *The 2017 Gulf crisis: An interdisciplinary approach*. Springer.

Rodrigue, J.-P. (2020). Transport, energy and environment. In J.-P Rodrigue (Ed.), *The geography of transport systems* (pp. 124–150). Routledge.

Skelhorn, C. (2019). Planning and design for sustainable cities in the MENA region. *Smart and Sustainable Built Environment, 8*(2), 98–102.

Leslie A. Pal is Founding Dean of the College of Public Policy at Hamad Bin Khalifa University, Doha, Qatar. He is the author, co-author or editor of over thirty books, the most recent of which are Informing Action: Higher Education Countering Violent Extremism (2021), The Future of the Policy Sciences (2021), Global Governance and Muslim Organizations (2019), Policy Transfer: Micro-Dynamics and Macro-Effects (2017), Policy Making in a Transformative State: The Case of Qatar (2016), Frontiers of Governance: The OECD and Global Public Management Reform (2012), and Beyond Policy Analysis: Public Issue Management in Turbulent Times (6th edition, 2020). He has published over ninety articles and book chapters in a wide variety of areas of public policy and administration, and international public management reform. His articles have appeared in the journals Governance, Policy Sciences, Journal of Comparative Policy Analysis, Canadian Political Science Review, Political Studies Review, and Policy and Society. He was the Executive Editor of the Journal of Comparative Policy Analysis from 2015–2020, and is currently a member of the editorial boards of the International Review of Public Policy and the Journal of Economic Policy Reform. In January 2020 he was elected as one of the Vice-Presidents of the International Public Policy Association. He is also on the Board of the Association for Middle Eastern Public Policy and Administration (AMEPPA). He has served as a consultant to the Open Society Foundations, the World Bank and the OECD, as well as several other international organizations.

Reem Al-Hababi is a PhD student in the Gulf Studies Program at Qatar University. She holds a Master of Public Policy from HBKU.

Logan Cochrane is an Associate Professor in the College of Public Policy at HBKU and is Adjunct Associate Professor at Hawassa University. He has worked in non-governmental organizations internationally, including in Afghanistan, Benin, Burundi, the Democratic Republic of the Congo, Egypt, Ethiopia, Qatar, Saudi Arabia, South Africa, South Sudan, Tanzania, and Uganda.

Correction to: Sustainable Qatar

Logan Cochrane and Reem Al-Hababi

Correction to: Chapter 1 in: L. Cochrane and R. Al-Hababi (eds.), *Sustainable Qatar*, Gulf Studies 9, https://doi.org/10.1007/978-981-19-7398-7_1

The original version of the book was inadvertently published with the below errors in Chapter 1. Page 6: "the Doha Institute" is corrected to read as "the Doha Institute for Graduate Studies" Page 9: "Chapter 9 begins to look at the impacts of climate change on Qatar, and specifically how sea-level rise will affect the region in the coming decades" is corrected to read as "Chapter 9 begins to look at the impacts of climate change on Qatar, and specifically how sea-level rise will affect the region in the coming decades, highlighting how this poses a national security risk that needs to be on top of policy agendas across the region." The correction to the chapter has been updated with the changes.

The updated original version of this chapter can be found at https://doi.org/10.1007/978-981-19-7398-7_1

L. Cochrane and R. Al-Hababi (eds.), *Sustainable Qatar*, Gulf Studies 9,
https://doi.org/10.1007/978-981-19-7398-7_20

GPSR Compliance
The European Union's (EU) General Product Safety Regulation (GPSR) is a set of rules that requires consumer products to be safe and our obligations to ensure this.

If you have any concerns about our products, you can contact us on

ProductSafety@springernature.com

In case Publisher is established outside the EU, the EU authorized representative is:

Springer Nature Customer Service Center GmbH
Europaplatz 3
69115 Heidelberg, Germany

www.ingramcontent.com/pod-product-compliance
Ingram Content Group UK Ltd.
Pitfield, Milton Keynes, MK11 3LW, UK
UKHW021845060525
458290UK00001B/4
* 9 7 8 9 8 1 1 9 7 3 9 7 0 *